AF580226

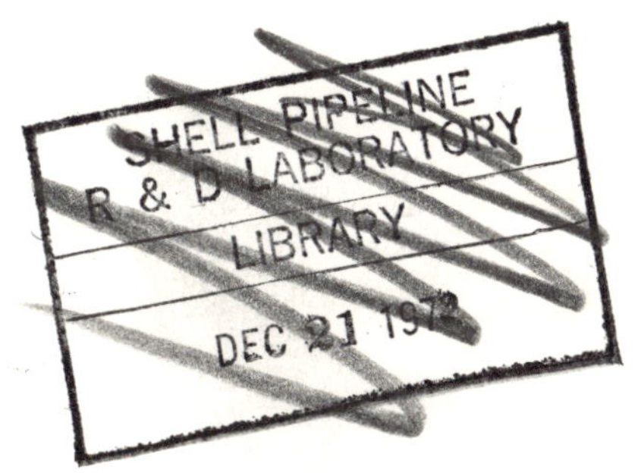
SHELL PIPELINE
R & D LABORATORY
LIBRARY
DEC 21 1972

NONLINEAR PARTIAL DIFFERENTIAL EQUATIONS IN ENGINEERING

Volume II

This is Volume 18-II in
MATHEMATICS IN SCIENCE AND ENGINEERING
A series of monographs and textbooks
Edited by RICHARD BELLMAN, *University of Southern California*

The complete listing of the books in this series is available from the Publisher upon request.

NONLINEAR PARTIAL DIFFERENTIAL EQUATIONS IN ENGINEERING

W. F. AMES

DEPARTMENT OF MECHANICS AND HYDRAULICS
THE UNIVERSITY OF IOWA
IOWA CITY, IOWA

Volume II

1972

ACADEMIC PRESS • New York • London

ACADEMIC PRESS, INC.
111 Fifth Avenue, New York, New York 10003

United Kingdom Edition published by
ACADEMIC PRESS, INC. (LONDON) LTD.
24/28 Oval Road, London NW1 7DD

LIBRARY OF CONGRESS CATALOG CARD NUMBER: 65-22767

AMS (MOS) 1970 Subject Classification: 35-02

PRINTED IN THE UNITED STATES OF AMERICA

IN MEMORY OF

KEVORK M. DANIELSON

Contents

CHAPTER 3. Approximate Methods

CHAPTER 4. Numerical Methods

Preface

During the seven years since the preparation of Volume I significant advances have occurred in several areas pertinent to this series. In particular, the rapid evolution of the *finite element method* is noteworthy, but surely as important is the development of a *deductive similarity theory* based on finite and infinitesimal continuous transformation groups. In all areas of study, endless sifting and winnowing continues, whereby man improves the present with the optimistic hope of a better future.

This book builds on the first volume, maintaining the goal of unifying as much of the scattered literature as possible. Analytic methods occupy fully one-half of the work with an emphasis on the application of modern algebra to nonlinear problems. The remainder is about equally divided between approximate procedures and numerical methods. While portions of the book have been used in several advanced graduate courses, it is primarily intended as a reference work for all those bedeviled scientists and engineers faced with the solution of nonlinear problems. No finite volume can be all encompassing. Over 500 references serve to extend and supplement the text.

I am indebted to many for their ingenuity, creativity, and resourcefulness. It is indeed difficult to build effective and general methods in an area which like Bunyan's road has a "deep ditch on one side, a quagmire on the other and . . . ends in a wilderness." I also wish to express my appreciation to Joseph Howe and Hunter Rouse for creating and encouraging a spirit of research and inquiry at Iowa. Mrs. Robert Panchyshyn was the lady behind the typewriter. Throughout a difficult manuscript she was always smiling. Lastly, my wife Terry was, for the sixth time, a "book widow." I am continually in her debt for many things but especially for understanding.

Contents of Volume I

CHAPTER **1**

Analytic Techniques and Solutions

1.0 INTRODUCTION

A considerable portion of the progress in the theory of linear partial differential equations has resulted from the use of *ad hoc*[†] methods. Examples of *ad-hoc* procedures include *separation of variables*, *integral transforms*, and *finite difference methods*. On the other hand, the method of characteristics and those procedures which develop *general solutions* are not *ad hoc* but arise *naturally* from the specific system under consideration. Such analyses will be said to develop *natural* methods.

Specific nonlinear problems have been shown to yield useful solutions when the *ad-hoc* processes of the linear theory are employed, but their great utility, in the linear theory, rests primarily upon the principle of superposition. In accordance with this *principle*, elementary solutions can be combined to yield more flexible ones, namely those which can satisfy the auxiliary conditions that arise from particular phenomena. The loss of this principle in nonlinear problems and the lack of effective replacements constitutes the major barrier to reasonable understanding of the present chaotic state.

Nevertheless, a considerable body of methodology is extant on nonlinear partial differential equations. The discussion in this chapter supplements the corresponding material in the author's *Nonlinear Partial Differential Equations in Engineering* (1965) which will be referred to as Volume I throughout the text.

[†] From the Latin, "for this case alone."

1.1 NONLINEAR SUPERPOSITION PRINCIPLES

Many significant successes in constructing effective theories for physical phenomena can be traced to the linear principle of superposition. This concept is an immediate consequence of the definition of a linear operator L:

$$L(af + bg) = aL(f) + bL(g). \tag{1.1}$$

Several questions immediately come to mind. While all linear equations have a linear superposition principle, can they also possess more than one? Second, are superposition principles possible for nonlinear equations?† These questions were considered by Jones and Ames [1]‡ using elementary analysis.

For simplicity, we restrict our attention to cases involving two independent variables. Let $u_i = u_i(x, y)$, $i = 1, 2,..., k$, be solutions of some equation $L(u) = 0$. A function $F = F(u_1, u_2, ..., u_k, x, y)$ is called a *connecting function* for $L(u) = 0$, if F is also a solution. This constitutes a *nonlinear superposition principle.*

Connecting functions for a class of linear and quasi-linear equations may be developed from a linear equation by dependent variable transformations. Consider the linear equation

$$a(x, y)u_x + b(x, y)u_y = c(x, y)u. \tag{1.2}$$

Setting $u = f[v(x, y)]$, Eq. (1.2) becomes

$$av_x + bv_y = cf/f'. \tag{1.3}$$

Since Eq. (1.2) is linear, $U = \sum_{i=1}^{k} u_i$ is a solution if the u_i, $i = 1,..., k$, are solutions. However, if $u_i = f(v_i)$, $i = 1,..., k$, and $V = f^{-1}(U)$, then

$$V = f^{-1}[f(v_1) + f(v_2) + \cdots + f(v_k)] \tag{1.4}$$

is a solution to Eq. (1.3). The basic Eq. (1.2) need not be first order or even linear. We could reason from a nonlinear equation for which a connecting function has been found.

Some very interesting examples follow from the previous results.

---

† Throughout the text all of our equations will be nonlinear partial differential equations. Consequently, all the descriptive adjectives can be omitted without loss of clarity.

‡ Numbers in brackets refer to References at the end of each chapter.

For example if $a = b = 1$, $c = n \neq 0$, and $f = v^n$, then Eq. (1.3) becomes the linear equation

$$v_x + v_y = v, \tag{1.5}$$

and Eq. (1.4) gives us the *nonlinear* superposition principle,

$$F = \left[\sum_{i=1}^{k} v_i^{\,n}\right]^{1/n}, \tag{1.6}$$

where $n \neq 0$ is any real number.† *Consequently, a linear equation can have a noncountable infinity of superposition principles*; when $n = 1$, the classical principle is recovered.

A more complicated example results from the transformation

$$f = \exp[v^{(1-n)}/(1-n)], \qquad n \neq 1,$$

which yields the equation

$$av_x + bv_y = cv^n. \tag{1.7}$$

For this equation

$$F^{1-n} = (1-n)\log\left\{\sum_{i=1}^{k} \exp[v_i^{1-n}/(1-n)]\right\} \tag{1.8}$$

is a connecting function.

The equation

$$u_t + uu_x = \lambda u_{xx}, \tag{1.9}$$

where λ is a parameter, has been utilized by Burgers [2] as a mathematical model of turbulence. By setting $u = v_x$, integrating once with respect to x, and discarding an arbitrary function of t, Eq. (1.9) becomes

$$v_t + (\tfrac{1}{2})v_x^{\,2} = \lambda v_{xx}, \tag{1.10}$$

an equation which has some significance in the burning of a gas in a rocket, as discussed by Forsythe and Wasow [3]. If we now set $v = -2\lambda \log(w)$, Eq. (1.10) transforms to

$$w_t = \lambda w_{xx}, \tag{1.11}$$

the one-dimensional diffusion equation! These transformations are

† Upon letting $n \to 0$, the alternate superposition principle $F = (\prod_{i=1}^{k} v_i)^{1/k}$ is obtained. This was observed by J. R. Ferron in a private discussion.

already well known. A complete discussion of them with proper priorities may be found in Volume I of Ames [4a].[†]

Reasoning as before, a connecting function for Eq. (1.10) is found to be

$$F = -2\lambda \log \left[\exp\left(-\frac{v_1}{2\lambda}\right) + \exp\left(-\frac{v_2}{2\lambda}\right)\right]. \tag{1.12}$$

Now, this can be used to find a connecting function for Eq. (1.9). This is seen to be

$$F = -2\lambda \frac{\partial}{\partial x}\left\{\log \left[\exp\left(-\frac{1}{2\lambda}\int_{x_0}^{x} u_1(\xi, t)\, d\xi\right) + \exp\left(-\frac{1}{2\lambda}\int_{x_0}^{x} u_2(\xi, t)\, d\xi\right)\right]\right\}, \tag{1.13}$$

where $x_0 < x$ is some lower limit.

Extensions to a larger class of problems, including the Navier–Stokes equations will be given later in this chapter (Section 1.2c).[‡]

1.2 GENERATION OF NONLINEAR EQUATIONS WITH BUILT-IN SOLUTIONS

In Volume I, Chapter 2, dependent variable transformations were shown to provide a vehicle for the generation of nonlinear equations with built-in solutions. In particular, Burgers' equation $u_t + uu_x = \nu u_{xx}$ could be transformed into the linear diffusion equation. Later papers employing this concept and its generalizations are due to Chu [5], Montroll [6], and Ames and Vicario [7]. Herein several examples drawn from these references are discussed followed by some general remarks.

a. In Wave Mechanics

The equations for wave propagation and vibration of a traveling threadline, derived by Ames *et al.* [8], are

$$V^2 u_{xx} + 2V u_{xt} + u_{tt} = \frac{T u_{xx}}{m(1 + u_x^2)}, \tag{1.14}$$

$$VV_x + V_t = \frac{1}{m(1 + u_x^2)^{1/2}} \left\{\frac{T}{(1 + u_x^2)^{1/2}}\right\}_x, \tag{1.15}$$

$$[mV(1 + u_x^2)^{1/2}]_x + [m(1 + u_x^2)^{1/2}]_t = 0, \tag{1.16}$$

$$m(T + N) = BN, \tag{1.17}$$

[†] The complete details for this work will be listed at this point only. We shall refer to this work hereafter simply as Volume I.

[‡] See also Levin [4b] and Kečkić [4c].

where u, v, m, T, x, t, B, and N are, respectively, transverse displacement, velocity, density per unit length, tension, distance, time (all dimensionless), and two physical constants. When the transverse vibrations are small, that is, when u_x is small compared with 1, Eqs. (1.15) and (1.16) are uncoupled from Eq. (1.14), becoming

$$VV_x + V_t = T_x/m, \tag{1.18}$$

$$m_t + (mV)_x = 0, \tag{1.19}$$

together with the (linear) constitutive relation, Eq. (1.17). We shall now demonstrate that Eqs. (1.18) and (1.19) can be transformed into a linear wave equation.

By defining a particle function ψ such that

$$m = \partial\psi/\partial x, \qquad -mV = \partial\psi/\partial t, \tag{1.20}$$

Eq. (1.19) is identically satisfied. However, we do not proceed further to develop an equation for ψ. Instead a Von-Mises (see Volume I) type of transformation, that is a transformation from the (x, t) to (ψ, t) plane, will be utilized. Upon considering V, m, and T as functions of ψ and t, the basic transformation relations are found to be

$$(\ )_x|_t = (\ )_\psi \psi_x, \qquad (\ )_t|_x = (\ )_\psi \psi_t + (\ )_t|_\psi, \tag{1.21}$$

where $|_t$ indicates operations at constant t. Under these transformations Eqs. (1.18) and (1.19) become

$$V_t = T_\psi \tag{1.22}$$

and

$$-m^{-2}m_t = V_\psi, \tag{1.23}$$

respectively.

Upon equating the ψ-derivative of Eq. (1.22) to the t-derivative of Eq. (1.23), that is, assuming $V_{t\psi} = V_{\psi t}$, we obtain $(m_t/m^2)_t = -T_{\psi\psi}$. When the constitutive equation (1.17) is introduced, the equation for m is

$$m_{tt} - (BN)m_{\psi\psi} + 2m^{-1}[(BN)m_\psi^2 - m_t^2] = 0. \tag{1.24}$$

To be especially noted are the *quadratic nonlinearities* of this equation. As was observed in Volume I, and as will be seen in our subsequent examples, equations with quadratic nonlinearities are often transformable into linear equations.

With the foregoing remark in mind, let us subject the linear wave equation

$$w_{tt} - k^2 w_{\psi\psi} = 0 \tag{1.25}$$

to the transformation

$$w = F(\phi), \tag{1.26}$$

for arbitrary but differentiable F. The resulting equation for ϕ is

$$\phi_{tt} - k^2\phi_{\psi\psi} - F''F'^{-1}[k^2\phi_\psi{}^2 - \phi_t{}^2]. \tag{1.27}$$

Comparing Eqs. (1.24) and (1.27), equivalence is obtained if we set $\phi = m$, $k^2 = BN$, and take $F''/F' = -2/m$, that is, $w = F(\phi) = m^{-1}$.

Since the general solution of Eq. (1.25) is

$$w(\psi, t) = f(\psi + kt) + g(\psi - kt), \tag{1.28}$$

it follows immediately that the *general solution* for m is

$$m(\psi, t) = [f(\psi + kt) + g(\psi - kt)]^{-1}. \tag{1.29}$$

Corresponding solutions for V and T follow from elementary analyses. A solution for $x(\psi, t)$ is also obtained to provide for return to the physical plane.

b. Diffusion and Reaction Problems

A number of typical equations possessing quadratic nonlinearities have been discussed by Montroll [6].

1. POPULATION GROWTH AND DIFFUSION[†]

The equation which describes the combination of population (n) growth and diffusion is

$$n_t = Dn_{xx} + kn(\theta - n)/\theta, \tag{1.30}$$

where D and k are constants, and θ represents a saturation population per unit length. The multidimensional generalization of Eq. (1.30) is

$$n_t = D\,\nabla^2 n + kn(\theta - n)/\theta, \tag{1.31}$$

where θ is the saturation population per unit area or volume.

[†] See Fisher [9] and Kolmogorov *et al.* [10].

2. CLANNISH RANDOM WALKERS

An equation similar to Eq. (1.30) is derived for the motion of two interacting populations which tend to be *clannish*—that is they wish to live near those of their own kind. In the one-dimensional case, the density function $f(x, T)$ for one species at point x and time T has the equation

$$f_T = \{Df_x + \alpha f(f - 1)\}_x . \tag{1.32}$$

This equation is reduced to one which is parameter free by setting

$$t = T\alpha^2/D, \qquad y = x\alpha/D, \qquad g = f - 1/2, \tag{1.33}$$

so that

$$g_t = [g_y + g^2]_y . \tag{1.34}$$

As has been previously observed in Section 1.1, the main equation [Eq. (1.9)] of Burgers' model of turbulence also has the form of Eq. (1.34). To obtain that form, set $x = \nu^{1/2} y$ and $u = -2g\nu^{1/2}$.

3. SEPARATION CASCADES†

Many of the separation processes in the chemical industry are staged cascades such as membrane gas separation cascades and distillation columns. If the separation factor β is very small so that many stages are required, then the length of a stage becomes short compared with the total length of the cascade. Consequently the concentration difference $(c_n - c_{n-1})/a \to \partial c(x, T)/\partial x$, as $a \to 0$, whereupon the equation for c takes the form

$$c_T = D\{c_x - \beta' c(1 - c) - pc\}_x . \tag{1.35}$$

Equation (1.35) is transformed into Eq. (1.32) by making the substitutions

$$c = f(p + \beta')/\beta', \qquad \alpha = D(\beta' + p).$$

Equation (1.32) is also valid for continuous separation processes such as thermal diffusion and distillation. Similar equations exist in centrifugal separation and chromatographic analyses.

4. MOLECULAR RECOMBINATION‡

The "growth" Eq. (1.31) can be converted into an equation for particle spreading by diffusion and loss by recombination (reaction). With

† See Benedict [11], Cohen [12], and Montroll and Newell [13].

‡ See Gray and Kerr [14].

$\gamma = k/\theta$, we take the limits $k \to 0$ and $\theta \to 0$, such that the ratio remains γ. Then Eq. (1.31) becomes

$$n_t = D \nabla^2 n - \gamma n^2. \tag{1.36}$$

If an electric arc is passed through a gas such as N_2, the free radicals N are formed which diffuse away from the arc. But they also recombine by collision. This recombination of free radicals, governed by Eq. (1.36), where $n(r, t)$ is the free radical density at point r and time t, is a typical example of this process.

In Volume I, page 23, it was observed that the dependent variable transformation $\psi = F(u)$ of the linear parabolic equation $\psi_t = \lambda \nabla^2 \psi$ generated the equation

$$u_t = \lambda \nabla^2 u + \lambda(F''/F')(\nabla u)^2 \tag{1.37}$$

with quadratic nonlinearities. If we wished to solve the equation

$$u_t = \lambda u_{xx} + u_x{}^2 G(u), \tag{1.38}$$

we would set

$$\lambda F'' = F' G(u) \tag{1.39}$$

and solve for F. For example, if $G(u) = \alpha$, then

$$\psi = F(u) = (\lambda/\alpha) \exp(\alpha u/\lambda), \tag{1.40}$$

so that

$$u = (\lambda/\alpha) \ln(\alpha\psi/\lambda). \tag{1.41}$$

Generally,

$$\psi = F(u) = \int^u \exp\left\{\int^{u_1} \lambda^{-1} G(u_2)\, du_2\right\} du_1 . \tag{1.42}$$

The solution of the nonlinear equation

$$u_t = \lambda u_{xx} + \alpha u_x{}^2 \tag{1.43}$$

is given by Eq. (1.41), where ψ is a solution of the diffusion equation. For the pure initial-value problem with $u(x, 0)$ known in an unbounded space,

$$\psi(x, 0) = (\lambda/\alpha) \exp[\alpha u(x, 0)/\lambda]$$

and

$$\psi(x, t) = [2(\pi t\lambda)^{1/2}]^{-1} \int_{-\infty}^{\infty} (\lambda/\alpha) \exp[\alpha u(x', 0)/\lambda - (x - x')^2/4\lambda t]\, dx',$$

whereupon

$$u(x, t) = (\lambda/\alpha) \ln(\alpha\psi/\lambda).$$

Another class of equations can be generated from the diffusion equation $\psi_t = \lambda\psi_{xx}$ by the transformation

$$\psi = \exp\left[\int^x F(u)\,dx\right], \tag{1.44}$$

where $F(u)$ is again arbitrary. One finds the resulting equation to be

$$u_t = \lambda\{u_{xx} + F''u_x^2/F'\} + 2\lambda F(u)u_x\,. \tag{1.45}$$

Still another class follows from transforming the diffusion equation with

$$\psi = f(t)F(u) \tag{1.46}$$

so that

$$u_t = \lambda\{u_{xx} + F''u_x^2/F'\} - f'F/fF'. \tag{1.47}$$

Montroll [15] found that by choosing $f = \exp[-kt]$ and $F/F' = G(u)$, then

$$u_t = \lambda\{u_{xx} + [(1 - G')/G]u_x^2\} + kG(u), \tag{1.48}$$

with

$$F(u) = \int^u du/G(u).$$

If we select

$$G(u) = u(\theta - u)/\theta, \tag{1.49}$$

the right-hand term of Eq. (1.48) becomes the same as that of Eq. (1.30), but the full equation has the form

$$u_t = \lambda\{u_{xx} + 2u_x^2/(\theta - u)\} + ku(\theta - u)/\theta. \tag{1.50}$$

Clearly $F(u) = u/(\theta - u)$, and u is related to the function ψ, which satisfies the diffusion equation, through

$$u(x, t) = \theta\psi/[e^{-kt} + \psi]. \tag{1.51}$$

c. A Class of Reducible Equations

A class of equations reducible to a single linear-diffusion equation was discovered by Chu [5]. The system of n equations is

$$\frac{\partial u_i}{\partial t} + F_j\frac{\partial u_i}{\partial x_j} = G_i\frac{\partial u_i}{\partial x_j}\frac{\partial u_i}{\partial x_j} + k\frac{\partial^2 u_i}{\partial x_j\,\partial x_j} + H_iR_i\,, \qquad i, j = 1, 2,\ldots, n, \tag{1.52}$$

where the summation convention is adopted with the *index i not summed* throughout this section. Here F_i, G_i, and H_i are functions of u_i, which are at least twice continuously differentiable; k is a constant; and R_i is a continuously differentiable function of t, $x_1, \ldots, x_n$. With certain restrictions on F_i, G_i, H_i, and R_i, Eq. (1.52) can be transformed into a diffusion equation in n dimensions.

Motivated by the technique applied to the Burgers' equation, we consider the transformation

$$F_i(u_i) = -\frac{2k}{\phi}\frac{\partial\phi}{\partial x_i}, \tag{1.53}$$

corresponding to which we have

$$F_i' \frac{\partial u_i}{\partial t} = \frac{2k}{\phi^2}\frac{\partial\phi}{\partial t}\frac{\partial\phi}{\partial x_i} - \frac{2k}{\phi}\frac{\partial^2\phi}{\partial t\,\partial x_i}$$

$$F_i' \frac{\partial u_i}{\partial x_j} = \frac{2k}{\phi^2}\frac{\partial\phi}{\partial x_j}\frac{\partial\phi}{\partial x_i} - \frac{2k}{\phi}\frac{\partial^2\phi}{\partial x_j\,\partial x_i}$$

and

$$\begin{aligned} F_i' \frac{\partial^2 u_i}{\partial x_j\,\partial x_j} = & -\frac{4k}{\phi^3}\frac{\partial\phi}{\partial x_j}\frac{\partial\phi}{\partial x_j}\frac{\partial\phi}{\partial x_i} + \frac{2k}{\phi^2}\frac{\partial^2\phi}{\partial x_j\,\partial x_j}\frac{\partial\phi}{\partial x_i} \\ & + \frac{4k}{\phi^2}\frac{\partial\phi}{\partial x_j}\frac{\partial^2\phi}{\partial x_j\,\partial x_i} - \frac{2k}{\phi}\frac{\partial^3\phi}{\partial x_j\,\partial x_j\,\partial x_i} \\ & - \frac{4k^2}{\phi^2}\frac{F_i''}{(F_i')^2}\left[\frac{1}{\phi}\frac{\partial\phi}{\partial x_j}\frac{\partial\phi}{\partial x_i} - \frac{\partial^2\phi}{\partial x_j\,\partial x_i}\right] \\ & \times \left[\frac{1}{\phi}\frac{\partial\phi}{\partial x_j}\frac{\partial\phi}{\partial x_i} - \frac{\partial^2\phi}{\partial x_j\,\partial x_i}\right]. \end{aligned}$$

Setting these into Eq. (1.52), there results

$$\begin{aligned} & \frac{2k}{\phi^2}\left[\frac{\partial\phi}{\partial x_i} - \phi\frac{\partial}{\partial x_i}\right]\left[\frac{\partial\phi}{\partial t} - k\frac{\partial^2\phi}{\partial x_j\,\partial x_j}\right] \\ & = F_i' H_i R_i + \frac{4k^2}{\phi^2 F_i'^2}[G_i F_i' - kF_i''] \\ & \times \left[\frac{1}{\phi}\frac{\partial\phi}{\partial x_j}\frac{\partial\phi}{\partial x_i} - \frac{\partial^2\phi}{\partial x_j\,\partial x_i}\right]\left[\frac{1}{\phi}\frac{\partial\phi}{\partial x_j}\frac{\partial\phi}{\partial x_i} - \frac{\partial^2\phi}{\partial x_j\,\partial x_i}\right]. \end{aligned} \tag{1.54}$$

The left-hand side of Eq. (1.54) is seen to be

$$-\frac{\partial}{\partial x_i}\left[\frac{2k}{\phi}\left(\frac{\partial\phi}{\partial t} - k\frac{\partial^2\phi}{\partial x_j\,\partial x_j}\right)\right].$$

Consequently, if we set

$$G_i F_i' - k F_i'' = 0, \qquad F_i' H_i = 1, \qquad R_i = -\partial P/\partial x_i\,, \tag{1.55}$$

it then follows that Eq. (1.54) becomes

$$\frac{\partial}{\partial x_i}\left[\frac{2k}{\phi}\left(\frac{\partial \phi}{\partial t} - k\frac{\partial^2 \phi}{\partial x_j\,\partial x_j}\right) - P\right] = 0. \tag{1.56}$$

This integrates to

$$\frac{\partial \phi}{\partial t} = k\frac{\partial^2 \phi}{\partial x_j\,\partial x_j} + \left[c(t) + \frac{P}{2k}\right]\phi, \tag{1.57}$$

which is a linear diffusion equation. Consequently, solutions of a system of n quasi-linear equations (1.52) can be obtained from the solutions of a linear equation (1.57) whenever the "reducibility conditions," Eq. (1.55), are satisfied. A necessary and sufficient condition for the first two of those is that F_i, G_i, and H_i are obtained from a generating function $f_i(u_i)$ by means of

$$\begin{aligned} F_i &= \int^{u_i} f_i(r)\,dr, \\ G_i &= k\,d(\ln f_i)/du_i\,, \\ H_i &= f_i^{-1}, \end{aligned} \tag{1.58}$$

where i is not summed. For the last condition of Eq. (1.55), a necessary and sufficient condition is that the Stokes tensor S, for R_i, vanishes identically; that is,

$$S_{ij} = \frac{\partial R_j}{\partial x_i} - \frac{\partial R_i}{\partial x_j} = 0. \tag{1.59}$$

Of course Eq. (1.57) will not yield all solutions of Eq. (1.52), because of the restrictions imposed by the transformation, Eq. (1.53). In actual applications, difficulties may occur in transforming the boundary and initial conditions.

As an example let us consider the Navier–Stokes equations for incompressible fluid flow

$$\frac{\partial u_i}{\partial t} + u_j\frac{\partial u_i}{\partial x_j} = -\frac{1}{\rho}\frac{\partial p}{\partial x_i} + \nu\frac{\partial^2 u_i}{\partial x_i\,\partial x_j}\,, \qquad i, j = 1, 2, 3 \tag{1.60}$$

(sum on j), where u_i is the velocity component in the x_i direction, p

is pressure, ρ is (constant) density, and ν is (constant) kinematic viscosity. In the notation of the general theory

$$F_i(u_i) = u_i = -\frac{2\nu}{\phi}\frac{\partial \phi}{\partial x_i}, \tag{1.61}$$

and the reducibility conditions are easily shown to hold. Thus through transformation Eq. (1.61) the Navier–Stokes equations reduce to a linear diffusion equation

$$\frac{\partial \phi}{\partial t} = \nu \frac{\partial^2 \phi}{\partial x_j \, \partial x_j} + \frac{p(t, x_1, x_2, x_3)}{2\rho\nu}\phi. \tag{1.62}$$

This result, in the case of zero pressure gradient, was also obtained by Cole [16]. One way to view Eq. (1.62) is as the equation for a viscous flow in a pure initial-value problem with a prescribed pressure p. The velocity field so obtained requires a corresponding source distribution given by

$$Q(t, x_1, x_2, x_3) = -2\nu(\partial^2 \ln \phi/\partial x_j \, \partial x_j) \tag{1.63}$$

to satisfy the continuity equation (conservation of mass).

Perhaps the more physical case wherein the source distribution is prescribed should be examined. If for instance $Q \equiv 0$, Eq. (1.62), in combination with the continuity equation, transforms into a Bernoulli equation. Conversely, the nonlinear Bernoulli equation

$$\frac{\partial \theta}{\partial t} + \frac{1}{2}\frac{\partial \theta}{\partial x_j}\frac{\partial \theta}{\partial x_j} + \frac{p}{\rho} = 0 \tag{1.64}$$

is converted into a linear diffusion equation by means of the equation of continuity and the change of variable $\theta = \ln \phi$.

Table 1-1 lists a few reducible equations in one dimension. In the

TABLE 1-1

Generating function f	Equation
0	$u_t = ku_{xx}$
1	$u_t + uu_x = ku_{xx} + R(t, x)$
nu^{n-1}	$u_t + u^n u_x = \frac{kn(n-1)}{u}(u_x)^2 + ku_{xx} + \frac{R(t, x)}{nu^{n-1}}$
e^u	$u_t + e^u u_x = k(u_x)^2 + ku_{xx} + e^{-u}R(t, x)$
$\cos u$	$u_t + (\sin u)\, u_x = -k \tan u(u_x)^2 + ku_{xx} + \sec uR(t, x)$

one-dimensional case, the transformation specified by Eq. (1.53) imposes no restrictions on the solution u (other than that of existence).

d. The Inverse Transformation

In his study of nonlinear ordinary and partial differential equations associated with Appell functions, Vein [17] developed an inverse technique for finding parabolic equations transformable into the linear diffusion equation. Let $z(x, y)$ be any solution of

$$z_y = z_{xx} \tag{1.65}$$

and set

$$\theta = z_x , \qquad \psi = z_y , \tag{1.66}$$

whereupon it follows from Eq. (1.65) that

$$\psi = \theta_x . \tag{1.67}$$

If $\phi(x, y)$ is the *inverse function* of $z(x, y)$ with respect to x, that is

$$x = z(\phi, Y) \qquad (Y = y), \tag{1.68}$$

then, from Eq. (1.66), we have

$$\begin{aligned} z_\phi(\phi, Y) &= \theta(\phi, Y) = u, \\ z_Y(\phi, Y) &= \psi(\phi, Y) = v, \\ \theta_\phi(\phi, Y) &= \psi(\phi, Y) = v. \end{aligned} \tag{1.69}$$

The quantities u and v are defined by Eqs. (1.69).

From Eq. (1.68) we find, by differentiating with respect to x and y, that

$$1 = u\phi_x , \qquad 0 = u\phi_y + v, \tag{1.70}$$

and from Eqs. (1.69)

$$u_x = v\phi_x . \tag{1.71}$$

As a consequence of Eqs. (1.70), it is obvious that $u = \phi_x^{-1}$ and $v = -\phi_y/\phi_x$. When these are substituted into Eq. (1.71), we find that ϕ satisfies

$$\phi_{xx} = \phi_x{}^2\phi_y . \tag{1.72}$$

Thus a solution of Eq. (1.72) is the inverse function with respect to

x of any solution of the diffusion equation. To illustrate these solutions, consider the function

$$z = y^{-1/2} \exp[-x^2/4y],$$

which satisfies Eq. (1.65). The inverse function ϕ, with respect to x, is obtained by replacing z by x and x by ϕ,

$$x = y^{-1/2} \exp[-\phi^2/4y],$$

that is,

$$\phi^2 = -4y \log(xy^{1/2}). \tag{1.73}$$

It follows from an elementary demonstration, that Eq. (1.73) satisfies Eq. (1.72).

The versatility of the inverse transformation is further demonstrated by noting that Eqs. (1.70) and (1.71) imply $v = uu_x$. Upon eliminating v there results $\phi_y = -u_x$, which becomes

$$u^2 u_{xx} = u_y \tag{1.74}$$

when ϕ is eliminated, using $1 = u\phi_x$. Thus the solution of Eq. (1.74) is

$$u = \theta(\phi, y) = [z_x(x, y)]_{x=\phi} = z_\phi(\phi, y), \tag{1.75}$$

where z is any solution of the diffusion equation.

Similar operations verify the following:

(i) A solution of $vv_{xxx} = v_{xy}$ is $z_\phi^2(\phi, y)$.

(ii) A solution of $x^2 v_{xx} = v_x^2 v_y$ is the inverse function with respect to x of $z_\phi(\phi, y)$.

1.3 EMPLOYING THE WRONG EQUATION TO FIND THE RIGHT SOLUTION

Very often a transformation cannot be found to linearize a nonlinear equation. However, even in such cases, one can sometimes find another nonlinear equation possessing an extra term or terms not appearing in the equation of interest but which can be transformed to a linear one. Now the new equation may have the feature that, for certain initial conditions, the extra term is always very small. Then the solution of the new equation, under those initial conditions, may provide a good approximation to the original problem. Montroll [15] has obtained approximate solutions for the Fisher equation utilizing this concept.

For ease in reference, the pertinent equations are reproduced here. The Fisher equation (1.30) is

$$u_t = Du_{xx} + ku(\theta - u)/\theta,$$

while the equation (1.50)

$$u_t = D\{u_{xx} + 2u_x^2/(\theta - u)\} + ku(\theta - u)/\theta$$

is linearizable, and u is related to ψ, which satisfies the diffusion equation, through Eq. (1.51):

$$u(x, t) = \theta\psi/[e^{-kt} + \psi].$$

Suppose that we consider the pure initial-value problem in which $u(x, 0) \ll \theta$ for all x. Then as $\theta \to \infty$, Eqs. (1.30) and (1.50) have the same form, $u_t = Du_{xx} + ku$, corresponding to unlimited growth. On the other hand as $t \to \infty$, we see from Eq. (1.51) that $u(x, t) \to \theta$, that is, saturation at all points, which would be the case with Eq. (1.30). Thus at both early and late times, Eqs. (1.50) and (1.30) are equivalent under the above initial condition. We now solve Eq. (1.50) subject to several initial conditions and use these results to show that, for certain initial conditions, the solution of Eq. (1.50) is essentially the same as that for Eq. (1.30).

If $u(x, 0) = \theta/(1 + e^{\beta x})$[†] then $\psi(x, 0) = e^{-\beta x}$, whereupon the linear diffusion equation has the solution

$$\psi(x, t) = \exp[-\beta(x - \beta Dt)] \tag{1.76}$$

and, from Eq. (1.51),

$$u(x, t) = \theta/\{1 + \exp \beta[x - (\beta D + k/\beta)t]\}. \tag{1.77}$$

The rate at which the (mutant) population front propagates[‡] is

$$dx/dt = v = \beta D + (k/\beta);$$

saturation exists behind the front, and the shape of the diffusion front remains invariant as the wave propagates. The propagation rate is determined by the initial slope $(-\theta\beta/2)$ at $x = 0$. If the initial front is steep (β large), the propagation is dominated by diffusion. If it is broad (β

† For the physical interpretation of these problems we refer to the original papers of Montroll [6, 15].

‡ The solution is in the form of a traveling wave $h(x - vt)$.

small), it is dominated by saturation development. When both processes contribute equally, that is $\beta = (k/D)^{1/2}$, the *minimum propagation velocity* is $v_{\min} = 2(kD)^{1/2}$. Thus, no solution of the form $h(x - vt)$ can be found for Eq. (1.50)—for which $|v| < 2(kD)^{1/2}$. A similar result was noted by Fisher [9] for Eq. (1.30).

As a second example, suppose the initial distribution extends over a length a, peaking at the origin at a fraction η ($\eta < 1$) of the saturation value θ. Then

$$u(x, 0) = \eta\theta/[\eta + (1 - \eta) \exp(x^2/2a^2)] \tag{1.78}$$

and, from Eq. (1.51), it follows that

$$\psi(x, 0) = [\eta/(1 - \eta)] \exp[-x^2/2a^2]. \tag{1.79}$$

The solution of the diffusion equation, subject to the initial condition given by Eq. (1.79), is

$$\psi(x, t) = \frac{\eta a \exp[-x^2/2(a^2 + 2Dt)]}{(1 - \eta)(a^2 + 2Dt)^{1/2}} .$$

Consequently, from Eq. (1.51),

$$u(x, t) = \theta\eta a\{\eta a + (1 - \eta)(a^2 + 2Dt)^{1/2} \exp[-kt + x^2/2(a^2 + 2Dt)]\}^{-1}. \tag{1.80}$$

When $| x | \ll [2kt(a^2 + 2Dt)]^{1/2}$, the exponential term in Eq. (1.80) can be neglected and saturation is achieved. The propagating (population) front at any time can be identified with the value of x which makes the argument of the exponential vanish, that is,

$$x = \pm\{2kt(a^2 + 2Dt)\}^{1/2}. \tag{1.81}$$

Thus as t increases, the velocity of propagation $v = dx/dt$, approaches $2(kD)^{1/2}$, which is independent of a. Thus, in this (unsaturated) case, the propagation velocity becomes independent of the initial distribution. In fact, $2(kD)^{1/2}$ is just the minimum propagation velocity that can be achieved by an initial distribution which has a saturated region (our first example). In our first example, which contains a saturation region, the propagation velocity v depends upon the initial wave shape!

Apparently no one has succeeded in linearizing Fisher's equation (1.30), but we shall show that the solution of the more complicated Eq. (1.50) acts like that of Eq. (1.30) when the initial condition is such that no region is near saturation θ. The analysis demonstrating this is carried out through a detailed analysis of Eq. (1.80) which resulted from employment of the near-Gaussian initial condition, Eq. (1.79).

First, at early times in the growth process—before saturation occurs — the linearized versions of both equations are valid. Since these are identical, so, therefore, are their solutions.

Second, we consider the later stages of the process when $2Dt \gg a^2$ by substituting the solution, given by Eq. (1.80), into Eq. (1.50). The contribution to $\partial u/\partial t$ of each of the three terms on the right-hand side of Eq. (1.50) at all times is

$$\text{(i)}\quad u_{xx} = -F\{\eta a\{1 + [x^2/(a^2 + 2Dt)]\} + (1-\eta)(a^2 + 2Dt)^{1/2}\{1 - [x^2/(a^2 + 2Dt)]\}E\}, \tag{1.82}$$

$$\text{(ii)}\quad 2u_x^2/(\theta - u) = F\{2a\eta x^2/(a^2 + 2Dt)\}, \tag{1.83}$$

$$\text{(iii)}\quad ku(\theta - u)/\theta = F\{(a^2 + 2Dt)[\eta a + (1-\eta)(a^2 + 2Dt)^{1/2}E]k/D\}, \tag{1.84}$$

where E and F are common factors defined by

$$E = \exp\{-kt + x^2/2(a^2 + 2Dt)\} \tag{1.85}$$

and

$$F = [a\theta\eta(1-\eta)E/(a^2 + 2Dt)^{1/2}]\{\eta a + (1-\eta)(a^2 + 2Dt)^{1/2}E\}^{-3}. \tag{1.86}$$

Each of the three terms will be examined in the three regimes $x^2 - 4Dk^2t \gg 4Dt$; $|x^2 - 4Dk^2t| \ll 4Dt$; $x^2 \ll 4Dkt^2$ as suggested from Eq. (1.81). Since we assumed $2Dt \gg a^2$, the a^2 can be neglected compared with $2Dt$ in all the above equations. Thus, in the first regime, with $x^2 - 4Dk^2t \gg 4Dt$,

$$E \sim \exp\{(x^2 - 4Dkt^2)/4Dt\} \tag{1.87}$$

is very large, which implies that all the nonexponential terms in Eqs. (1.82)–(1.84) can be neglected. Our "extra" diffusion term does not contain E other than in the term F, which is common to all expressions. Thus (ii) is negligible when compared with the regular diffusion term (i). The exponential term in (i) has a coefficient $x^2/(2Dt)^{1/2}$ to be compared with $(2Dt)^{3/2}k/D$ of (iii). Thus in our first regime the regular diffusion term is the most important contributor to $\partial u/\partial t$. Actually $\partial u/\partial t$ is very small in this regime, since F is proportional to $E^{-2}/(2Dt)^2$ there.

Toward the center of the wave, $|x^2 - 4Dkt^2| \ll 4Dt$, whereupon $E \sim 1$, so that

$$u_{xx} \sim F(1-\eta)(2t)^{3/2}kD^{1/2},$$

$$2u_x^2/(\theta - u) \sim 4\eta aktF,$$

$$ku(\theta - u)/\theta \sim (1-\eta)Fk(2t)^{3/2}D^{1/2}.$$

Consequently in the second regime, where the action is, (i) and (iii) contribute the same amount to $\partial u/\partial t$, and both exceed our extra term by a factor $t^{1/2}$. Therefore the additional diffusion term becomes less and less significant as time increases.

Lastly, in the third regime, that of the saturation region behind the wave, $x^2 \ll 4Dkt^2$. The exponential term E is negligible, and the main contribution comes from the growth term (iii) which is of order $2F\eta akt$, compared with $F\eta a$, the main term of (i), and $Fa\eta x^2/Dt$ in the extra diffusion term. However, because of the form of F, all terms in this regime are small.

In each of the three regimes, the additional diffusion term $2u_x{}^2/(\theta - u)$ contributes insignificantly to u_t compared with the other two terms of diffusion and generation. Thus it appears that so long as the region is not initially saturated, the generation and diffusion of an initial disturbance described by Montroll's equation (1.50) should be essentially the same as that obtained by solving Fisher's equation (1.30).

1.4 APPLICATION OF THE QUASI-LINEAR THEORY

The general form of the theory of characteristics for simultaneous quasi-linear first-order partial differential equations, herein called the quasi-linear theory, is given in Volume I, pages 72–84. Here we shall present an application of that theory to the problem of longitudinal wave propagation on a traveling threadline. The mathematical model for this problem, specified by Ames and Vicario [7], is discussed in Section 1.2. The dimensional equations are Eqs. (1.18), (1.19), and the constitutive relation $m(T + EA_0) = EA_0 m_0$. These can be rendered dimensionless and constant free by setting

$$\begin{aligned} T' &= T/EA_0, \qquad & m' &= m/m_0, \qquad & x' &= x/L, \\ V' &= V/c_0, \qquad & t' &= c_0 t/L, \qquad & c_0{}^2 &= EA_0/m_0. \end{aligned} \tag{1.88}$$

Rewriting and discarding the primes, the dimensionless form in m and V becomes

$$V_t + VV_x = -m^{-3} m_x, \qquad m_t + (mV)_x = 0, \qquad m(T + 1) = 1. \tag{1.89}$$

Upon applying the quasi-linear theory to Eqs. (1.89), we find the characteristics to be

$$(dx/dt)\,|_{1,2} = V \pm m^{-1},$$

while the canonical characteristic equations are

$$m^2 V_\alpha + m_\alpha = 0,$$

$$-m^2 V_\beta + m_\beta = 0,$$

$$x_\alpha = (V + m^{-1}) t_\alpha\,, \tag{1.90}$$

$$x_\beta = (V - m^{-1}) t_\beta\,. \tag{1.91}$$

The Riemann invariants r and s are

$$-r = \tfrac{1}{2}(V - m^{-1}), \qquad \text{invariant along } \beta \text{ characteristics,}$$

$$s = \tfrac{1}{2}(V + m^{-1}), \qquad \text{invariant along } \alpha \text{ characteristics.}$$

In terms of the invariants, Eqs. (1.90) and (1.91) become

$$x_s = 2st_s\,, \qquad x_r = -2rt_r\,,$$

and when x is eliminated there remains $(s + r)t_{sr} = 0$. Since $s + r = m^{-1} \neq 0$, $t_{sr} = 0$, whereupon

$$t(r, s) = f(r) + g(s)$$

or

$$t(m, V) = f[\tfrac{1}{2}(m^{-1} - V)] + g[\tfrac{1}{2}(m^{-1} + V)], \tag{1.92}$$

where f and g are arbitrary. Similarly, $x_{rs} = 0$, whose general solution is

$$x(r, s) = F(r) + G(s)$$

or

$$x(m, V) = F[\tfrac{1}{2}(m^{-1} - V)] + G[\tfrac{1}{2}(m^{-1} + V)].$$

The functions f, g, F, and G are determined by employing auxiliary conditions. To illustrate the approach, consider the pure initial-value problem, which specifies

$$m(x, 0) = 1, \qquad V(x, 0) = V_0(x)$$

as initial conditions in the physical plane. In the transformed plane, these conditions become

$$t(V, 1) = 0, \qquad x(V, 1) = x_0(V),$$

where x_0 and V_0 are inverse functions. When these data are applied to Eq. (1.92), we have

$$0 = f[\tfrac{1}{2}(1 - V)] + g[\tfrac{1}{2}(1 + V)],$$

and upon setting $\eta = \frac{1}{2}(1 + V)$, it follows that

$$g(\eta) = -f(1 - \eta).$$

Thus the solution for t becomes

$$t(V, m) = f[\tfrac{1}{2}(m^{-1} - V)] - f[1 - \tfrac{1}{2}(m^{-1} + V)]. \tag{1.93}$$

To complete the evaluation of t, we utilize the reducible nature of the first two of Eqs. (1.89).

Since Eqs. (1.89) are reducible, the dependent and independent variables can be interchanged, thereby generating the pair

$$x_m - Vt_m + m^{-3}t_V = 0, \qquad x_V + mt_m - Vt_V = 0.$$

From the first of these, we find

$$(\partial t/\partial V)\Big|_{\substack{m=1\\ V=V_0}} = [V(\partial t/\partial m) - (\partial x/\partial m)]\Big|_{\substack{m=1\\ V=V_0}}, \tag{1.94}$$

and from the second

$$(\partial t/\partial V)\Big|_{\substack{m=1\\ V=V_0}} = (\partial x/\partial V + \partial t/\partial m)/V\Big|_{\substack{m=1\\ V=V_0}}. \tag{1.95}$$

Eliminating $\partial t/\partial V$ between Eqs. (1.94) and (1.95), we find

$$(\partial t/\partial m)\Big|_{\substack{m=1\\ V=V_0}} = [\partial x/\partial V + V(\partial x/\partial m)]/(V^2 - 1)\Big|_{\substack{m=1\\ V=V_0}}.$$

Under the assumption that

$$(\partial x/\partial m)\Big|_{\substack{m=1\\ V=V_0}} = (\partial/\partial m)[x(V, 1)] = 0,$$

we finally obtain

$$(\partial t/\partial m)\Big|_{\substack{m=1\\ V=V_0}} = \frac{(dx/dV)\Big|_{\substack{m=1\\ V=V_0}}}{(V_0^2 - 1)}. \tag{1.96}$$

Alternatively, if Eq. (1.93) is differentiated with respect to m, and evaluated at $m = 1$, $V = V_0$, we get

$$(\partial t/\partial m)\Big|_{\substack{m=1\\ V=V_0}} = -f'[\tfrac{1}{2}(1 - V_0)]. \tag{1.97}$$

Upon equating Eqs. (1.96) and (1.97), setting $\xi = \frac{1}{2}(1 - V_0)$, the form of f is determined by solving the simple equation

$$f'(\xi) = \frac{(dx/d\xi)_{V=V_0}}{8\xi(\xi - 1)}.$$

From Eq. (1.93) we may now determine $t(V, m)$. Employing similar arguments, it follows that

$$x(V, m) = F[\tfrac{1}{2}(m^{-1} - V)] - F[1 - \tfrac{1}{2}(m^{-1} + V)] - x_0(V),$$

where

$$F'[\tfrac{1}{2}(m^{-1} - V)] = (V - m^{-1})f'[\tfrac{1}{2}(m^{-1} - V)].$$

In his study of wake collapse in density-stratified fluids, Mei [18] has employed a similar analysis to investigate the early stages of the collapse process.

1.5 EARNSHAW'S PROCEDURE

A simple, but nevertheless sometimes fruitful, *ad-hoc* procedure was introduced by Earnshaw [19] in his study of sound waves. To introduce this concept and also to provide the nonlinear equations for a subsequent discussion, we herein sketch the development of the equations for the propagation of finite disturbances in bars of rubberlike materials (Nowinski [20], Ames [21]).

Let a perfectly elastic incompressible straight bar or wire of uniform, finite-area cross section have negligible transverse dimensions. Furthermore, assume the following:

(i) The bar is infinitely long, so that no reflections of waves occur and other possible wave interferences are discarded.
(ii) Transverse inertia during the bar motion is neglected.
(iii) In compression and tension zones, the bar does not experience instability.
(iv) The bar is subjected to simple unidirectional strain, in the sense that the only identically nonvanishing stress component is the longitudinal, normal stress component, which is uniformly distributed over the cross section.
(v) The effect of strain rate on the constitutive equations is neglected, and the static stress–strain relations are extended to the dynamic case.

Adopting the Lagrangian formulation, let both the material coordinate X and spatial coordinate x be referred to the same fixed Cartesian system, one of whose axes coincides with the axis of the bar as in Fig. 1-1. Let

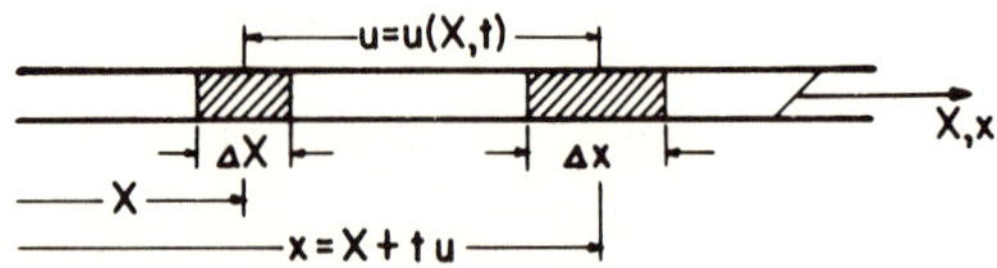

FIG. 1-1. Coordinate systems for nonlinear wave propagation.

ρ_0 and ρ be the mass densities in the stress free configuration (associated with the X coordinate) and deformed configuration [associated with $x = x(X, t)$]. If t is time, σ_0 the normal stress referred to the undeformed cross section of the rod, and u the particle displacement, then Cauchy's law of motion becomes (James and Guth [22])

$$\partial\sigma_0/\partial X = \rho_0(\partial^2 u/\partial t^2). \tag{1.98}$$

Since $x = X + u$, the stretch (extension ratio) $\lambda = \partial x/\partial X$ can be written as

$$\lambda = 1 + (\partial u/\partial X). \tag{1.99}$$

Consequently, Eq. (1.98) becomes

$$\partial^2 x/\partial t^2 = C^2(\partial^2 x/\partial X^2) \tag{1.100}$$

or, in terms of the stretch,

$$(\partial^2\lambda/\partial t^2) = (\partial/\partial X)\{C^2(\partial\lambda/\partial X)\}, \tag{1.101}$$

where

$$C^2 = (1/\rho_0)(d\sigma_0/d\lambda), \qquad \sigma_0 = \sigma_0(\lambda). \tag{1.102}$$

We shall assume throughout that $d\sigma_0/d\lambda > 0$.

It is instructive to note that Eq. (1.100) can easily be split into a system of two first-order hyperbolic equations. For this purpose, we denote by $V = \partial x/\partial t$ the (absolute) particle velocity, whereupon Eq. (1.100) transforms into

$$(\partial V/\partial t) - C^2(\partial\lambda/\partial X) = 0, \qquad (\partial V/\partial X) - (\partial\lambda/\partial t) = 0, \tag{1.103}$$

where the last equation represents the identity of the two cross partial derivatives (integrability condition). Upon applying the theory of characteristics (see Volume I) to Eqs. (1.103), we readily find the equations of the characteristics to be $dX/dt = \pm C(\lambda)$, and the equations along the characteristics are $dV/d\lambda = \pm C(\lambda)$.

Similar reasoning may be applied to establish the Eulerian form of the equations (see, e.g., Taylor [23]). In our notation, they take the form

$$\begin{aligned} VV_x + V_t - \lambda C^2 \lambda_x &= 0, \\ -\lambda V_x + V\lambda_x + \lambda_t &= 0. \end{aligned} \tag{1.104}$$

Under the assumption that the strain energy exists, the theory of finite elastic deformations (Truesdell [24, Eq. (42.11)]) furnishes the stress–stretch relation

$$\sigma_0 = 2\left[\frac{\partial W}{\partial \mathrm{I}} + \frac{1}{\lambda}\frac{\partial W}{\partial \mathrm{II}}\right]\left(\lambda - \frac{1}{\lambda^2}\right) \tag{1.105}$$

for an incompressible body in simple extension. Here, W is the elastic strain-energy function, and for an incompressible material, the strain invariants are

$$\mathrm{I} = 2\lambda^{-1} + \lambda^2, \qquad \mathrm{II} = 2\lambda + \lambda^{-2}, \qquad \mathrm{III} = 1. \tag{1.106}$$

From the experimental results of Rivlin and Saunders (see Truesdell [24, p. 214]), data are well approximated by

$$W = \alpha(\mathrm{I} - 3) + f(\mathrm{II} - 3), \tag{1.107}$$

where α is a constant and f is an arbitrary function to be obtained. The expanded form

$$W = \alpha(\mathrm{I} - 3) + \sum_{k=1}^{n} \beta_k(\mathrm{II} - 3)^k,$$

with β_k constants, has been employed. Retention of only the linear term leads to

$$W = \alpha(\mathrm{I} - 3) + \beta(\mathrm{II} - 3), \qquad \alpha, \beta > 0,$$

corresponding to the so called Mooney–Rivlin material. If $\beta = 0$, we obtain Rivlin's neo-Hookean material.

If W takes the general form Eq. (1.107), then Eq. (1.101) becomes

$$\frac{\partial^2 \lambda}{\partial t^2} = \frac{\partial}{\partial X}\left\{\frac{2}{\rho_0}\left[\alpha(1 + 2\lambda^{-3}) + 3\lambda^{-4}\frac{\partial f}{\partial \mathrm{II}} + 2(1 - \lambda^{-3})^2\frac{\partial^2 f}{\partial \mathrm{II}^2}\right]\frac{\partial \lambda}{\partial X}\right\}. \tag{1.108}$$

Alternately, we may write

$$\frac{\partial^2 x}{\partial t^2} = \frac{2}{\rho_0}\left\{\alpha\left[1 + 2\left(\frac{\partial x}{\partial X}\right)^{-3}\right] + 3\left(\frac{\partial x}{\partial X}\right)^{-4}\frac{\partial f}{\partial \mathrm{II}} + 2\left[1 - \left(\frac{\partial x}{\partial X}\right)^{-3}\right]^2 \frac{\partial^2 f}{\partial \mathrm{II}^2}\right\}\frac{\partial^2 x}{\partial X^2}. \tag{1.109}$$

For the Mooney–Rivlin material, Eq. (1.109) becomes

$$\frac{\partial^2 x}{\partial t^2} = \frac{2}{\rho_0}\left\{\alpha\left[1 + 2\left(\frac{\partial x}{\partial X}\right)^{-3}\right] + 3\beta\left(\frac{\partial x}{\partial X}\right)^{-4}\right\}\frac{\partial^2 x}{\partial X^2}$$

and for the neo-Hookean material, Eq. (1.109) becomes

$$\frac{E}{3\rho_0}\left\{1 + 2\left(\frac{\partial x}{\partial X}\right)^{-3}\right\}\frac{\partial^2 x}{\partial X^2} = \frac{\partial^2 x}{\partial t^2}. \tag{1.110}$$

In Eq. (1.110), 2α has been replaced by $E/3$, a value that is suggested by the desirability of obtaining the familiar infinitesimal strain relation $\sigma_0 = E\epsilon$ from Eq. (1.105).

From his own theory of finite elasticity, Seth [25] obtains the corresponding equation

$$\frac{E}{\rho_0}\left\{1 + \frac{\partial x}{\partial X}\right\}^{-3}\frac{\partial^2 x}{\partial X^2} = \frac{\partial^2 x}{\partial t^2},$$

which differs fundamentally from Eq. (1.110).

As suggested by Earnshaw, we now assume that the absolute particle velocity $V = \partial x/\partial t$ is a yet to be determined function of the stretch λ, that is,

$$V = F(\lambda). \tag{1.111}$$

Upon setting this into Eq. (1.103), there results

$$F'(\lambda)\lambda_t - C^2\lambda_X = 0, \qquad \lambda_t - F'(\lambda)\lambda_X = 0. \tag{1.112}$$

These are identical if

$$\frac{dF}{d\lambda} = \pm C = \pm\left(\frac{1}{\rho_0}\frac{d\sigma_0}{d\lambda}\right)^{1/2}. \tag{1.113}$$

If we suppose that the particle velocity is zero for the undisturbed medium, for which $\lambda = 1$, then upon integration we find

$$V = \pm\int_1^\lambda \rho_0^{-1}\left(\frac{d\sigma_0}{d\lambda}\right)^{1/2} d\lambda. \tag{1.114}$$

In addition to Nowinski, this method has been employed by Ames *et al.* [26] in their study of wave propagation on a traveling threadline.

1.6 TRAVELING-WAVE SOLUTIONS

A wide variety of physical problems (see Volume I) are governed by the "power law" diffusion equation

$$u_t = (u^n)_{xx} \,. \tag{1.115}$$

If λ is a constant, what form must f have so that $f(x - \lambda t)$ is a solution of Eq. (1.115)? Upon substitution we find that f must satisfy the differential equation

$$(f^n)'' + \lambda f' = 0, \tag{1.116}$$

where the prime indicates differentiation with respect to $\eta = x - \lambda t$. The first integration is immediate, so that $(f^n)' + \lambda f = A$. The final integration is easily accomplished if n is a positive integer, yielding the implicit solution

$$\sum_{j=0}^{n-2} \frac{(A/\lambda)^j u^{n-1-j}}{n-1-j} + \left(\frac{A}{\lambda}\right)^{n-1} \ln\left(u - \frac{A}{\lambda}\right) = \frac{\lambda}{n}(\lambda t - x + B), \tag{1.117}$$

where $A \neq 0$ and B are constants. If $A = 0$, the integration generates the explicit form

$$u(x, t) = \left[\frac{\lambda(n-1)}{n}(\lambda t - x + B)\right]^{1/(n-1)} \qquad \text{for} \quad n \neq 1. \tag{1.118}$$

If $n = 1$, then

$$u(x, t) = B \exp[-\lambda(x - \lambda t)].$$

Thus it is possible for a diffusion equation to have a traveling-wave solution propagating undisturbed through the medium.

A similar question can and has been asked in wave mechanics. This (traveling-wave) class of special solutions represents waves of permanent profile that propagate with constant velocity and unchanging shape. For steady propagating waves, the dynamic variables will be functions only of $\eta = x - \lambda t$, where λ is an *assumed* velocity of propagation. Clearly, the partial derivatives will be related by

$$\frac{\partial}{\partial x} = \frac{d}{d\eta} = -\frac{1}{\lambda}\frac{\partial}{\partial t}\,. \tag{1.119}$$

In the event that the original partial-differential equations, in x and t, can be reduced to ordinary differential equations in η, then the solutions to these, if they exist, will be possible modes of steady propagation. These solutions depend parametrically upon the assumed velocity λ. The relation $\eta = x - \lambda t$ transforms our spatial coordinate system to one which is moving with respect to the wave medium. The solutions sought by this *dynamic steady-state* method are stationary in the moving coordinate system.

In transmission lines (see Scott [27]) it sometimes happens that steady propagation does occur on dispersive,† lossless, nonlinear lines. This is in contrast to the known results for lossless linear lines. A linear non-dispersive line can support steady propagation with arbitrary wave shape, but a linear dispersive line cannot!

A mechanical transmission line treated by Scott [28] is modeled by the equation

$$\phi_{xx} - \phi_{tt} = G(\phi). \tag{1.120}$$

For his special case of the pendula rotation, $G(\phi) = \sin \phi$. If we ask for solutions of the form $\phi = \phi(\eta)$, $\eta = x - \lambda t$, Eq. (1.120) becomes

$$d^2\phi/d\eta^2 = G(\phi)/(1 - \lambda^2). \tag{1.121}$$

In the special case, $G(\phi) = \sin \phi$, two distinct pulse solutions are obtained, namely,

$$\phi = 4 \tan^{-1}\{\exp \pm [(x - \lambda t)/(1 - \lambda^2)^{1/2}]\}, \qquad \lambda < 1,$$

and

$$\phi = 4 \tan^{-1}\{\exp \pm [(x - \lambda t)/(\lambda^2 - 1)^{1/2}]\} + \pi, \qquad \lambda > 1.$$

In the first case, the constant of integration is set equal to $+1$ and in the second -1.

If the integration constant (say, E) is not equal to ± 1, ϕ can still be written as an implicit function of η, given by

$$(1 - \lambda^2)^{1/2} \int_0^\phi \frac{d\phi}{[2(E - \cos \phi)]^{1/2}} = \eta. \tag{1.122}$$

† Dispersion essentially implies that different harmonic components of the waveform travel at different velocities. Consequently, dispersive effects are expected to greatly influence shock formation.

In the case that $E > 1$, $\lambda < 1$, ϕ is a *monotonically increasing function* of η

$$\phi = \cos^{-1}\left\{2\,\mathrm{cd}^2\left(\frac{\eta}{\gamma(1-\lambda^2)^{1/2}}\right) - 1\right\},$$

where $\mathrm{cd}\,x = \mathrm{cn}\,x/\mathrm{dn}\,x$ is an elliptic function (see Abramowitz and Stegun [29]) of modulus $\gamma = 2/(E+1)$. The second case of interest is that for which $-1 < E < 1$ and $\lambda > 1$. Here we find that ϕ is a periodic function of η

$$\phi = 2\sin^{-1}\left\{\gamma\,\mathrm{sn}\left[\frac{\eta}{(\lambda^2-1)^{1/2}}\right]\right\},$$

where sn is an elliptic function of modulus $2\gamma^2 = 1 - E$.

The effects discussed in the previous three paragraphs are not new, having been observed as early as 1844 by Scott–Russell and discussed in some detail by Korteweg and deVries [30]. A review of the theory of water waves of permanent profile is found in the work of Stoker [31].

Korteweg and deVries studied the equation

$$u_t + uu_x + u_{xxx} = 0, \tag{1.123}$$

which is a special case of the equation

$$u_t \pm u^p u_x + \delta^2 u_{xxx} = 0 \tag{1.124}$$

studied by Zabusky [32]. Equation (1.123) is found to have solitary-wave pulse solutions $u = \phi(\eta)$, $\eta = x - \lambda t$, called *solitons*, of the form

$$\phi(\eta) = A_0 + A\,\mathrm{sech}^2(\eta/\Delta),$$

where A_0 is the value u takes as $|x| \to \infty$, $A = 3\lambda$, and $\Delta = 2\lambda^{-1/2}$. Clearly, larger velocity λ implies larger amplitude and smaller pulse width.

Extensive studies of Eqs. (1.123) and (1.124) have been carried out by Kruskal, Zabusky, Miura *et al.* (see Zabusky [32] for a bibliography to 1966), Miura [33], Miura *et al.* [34], and Lax [35].

1.7 ARBITRARY FUNCTIONS

The freedom of analysis provided by the general solutions of certain equations (see Volume I) is primarily a result of the presence of arbitrary functions whose form is determined by employing boundary and initial conditions. In this section we shall use that suggestion and demonstrate how some solutions for the momentum equations of fluid mechanics can be constructed.

Burgers' equation (1.9), previously discussed in Section 1, was transformed into Eq. (1.10):

$$v_t + \tfrac{1}{2}v_x^2 = \lambda v_{xx}.$$

We now seek a solution to Eq. (1.10) in the form

$$v = f[t + g(x)], \tag{1.125}$$

where f and g are to be determined. Upon setting this into Eq. (1.10), we find

$$f'(\omega)[1 - \lambda g''(x)] = [g'(x)]^2 [\lambda f'' - \tfrac{1}{2}(f')^2],$$

which may be "separated" into

$$[1 - \lambda g''(x)]/(g')^2 = [\lambda f'' - \tfrac{1}{2}(f')^2]/f' = \text{constant} = c, \tag{1.126}$$

where $\omega = t + g(x)$. Thus f and g must satisfy

$$\begin{aligned} \lambda f'' - \tfrac{1}{2}(f')^2 - cf' &= 0, \\ \lambda g'' + c(g')^2 &= 1, \end{aligned} \tag{1.127}$$

both of which are easily integrated by reduction of order. Actual integrals will be recorded for the following example.

The successful use of Eq. (1.125) in developing a five-parameter class of solutions for the transformed Burgers' equation suggests that generalizations may be useful. For the pressure free two-dimensional Navier–Stokes equations

$$u_t + uu_x + vu_y = \nu(u_{xx} + u_{yy}), \tag{1.128}$$

$$v_t + uv_x + vv_y = \nu(v_{xx} + v_{yy}), \tag{1.129}$$

we introduce the auxiliary function ψ defined by $u = \psi_x$ and $v = \psi_y$ and obtain, respectively,

$$\begin{aligned} (\psi_t)_x + \tfrac{1}{2}[\psi_x^2 + \psi_y^2]_x &= \nu[\psi_{xx} + \psi_{yy}]_x, \\ (\psi_t)_y + \tfrac{1}{2}[\psi_x^2 + \psi_y^2]_y &= \nu[\psi_{xx} + \psi_{yy}]_y. \end{aligned}$$

After discarding arbitrary functions of y and x, respectively, both of these equations integrate to

$$\psi_t + \tfrac{1}{2}(\psi_x^2 + \psi_y^2) = \nu(\psi_{xx} + \psi_{yy}), \tag{1.130}$$

which is the two-dimensional form of Eq. (1.10).

A solution of the form

$$\psi = f[t + g(x) + h(y)] \tag{1.131}$$

will be sought for Eq. (1.130). With $\omega = t + g(x) + h(y)$, we find that

$$\nu f'' - \tfrac{1}{2}(f')^2 - cf' = 0,$$
$$\nu g'' + c(g')^2 = 1 - c_1,$$
$$\nu h'' + c(h')^2 = c_1,$$

whose explicit solutions are

$$f = -2\nu A_1 c \ln\left[1 - \tfrac{1}{2}\exp\left(\frac{2\omega A_2}{\nu}\right)\right],$$

$$g = \left[\frac{c_1 - 1}{c}\right]^{1/2} \ln\left\{\cos\left(\frac{\nu}{[c(c_1 - 1)]^{1/2}}[x + B_1]\right)\right\} + B_2,$$

$$h = \left[\frac{c_1}{c}\right]^{1/2} \ln\left\{\cosh\left(\frac{(c_1 c)^{1/2}}{\nu}[y + E_1]\right)\right\} + E_2.$$

This system constitutes an eight-parameter family of solutions, since c, c_1, A_1, A_2, B_1, B_2, E_1, and E_2 are not yet determined.

1.8 EQUATION SPLITTING

Past successes in the development of *particular solutions* for linear partial-differential equations have served to downgrade the position of the *general* solution. Research into techniques for the development of general solutions reached a zenith around 1890. The treatise by Forsyth [36], recently reprinted by Dover, bears witness to the extensive efforts put forth prior to that time. Since every solution can be put in the form of the general solution, it clearly gives the form of the broadest class of solutions. Furthermore, it is not *ad hoc* in character, and is not restricted in its utility by any linearity assumption or superposition principle. A resurgence of interest in these solutions appears to be necessary before any real depth of knowledge can develop on the difficult problems that concern us here. We shall employ the general solution in the development of the splitting concept which was introduced by Ames [37].

The splitting concept is fundamentally simple—although not always simple to execute. What we do is disregard the inviolate nature of the equation(s) to be solved. It is then decomposed into parts, which are equated to a common factor, in such a manner that a general solution (containing the appropriate number of arbitrary functions) can be constructed for at least one part. The form of the arbitrary function(s) is then obtained by means of the requirement that the other part be satisfied.

To fix the ideas, we consider the stream-function form of the boundary-layer equations

$$\psi_y \psi_{xy} - \psi_x \psi_{yy} = \nu \psi_{yyy} . \tag{1.132}$$

Upon splitting the right-hand side, the two equations

$$\psi_y \psi_{xy} - \psi_x \psi_{yy} = 0 \tag{1.133}$$

and

$$\psi_{yyy} = 0 \tag{1.134}$$

are obtained. Note here that alternative expressions such as

$$\psi_y \psi_{xy} - \psi_x \psi_{yy} = \psi_y^3$$

are possible and do generate solutions. The general solution of Eq. (1.133) is constructed by the Mongé method (see Volume I or Forsyth [36]) to be

$$\psi = F[y + G(x)] \tag{1.135}$$

with the functions F and G arbitrary. What forms must F and G have so that Eq. (1.134) is also satisfied? Considerable generality may sometimes be maintained if only one arbitrary function needs to be particularized. This is, in fact, the case here. For Eq. (1.135) to satisfy Eq. (1.134) it is obvious that $F'''(\eta) = 0$, $\eta = y + G(x)$, and upon integration we obtain

$$\psi = F[y + G(x)] = a[y + G(x)]^2 + b[y + G(x)] + c, \tag{1.136}$$

where a, b, c are constants. Equation (1.136) is a solution of the boundary-layer equation for arbitrary $G(x)$.

The more general problem specified by the Navier–Stokes equations in two dimensions will now be examined. Upon introducing the stream function ψ defined by means of $u = \psi_y$, $v = -\psi_x$, the dimensionless equations take the form

$$\psi_y \psi_{xy} - \psi_x \psi_{yy} = -p_x + \psi_{yyy} + \mathrm{Re}^{-1} \psi_{yxx} , \tag{1.137}$$

$$p_y + \mathrm{Re}^{-1}[\psi_x \psi_{xy} - \psi_y \psi_{xx} + \psi_{xyy}] = -\mathrm{Re}^{-2} \psi_{xxx} , \tag{1.138}$$

where

$$u = \bar{u}/U, \qquad v = \mathrm{Re}^{1/2}(\bar{v}/U), \qquad p = (\bar{p} - p_\infty)/\rho U^2,$$

$$x = \bar{x}/L, \qquad y = \mathrm{Re}^{1/2}(\bar{y}/L), \qquad \mathrm{Re} = UL/\nu,$$

are dimensionless variables.

We suppose $p = p(y)$ and split Eq. (1.137) into the two parts

$$\psi_y\psi_{xy} - \psi_x\psi_{yy} = F(x, y, \psi, \psi_x, \psi_y, \psi_{yy}, \psi_{xy}, \psi_{xx}), \tag{1.139}$$

$$\psi_{yyy} + \mathrm{Re}^{-1}\psi_{yxx} = F(x, y, \psi, \psi_x, \psi_y, \psi_{yy}, \psi_{xy}, \psi_{xx}), \tag{1.140}$$

in such a way that the general solution of at least one of these can be developed. The form of the arbitrary functions is then determined by requiring that the other equation is also satisfied. There will remain certain arbitrary constants which we select in such a way that the y-momentum Eq. (1.138) is satisfied—that is so that $p = p(y)$.

It is clear that the choice made for F strongly influences the general solution obtained, the labor involved, and the final result. For simplicity, it is herein again chosen as zero, although other forms have been used. With this choice, our system becomes

$$p_x = 0, \tag{1.141}$$

$$\psi_y\psi_{xy} - \psi_x\psi_{yy} = 0, \tag{1.142}$$

$$\psi_{yyy} + \mathrm{Re}^{-1}\psi_{yxx} = 0, \tag{1.143}$$

and p_y is defined by

$$p_y = -\mathrm{Re}^{-2}\psi_{xxx} - \mathrm{Re}^{-1}[\psi_x\psi_{yx} - \psi_y\psi_{xx} + \psi_{xyy}]. \tag{1.144}$$

The general solution of Eq. (1.142) has the explicit form, with arbitrary ϕ and η,

$$\psi = \phi[y + \eta(x)], \tag{1.145}$$

although it is the usual situation that the general solution is *implicit*. In such a case, the details of the computation are more complicated. Upon substituting Eq. (1.145) into Eq. (1.143) we find, with $\omega = y + \eta(x)$, that

$$\phi'''(\omega)[1 + \alpha^2(\eta')^2] + \alpha^2\phi''(\omega)\eta''(x) = 0,$$

where $\alpha^2 = \mathrm{Re}^{-1}$. To eliminate the dependence on x, and thus determine η, set

$$1 + \alpha^2(\eta')^2 = A\alpha^2\eta'', \tag{1.146}$$

where A is an arbitrary constant. Equation (1.146) has the solution

$$\eta(x) = -A \ln \cos[(\mathrm{Re}^{1/2}x/A) + C] + c_1 .$$

Then ϕ satisfies

$$\phi''' + A^{-1}\phi'' = 0$$

or

$$\phi(\omega) = \Gamma + \gamma A\omega + \epsilon \exp[-\omega/A].$$

Finally, we see that

$$\begin{aligned}\psi(x, y) = \Gamma + \gamma Ay - \gamma A^2 \ln \cos[(\mathrm{Re}^{1/2}x/A) + C] \\ + \epsilon\, e^{-y/A} \cos[(\mathrm{Re}^{1/2}x/A) + C],\end{aligned} \tag{1.147}$$

where Γ, γ, A, C, ϵ are arbitrary constants.

Lastly, are there any values of these constants for which $p = p(y)$? From Eq. (1.144) we find that this will be the case when $\gamma \equiv 0$. Hence

$$\begin{aligned}\psi &= \Gamma + \epsilon \exp[-y/A] \cos[(\mathrm{Re}^{1/2}\, x)/A + C],\\ u &= -\epsilon A^{-1} \exp[-y/A] \cos[(\mathrm{Re}^{1/2}\, x)/A + C],\\ v &= \epsilon A^{-1}\, \mathrm{Re}^{1/2} \exp[-y/A] \sin[(\mathrm{Re}^{1/2}\, x)/A + C],\\ p(y) &= -\tfrac{1}{2}(\epsilon A)^{-2} \exp[-2y/A] + D.\end{aligned}$$

With $C = 0$ this might be interpreted as the flow through a porous flate plate on the y-axis. Note especially the appearance of the $\mathrm{Re}^{1/2}$. In this case ($F = 0$) the solution can also be obtained by the *ad-hoc* assumption of separation.

1.9 INVERSION OF DEPENDENT AND INDEPENDENT VARIABLES

The technique of interchanging the roles of the dependent and independent variables has been found useful in the study of nonlinear equations. Examples are given in Volume I (pp. 35–38). Herein we discuss an extension of the scope of this inversion due to Dasarathy [38]. Certain classes of *coupled* nonlinear second-order partial-differential equations are shown to reduce to a pair of *uncoupled* linear Laplace equations under the interchange of dependent and independent variables.

When the quasi-linear coupled second-order equations

$$(a^2U_y^2 + b^2V_y^2)U_{xx} - 2(a^2U_xU_y + b^2V_xV_y)U_{xy} + (a^2U_x^2 + b^2V_x^2)U_{yy} = 0,$$
$$(a^2U_y^2 + b^2V_y^2)V_{xx} - 2(a^2U_xU_y + b^2V_xV_y)V_{xy} + (a^2U_x^2 + b^2V_x^2)V_{yy} = 0,$$

are subjected to the linear transformation

$$u(x, y) = aU, \qquad v(x, y) = bV,$$

they become

$$(u_y^2 + v_y^2)u_{xx} - 2(u_xu_y + v_xv_y)u_{xy} + (u_x^2 + v_x^2)u_{yy} = 0, \tag{1.148}$$
$$(u_y^2 + v_y^2)v_{xx} - 2(u_xu_y + v_xv_y)v_{xy} + (u_x^2 + v_x^2)v_{yy} = 0. \tag{1.149}$$

The interchange of dependent and independent variables is tantamount to assuming that $x = x(u, v)$, $y = y(u, v)$. When these are differentiated with respect to x, there results

$$1 = x_uu_x + x_vv_x, \qquad 0 = y_uu_x + y_vv_x,$$

thereby leading to the relations

$$u_x = Jy_v, \qquad v_x = -Jy_u. \tag{1.150}$$

In a similar fashion, we are lead to

$$u_y = -Jx_v, \qquad v_y = Jx_u, \tag{1.151}$$

where the Jacobian J defined by

$$J = u_xv_y - u_yv_x = (x_uy_v - x_vy_u)^{-1} \tag{1.152}$$

cannot be zero for a valid transformation. Further differentiation of Eqs. (1.150)–(1.152) with respect to x and y generates the second derivatives of u and v. When these are substituted into Eqs. (1.148) and (1.149), the *linear uncoupled Laplace equations*,

$$x_{uu} + x_{vv} = 0, \qquad y_{uu} + y_{vv} = 0, \tag{1.153}$$

are obtained, after extensive algebraic simplification.

Since Eqs. (1.148) and (1.149) are transformable into Eqs. (1.153) by this interchange, it is clear that this method has much wider horizons than previously observed. For example, if we set

$$u = F(U), \qquad v = G(V), \tag{1.154}$$

then Eqs. (1.148) and (1.149) become

$$[(F'U_y)^2 + (G'V_y)^2]U_{xx} - 2[(F')^2U_xU_y + (G')^2V_xV_y]U_{xy} + [(F'U_x)^2 + (G'V_x)^2]U_{yy} + (F''/F')(G')^2(U_xV_y - V_xU_y)^2 = 0, \quad (1.155)$$

and

$$[(F'U_y)^2 + (G'V_y)^2]V_{xx} - 2[(F')^2U_xU_y + (G')^2V_xV_y]V_{xy} + [(F'U_x)^2 + (G'V_x)^2]V_{yy} + (G''/G')(F')^2(U_xV_y - V_xU_y)^2 = 0. \quad (1.156)$$

The course of final solution is to obtain the results for $x(u, v)$, $y(u, v)$, determine the relationship $u(x, y)$, $v(x, y)$ graphically or otherwise, and finally invert the transformation specified by Eq. (1.154) to obtain U, V. These processes are often limited in their utility by the form of the boundary conditions.

1.10 CONTACT TRANSFORMATIONS

First employed by Lagrange (see Forsyth [36, p. 131]) and originally called a *tangential* transformation, the *contact* transformation is perhaps best known because of one of its kind, the Legendre transformation. A discussion and some applications of the Legendre transformation to quasi-linear equations are found in Volume I, pages 37–40. In this section we shall present a more general discussion and describe further applications to nonlinear equations.

We are often in a position to use coordinate transformations

$$\alpha^i = \alpha^i(x^1, \ldots, x^n), \qquad i = 1, 2, \ldots, n, \quad (1.157)$$

which map a hypersurface S defined by $U = U(x^1, \ldots, x^n)$, in the n independent variables x^i, into a corresponding hypersurface Σ in the n independent variables α^i. Since this takes a point P of S into a point Q of Σ it is called a *point transformation.*

A more general type of mapping follows from the establishment of a correspondence between a surface element of S and a *surface* element of Σ. The hypersurface S may be regarded as the *envelope* of a family of hypersurfaces S^1 all of which make tangential contact with S. The one-dimensional case is illustrated in Fig. 1-2. Consequently, a description of S can then be given in terms of the parameters defining the family S^1. From among all such transformations, consider that class having the property that two hypersurfaces S_1 and S_2 tangent at P are mapped

FIG. 1-2. One-dimensional hypersurface as an envelope of tangent lines.

into two hypersurfaces Σ_1 and Σ_2 tangent at P^1, the map of P. These are *contact transformations*.

To illustrate the idea, consider a simple curve C, with equation $y = f(x)$, and a mapping of points P of C by means of a transformation which depends *both upon the point P and the tangent to C at P*. Such a mapping, taking $C(x, y) \rightarrow \Gamma(\alpha, \beta)$, is describable by

$$\alpha = \alpha(x, y, p), \qquad \beta = \beta(x, y, p), \qquad p = dy/dx. \tag{1.158}$$

The slope of Γ at P^1 may be written in terms of the curve C as

$$\left.\frac{d\beta}{d\alpha}\right|_{\Gamma(P^1)} = \left.\frac{(\partial\beta/\partial x) + (\partial\beta/\partial y)p + (\partial\beta/\partial p)(dp/dx)}{(\partial\alpha/\partial x) + (\partial\alpha/\partial y)p + (\partial\alpha/\partial p)(dp/dx)}\right|_{C(P)}. \tag{1.159}$$

Now choose another curve $C^1 \rightarrow \Gamma^1$, such that C^1 is tangent to C at P. Clearly Γ and Γ^1 will have a common point P^1 in the (α, β) plane, but Γ^1 will not necessarily be tangent to Γ at P^1. This is a consequence of the fact that dp/dx at P is different for C and C^1; hence

$$\left.\frac{d\beta}{d\alpha}\right|_{\Gamma^1(P^1)} \neq \left.\frac{d\beta}{d\alpha}\right|_{\Gamma(P^1)}.$$

However, if the right-hand side of Eq. (1.159) is independent of dp/dx, the curves Γ and Γ^1 will always be tangent at P^1, as required for a contact transformation. The condition for independence of dp/dx is found by performing the division on the right-hand side of Eq. (1.159). Thus $d\beta/d\alpha$ is independent of dp/dx when

$$\frac{\partial\alpha}{\partial p}\left[\frac{\partial\beta}{\partial x} + \frac{\partial\beta}{\partial y}p\right] = \frac{\partial\beta}{\partial p}\left[\frac{\partial\alpha}{\partial x} + \frac{\partial\alpha}{\partial y}p\right], \tag{1.160}$$

i.e., this is the condition that must be satisfied in order for the transformation Eq. (1.158) to be a contact transformation.

We first consider the special case where the C^1 are taken to be straight lines in the x, y-plane. These are completely defined in terms of their slope p and their intercept on the y-axis which, for reasons of notation, we denote by $-Y(p)$. Consequently, the tangent line is

$$y = -Y + px$$

or

$$Y = px - y. \tag{1.161}$$

Since $dy/dp = p\,dx/dp$, it follows that $dY/dp = x$. Thus this contact transformation, called the *Legendre transformation*, is

$$x = dY/dX = P, \qquad y = XP - Y, \tag{1.162}$$

and its inverse is

$$X = p, \qquad Y = px - y. \tag{1.163}$$

In practice this involutory† contact transformation is employed to convert first-order ordinary differential equations $f(x, y, p) = 0$ into the dual equation $f(P, XP - Y, X) = 0$, which is also of first order. The solution of the dual equation, say $F(X, Y, C) = 0$ is converted to the original variables by means of Eq. (1.163). Thus $F(p, px - y, C) = 0$, and finally this together with $f(x, y, p) = 0$ provides a parametric solution for x and y, with parameter p.

Extensions to functions of more than one variable follow a similar pattern. Thus, if we write $\xi = z_x$ and $\eta = z_y$ and denote the intercept on the z-axis by $-Z(\xi, \eta)$, then the tangent plane at a point on the surface S has the equation

$$z(x, y) = \xi x + \eta y - Z(\xi, \eta). \tag{1.164}$$

From Eq. (1.164) it follows that $Z_\xi = x$, $Z_\eta = y$.

The symmetric character of the Legendre transformation is nicely summarized as follows:

$$\begin{gathered} Z(\xi, \eta) + z(x, y) = \xi x + \eta y, \\ \begin{array}{l|l} Z_\xi = x & z_x = \xi, \\ Z_\eta = y & z_y = \eta. \end{array} \end{gathered} \tag{1.165}$$

First-order equations $G(x, y, z, z_x, z_y) = 0$ are transformed into $G(Z_\xi, Z_\eta, \xi Z_\xi + \eta Z_\eta - Z, \xi, \eta) = 0$ by Eqs. (1.165). Thus, for example, $Z_x + ZZ_y = 0$ becomes the linear equation $\xi\eta Z_\xi + \eta^2 Z_\eta - Z\eta + \xi = 0$.

---

† A transformation T is involutory if $T^2 = I$, where I is the identity.

For functions of n independent variables x_i, $i = 1, 2,..., n$, the Legendre transformation becomes

$$Z(\xi_1,..., \xi_n) + z(x_1,..., x_n) = \sum_{i=1}^{n} x_i \xi_i, \qquad (1.166)$$

$$Z_{\xi_i} = x_i \quad | \quad z_{x_i} = \xi_i .$$

If transformations of second-order equations are desired (see Volume I, page 39), there results, in addition to Eq. (1.165), the second derivatives

$$z_{xx} = jZ_{\eta\eta}, \qquad z_{xy} = -jZ_{\xi\eta}, \qquad z_{yy} = jZ_{\xi\xi}, \qquad (1.167)$$

where

$$j = z_{xx}z_{yy} - z_{xy}^2 = (Z_{\xi\xi}Z_{\eta\eta} - Z_{\xi\eta}^2)^{-1} \neq 0.$$

In Section 1.5, there appeared wave equations of the form $u_{tt} = f(u_x)u_{xx}$. Upon applying Eqs. (1.165) and (1.167) (with $u = z$ and $t = y$), we get the linear equation $Z_{\xi\xi} = f(\xi)Z_{\eta\eta}$. Additional examples are presented in Volume I.

Varley [39] has studied the two-dimensional flow of a rheological (dilatant) fluid by means of the equations

$$F = u_x + v_y = 0, \qquad (1.168)$$

$$G = \tfrac{1}{2}[u_x{}^2 + v_y{}^2] + \tfrac{1}{4}[v_x + u_y]^2 - 1 = 0, \qquad (1.169)$$

which are to be solved in a neighborhood of the curve Γ: $x = x_0(l)$, $y = y_0(l)$, such that on Γ, $u = v = 0$. At this point, one is often tempted to introduce a stream function ψ defined through $\psi_y = u$, $\psi_x = -v$. Consequently, Eq. (1.168) is identically satisfied, whereas

$$G = \psi_{xy}^2 + \tfrac{1}{4}[\psi_{yy}^2 - \psi_{xx}^2] - 1 = 0.$$

When the Legendre transformation is applied, we find

$$j^2(Z_{\xi\eta})^2 + \tfrac{1}{4}j^2[Z_{\xi\xi}^2 - Z_{\eta\eta}^2] - 1 = 0.$$

Clearly, this is no improvement over our original equation. Thus it appears that equation consolidation, via the stream function, should be abandoned.

Nevertheless, the Legendre transformation may be a good choice if we investigate the problem in its primitive form, Eqs. (1.168) and (1.169). A parametrization due to Varley [39] is one fruitful way of reducing the complexities of this system.

1.11 PARAMETRIZATION AND THE LEGENDRE TRANSFORMATION

Let u and v satisfy the two, nonlinear first-order equations

$$F(u_x, u_y, v_x, v_y) = G(u_x, u_y, v_x, v_y) = 0, \tag{1.170}$$

and suppose F and G have continuous first partial derivatives in each of their arguments. Our goal is the replacement of the nonlinear equations (1.170) by two "equivalent" first-order linear equations. As a vehicle for this, we can think of Eqs. (1.170) parametrized with parameters α and β introduced, so that

$$u_x = p(\alpha, \beta), \qquad u_y = q(\alpha, \beta), \qquad v_x = r(\alpha, \beta), \qquad v_y = s(\alpha, \beta). \tag{1.171}$$

Then, Eqs. (1.170) can be regarded as those obtained by eliminating α and β from Eqs. (1.171).

The Legendre transformations (two, in this case) U and V are now introduced by means of the quantities

$$u = xp + yq + U, \qquad v = xr + ys + V. \tag{1.172}$$

Upon calculating u_x, u_y, v_x, and v_y, it follows that Eqs. (1.172) satisfy Eqs. (1.171) if[†]

$$x\,dp + y\,dq + dU = 0, \qquad x\,dr + y\,ds + dV = 0. \tag{1.173}$$

To obtain relations between (x, y) and the (α, β) partial derivatives of U and V, we find from Eqs. (1.173) that

$$(xp_\alpha + yq_\alpha + U_\alpha)\,d\alpha + (xp_\beta + yq_\beta + U_\beta)\,d\beta = 0,$$

and

$$(xr_\alpha + ys_\alpha + V_\alpha)\,d\alpha + (xr_\beta + ys_\beta + V_\beta)\,d\beta = 0.$$

If α and β are to be *independent*[‡] parameters, each of the four expressions in parentheses must vanish identically. Solving each pair for x and y there results

$$x = \partial_{\alpha\beta}(q, U)/\partial_{\alpha\beta}(p, q) = \partial_{\alpha\beta}(s, V)/\partial_{\alpha\beta}(r, s), \tag{1.174}$$

$$-y = \partial_{\alpha\beta}(p, U)/\partial_{\alpha\beta}(p, q) = \partial_{\alpha\beta}(r, V)/\partial_{\alpha\beta}(r, s), \tag{1.175}$$

[†] Of course, a trivial way of satisfying Eqs. (1.174) and (1.175) is to take α, β, U, and V constant.

[‡] If α is a function of β, the solutions can be easily obtained.

where

$$\partial_{\alpha\beta}(f, g) = \begin{vmatrix} f_\alpha & g_\alpha \\ f_\beta & g_\beta \end{vmatrix}.$$

Equations (1.174) and (1.175) are two linear equations for U and V as functions of α and β. They also provide the mapping from the α, β- to the x, y-plane in terms of the solution to these equations.

Returning to Eqs. (1.168) and (1.169), we first observe that a parametrization is *not unique*. One that works very well here is obtained by setting

$$\partial u/\partial x = -\sin 2\alpha, \qquad \partial v/\partial y = \sin 2\alpha, \tag{1.176}$$

whereupon Eq. (1.168) is satisfied. Then Eq. (1.169) becomes

$$(u_y + v_x)^2 = 4(1 - \sin^2 2\alpha) = 4 \cos^2 2\alpha,$$

which is satisfied if we choose

$$u_y = \beta + \cos 2\alpha, \qquad v_x = -\beta + \cos 2\alpha. \tag{1.177}$$

In our general notation we have now specified $p = -\sin 2\alpha$, $q = \beta + \cos 2\alpha$, $r = -\beta + \cos 2\alpha$, and $s = \sin 2\alpha$.

Having a parametrization at hand, we can now employ Eqs. (1.174) and (1.175). From the first of these there follows

$$x = [(2 \sin 2\alpha) U_\beta + U_\alpha]/(2 \cos 2\alpha) = V_\beta ,$$

so that one linear equation is

$$U_\alpha + 2 \sin 2\alpha U_\beta - 2 \cos 2\alpha V_\beta = 0. \tag{1.178}$$

From the second relation we obtain

$$V_\alpha - 2 \sin 2\alpha V_\beta - 2 \cos 2\alpha U_\beta = 0. \tag{1.179}$$

One can easily show that these form a hyperbolic system with characteristics $\alpha + \frac{1}{2}(\beta - 1) =$ constant, $\alpha \quad \frac{1}{2}(\beta - 1) =$ constant. In particular, if we set $\phi = (U + iV) \exp[-i\alpha]$, then $\phi_{\alpha\alpha} + \phi = 4\phi_{\beta\beta}$. The initial curve Γ maps into the curve $\beta = 1$ and $\alpha = \theta$, where $\tan \theta = y_0'(l)/x_0'(l)$.

Turning now to second-order equations, suppose

$$z_{xy} = f(z_{xx}, z_{yy}). \tag{1.180}$$

This can be replaced by a system of two first-order equations if we set $u = z_x$ and $v = z_y$, whereupon Eq. (1.180) becomes

$$u_y = v_x = f(u_x, v_y).$$

Such a form suggests a useful parametrization

$$p = \alpha, \qquad s = \beta, \qquad q = r = f(\alpha, \beta), \tag{1.181}$$

in the notation of Eq. (1.171).

Since the Mongé–Amperé equation with constant coefficients

$$z_{xx}z_{yy} - z_{xy}^2 + Az_{xx} + Bz_{xy} + Cz_{yy} = D \tag{1.182}$$

can be written in the form of Eq. (1.180), an example treating such a form will be briefly discussed. In Section 1.2a longitudinal wave propagation on a traveling threadline is shown to be governed by Eqs. (1.18) and (1.19):

$$VV_x + V_t = T_x/m, \qquad m_t + (mV)_x = 0,$$

and, in a (nonlinear) medium, by a constitutive law

$$T = F(m). \tag{1.183}$$

Upon multiplying Eqs. (1.18) and (1.19) by m and V, respectively, and adding, we obtain

$$(mV)_t + (mV^2 - T)_x = 0, \tag{1.184}$$

which will be used instead of Eq. (1.18). An auxiliary function $\theta(x, t)$ is introduced by defining

$$m = \theta_{xx}, \qquad -Vm = \theta_{xt}, \tag{1.185a}$$

so that Eq. (1.19) is satisfied identically. Equation (1.184) is also satisfied identically if the definition of θ is completed by setting

$$\theta_{tt} = mV^2 - T. \tag{1.185b}$$

From these it follows that

$$V = -\theta_{xt}/\theta_{xx}, \qquad m = \theta_{xx}, \qquad T = (\theta_{xt}^2 - \theta_{xx}\theta_{tt})/\theta_{xx},$$

whereupon the constitutive Eq. (1.183) generates the Mongé–Amperé equation

$$\theta_{xt}^2 - \theta_{xx}\theta_{tt} = \theta_{xx}F(\theta_{xx}). \tag{1.186}$$

One advantage of this form over others treated by Ames and Vicario [7] is that the independent variables are in the physical x, t-plane, a desirable feature for interpretation of the results.

Now set $u = \theta_x$, $v = \theta_t$, whereupon Eq. (1.186) becomes $u_t^2 = u_x v_t + u_x F(u_x)$. To parametrize, write $u_x = \alpha$, $v_t = \beta$, so that

$$u_t = v_x = \pm[\alpha\beta + \alpha F(\alpha)]^{1/2}.$$

Equations (1.174) and (1.175) are now employed to generate linear equations for U and V as functions of α and β.

Legendre and Amperé gave many examples of contact transformations. Lie developed the general theory in various works. The interested reader is referred to Forsyth [36]. In particular, we mention the contact transformation of Amperé (see Goursat [40]) $(x, y, z, p, q) \to (X, Y, Z, P, Q)$, namely,

$$X = x, \qquad Y = q, \qquad Z = qy - z, \qquad P = -p, \qquad Q = y,$$

with the converse

$$x = X, \qquad y = Q, \qquad z = QY - Z, \qquad p = -P, \qquad q = Y.$$

This transformation, like that of Legendre, is involutory.

1.12 BÄCKLUND TRANSFORMATIONS

The general theory of transformation of partial differential equations of order higher than the first is complete. Some of the classical work is available in Volume 6 of Forsyth [36] and in the work of Eisenhart [41]. In this section, our discussion relates to a transformation that had its origin in some investigations by Bäcklund [42, 43] concerning simultaneous equations of the first order arising in differential geometry (see also Eisenhart [41, page 284]).

From the foregoing sections we have seen how the contact transformation provides a method of constructing various classes of "equivalent" equations, thereby leading to the integrals of the original equation. The goal of the Bäcklund transformation is the same. In what follows, we shall denote by x, y, z, $p = z_x$, $q = z_y$ an element of any surface, and by x_1, y_1, z_1, p_1, q_1 an element of any other surface. To connect the two surface elements completely, though not uniquely, it is necessary to have five distinct equations relating the two sets of variables. However, since each set of five variables defines a surface element, they are related by the total differentials

$$dz = p\,dx + q\,dy, \qquad dz_1 = p_1\,dx_1 + q_1\,dy_1 . \tag{1.187}$$

Thus, if due regard is paid to Eqs. (1.187), only four independent equations

$$F_n(x, y, z, p, q; x_1, y_1, z_1, p_1, q_1) = 0, \qquad n = 1, 2, 3, 4, \tag{1.188}$$

are required to connect the surface elements. When these correspondences are used, there are certain cases when the variable z (or z_1) is an integral of an equation of Mongé–Amperé form.† When the variables z and z_1 thus separately satisfy equations of the second order, these equations can be regarded as transformable into one another by the four relations, Eqs. (1.188), usually called a *Bäcklund transformation.*

As the transformations change, it is important to ascertain the limitations or restrictions upon the allowable equations, other than that of belonging to Mongé–Amperé type. First one must ask whether a given Mongé–Amperé equation admits a Bäcklund transformation, and second how does one construct such a transformation. These are lengthy questions for the general case, and are beyond our scope. Nevertheless, the general treatment is available in the work of Forsyth [36, Volume 6, pp. 432–454]. Herein we shall demonstrate a development of the transformation equations, due to Clairin [44, 45], for a problem of Lamb [46–48] on the propagation of ultrashort optical pulses. The fundamental equation of his model is

$$\sigma_{\xi\tau} = \sin \sigma. \tag{1.189}$$

Probably the simplest Bäcklund case arises when there are two simultaneous equations involving two dependent variables z and z_1. We may take the questions in the form

$$f(x, y, z, z_1, p, q, p_1, q_1) = 0, \qquad g(x, y, z, z_1, p, q, p_1, q_1) = 0.$$

When these two equations can be solved algebraically for (say), p and q,

† The general Mongé–Amperé form is

$$Ru_{xx} + Su_{xy} + Tu_{yy} + U(u_{xx}u_{yy} - u_{xy}^2) = V,$$

where R, S, T, U, and V are functions of x, y, u, u_x, and u_y.

under the condition that $\partial_{pq}(f, g)$ does not vanish identically, we find

$$p = f_1(x, y, z, z_1, p_1, q_1), \qquad q = f_2(x, y, z, z_1, p_1, q_1). \tag{1.190}$$

The integrability condition $dp/dy = dq/dx$ generates the relationship

$$\frac{\partial f_1}{\partial y} + f_2 \frac{\partial f_1}{\partial z} + q_1 \frac{\partial f_1}{\partial z_1} + s_1 \frac{\partial f_1}{\partial p_1} + t_1 \frac{\partial f_1}{\partial q_1} = \frac{\partial f_2}{\partial x} + f_1 \frac{\partial f_2}{\partial z} + p_1 \frac{\partial f_2}{\partial z_1} + r_1 \frac{\partial f_2}{\partial p_1} + s_1 \frac{\partial f_2}{\partial q_1}, \tag{1.191}$$

where r_1, s_1, and t_1 represent $\partial^2 z_1/\partial x^2$, $\partial^2 z_1/\partial x\, \partial y$, and $\partial^2 z_1/\partial y^2$, respectively. Equation (1.191) is linear in r_1, s_1, and t_1 and, in general, depends upon x, y, z, z_1, p_1, and q_1.

Suppose our initial concern is with two simultaneous equations of first order in z and z_1. When z occurs in Eq. (1.191) we can think of that equation solved so as to express z in terms of $x, y, z_1, p_1, q_1, r_1, s_1$, and t_1. When the value of z so obtained is substituted into the given first-order equations, they become *two equations of the third order for the determination of* z_1. From the general theory, it is known that unless the original equations cannot be solved with respect to p and p_1, or with respect to q and q_1, they possess common integrals. Consequently, the two third-order equations, which are satisfied by z_1, must be compatible. They must therefore lead to a value or values of z_1 that involve arbitrary functions. To each such value of z_1, there corresponds a value of z as given earlier. (If the original equations are second order, then the equations for z_1 will be fourth order, in the preceding argument.)

Now it may happen that the integrability condition Eq. (1.191) is explicitly free of z. It then becomes a single, second-order equation for z_1. Upon solving this, the z_1 obtained is substituted into Eqs. (1.190) and a quadrature leads to a value of z containing an arbitrary constant.

Exceptional cases arise when Eq. (1.191) does not contain r_1, s_1, and t_1. This happens when

$$\partial f_1/\partial q_1 = 0, \qquad \partial f_2/\partial p_1 = 0, \qquad \partial f_1/\partial p_1 = \partial f_2/\partial q_1.$$

In this special case, if z is involved, then z_1 satisfies two equations of the second order. If z is not present, then z_1 satisfies a single equation of the first order.

The exceptional case in which $f = 0$, $g = 0$ and cannot be solved for p and q because the Jacobian is zero is discussed by Forsyth [36, p. 452].

1.13 AN EXAMPLE BÄCKLUND TRANSFORMATION

The problem of ultrashort optical pulses, discussed by Lamb [46–48], has the equation

$$\partial^2 z_1/\partial x\, \partial y = \sin z_1 . \tag{1.192}$$

Our attention will now be confined to explicitly solved transformations [see Eqs. (1.190)] of the form

$$p = f(p_1, z, z_1), \qquad q = \psi(q_1, z, z_1), \tag{1.193}$$

where z_1 is a solution of Eq. (1.192). The integrability requirement $dp/dy = dq/dx$ generates the relation

$$\Omega = \sin z_1 \left(\frac{\partial f}{\partial p_1} - \frac{\partial \psi}{\partial q_1}\right) - p_1 \frac{\partial \psi}{\partial z_1} + q_1 \frac{\partial f}{\partial z_1} + \psi \frac{\partial f}{\partial z_1} - f \frac{\partial \psi}{\partial z} = 0. \tag{1.194}$$

Calculation of successive derivatives of Ω to the point where z_1 is no longer explicitly present yields the following equations:

$$\frac{\partial \Omega}{\partial p_1} = \sin z_1 \frac{\partial^2 f}{\partial p_1{}^2} - \frac{\partial \psi}{\partial z_1} + q_1 \frac{\partial^2 f}{\partial p_1\, \partial z_1} + \psi \frac{\partial^2 f}{\partial p_1\, \partial z} - \frac{\partial f}{\partial p_1} \frac{\partial \psi}{\partial z} = 0, \tag{1.195}$$

$$\frac{\partial \Omega}{\partial q_1} = -\sin z_1 \frac{\partial^2 \psi}{\partial q_1{}^2} - p_1 \frac{\partial^2 \psi}{\partial q_1\, \partial z_1} + \frac{\partial f}{\partial z_1} + \frac{\partial \psi}{\partial q_1} \frac{\partial f}{\partial z} - f \frac{\partial^2 \psi}{\partial q_1\, \partial z} = 0, \tag{1.196}$$

$$\frac{\partial^2 \Omega}{\partial p_1\, \partial q_1} = -\frac{\partial^2 \psi}{\partial q_1\, \partial z_1} + \frac{\partial^2 f}{\partial p_1\, \partial z_1} + \frac{\partial \psi}{\partial q_1} \frac{\partial^2 f}{\partial p_1\, \partial z} - \frac{\partial f}{\partial p_1} \frac{\partial^2 \psi}{\partial q_1\, \partial z} = 0, \tag{1.197}$$

$$\frac{\partial^3 \Omega}{\partial p_1{}^2\, \partial q_1} = \frac{\partial^3 f}{\partial p_1{}^2\, \partial z_1} + \frac{\partial \psi}{\partial q_1} \frac{\partial^3 f}{\partial p_1{}^2\, \partial z} - \frac{\partial^2 f}{\partial p_1{}^2} \frac{\partial^2 \psi}{\partial q_1\, \partial z} = 0, \tag{1.198}$$

$$\frac{\partial^3 \Omega}{\partial p_1\, \partial q_1{}^2} = -\frac{\partial^3 \psi}{\partial q_1{}^2\, \partial z_1} + \frac{\partial^2 \psi}{\partial q_1{}^2} \frac{\partial^2 f}{\partial p_1\, \partial z} - \frac{\partial f}{\partial p_1} \frac{\partial^3 \psi}{\partial q_1{}^2\, \partial z} = 0, \tag{1.199}$$

$$\frac{\partial^4 \Omega}{\partial p_1{}^2\, \partial q_1{}^2} = \frac{\partial^2 \psi}{\partial q_1{}^2} \frac{\partial^3 f}{\partial p_1{}^2\, \partial z} - \frac{\partial^2 f}{\partial p_1{}^2} \frac{\partial^3 \psi}{\partial q_1{}^2\, \partial z} = 0. \tag{1.200}$$

Equation (1.200) is free of explicit dependence upon z_1 although, of course, solutions of that equation will depend parametrically upon z_1.

Upon integration of Eq. (1.200) arbitrary functions will appear which are determined by the requirement that Eqs. (1.195)–(1.199) must also be satisfied. It is possible to separate the p_1 and q_1 dependence of Eq. (1.200) and write it as

$$\frac{\partial^3\psi/\partial q_1^2\,\partial z}{\partial^2\psi/\partial q_1^2} = \frac{\partial^3 f/\partial p_1^2\,\partial z}{\partial^2 f/\partial p_1^2} = h(z, z_1). \tag{1.201}$$

From the usual separation argument it follows that the two left-hand terms of Eq. (1.201) do not depend upon p_1 and q_1 (as was expected). Hence the introduction of $h(z, z_1)$. Recalling that $f = f(p_1, z, z_1)$, $\psi = \psi(q_1, z, z_1)$, integration of Eqs. (1.201) gives

$$f = g(p_1, z_1)H(z, z_1) + l(z, z_1)p_1 + m(z, z_1), \tag{1.202}$$

$$\psi = \theta(q_1, z_1)H(z, z_1) + \lambda(z, z_1)q_1 + \mu(z, z_1), \tag{1.203}$$

where g, θ, l, λ, m, μ, and $H = \exp[\int h(z, z_1)\,dz]$ are arbitrary.

Choosing $H(z, z_1) = 1$ simplifies† the subsequent analysis, so we shall present only that case. Upon substituting Eqs. (1.202) and (1.203) into Eq. (1.199), we find

$$\frac{\partial^3\theta}{\partial z_1\,\partial q_1^2} - \frac{\partial^2\theta}{\partial q_1^2}\frac{\partial l}{\partial z} = 0. \tag{1.204}$$

Consequently $\partial l/\partial z$ is independent of z. Integration of Eq. (1.204) generates the solution for θ,

$$\theta(q_1, z_1) = v(q_1)\exp[u(z_1)] + r(z_1)q_1 + w(z_1), \tag{1.205}$$

where

$$z(du/dz_1) + \text{const} = l(z, z_1), \tag{1.206}$$

and v, r, w are arbitrary functions of their respective arguments. When θ and l are substituted into Eqs. (1.202) and (1.203), one finds that there is no loss in generality in setting $w(z_1)$ and the constant in Eq. (1.206) equal to zero.

Corresponding results are obtained from Eq. (1.198) for g and λ by similar analysis. At this point, the calculation is further restricted by setting $u = 0$ in Eq. (1.205) and similarly in the calculation for λ. Thus,

$$f(p_1, z, z_1) = \bar{v}(p_1) + \bar{r}(z_1)p_1 + m(z, z_1), \tag{1.207}$$

$$\psi(q_1, z, z_1) = v(q_1) + r(z_1)q_1 + \mu(z, z_1). \tag{1.208}$$

---

† Clearly, the transformation is not unique. Of course, we wish it to yield a simpler result than the initial system.

Next, by substituting Eqs. (1.207) and (1.208) into Eq. (1.197) we find that

$$(d/dz_1)(r - \bar{r}) = 0. \tag{1.209}$$

Turning now to Eq. (1.196), we find that it imposes the restriction

$$-\sin z_1 \frac{d^2 v}{dq_1{}^2} + \frac{\partial m}{\partial z}\frac{dv}{dq_1} = -\frac{\partial m}{\partial z_1} - r(z_1)\frac{\partial m}{\partial z}, \tag{1.210}$$

whose solution will yield the general form of $v(q_1)$. Since the right-hand side depends only upon z and z_1, so must the left-hand side. Thus, unless† $\partial m/\partial z = \sin z_1$, one must have $v(q_1) = c_1 q_1 + c_2$, where c_i are absolute constants. Following up this case, we see that $c_1 q_1$ can be absorbed into the term $r(z_1)q_1$ and c_2 into $\mu(z, z_1)$. Consequently, without loss of generality, we take

$$\psi(q_1, z, z_1) = r(z_1)q_1 + \mu(z, z_1) \tag{1.211}$$

and determine m by solving

$$\frac{\partial m}{\partial z_1} + r(z_1)\frac{\partial m}{\partial z} = 0, \tag{1.212}$$

which is the simplified form of Eq. (1.210). A similar procedure applied to Eq. (1.195) leads to

$$f(p_1, z, z_1) = \bar{r}(z_1)p_1 + m(z, z_1) \tag{1.213}$$

and the equation for μ,

$$\frac{\partial \mu}{\partial z_1} + \bar{r}(z_1)\frac{\partial \mu}{\partial z} = 0. \tag{1.214}$$

Finally, Eq. (1.194) yields

$$(\bar{r} - r)\sin z_1 + \mu\frac{\partial m}{\partial z} - m\frac{\partial \mu}{\partial z} = 0. \tag{1.215}$$

When one sets $\bar{r} = 1$, $r = -1$, Eqs. (1.212) and (1.214) have the simple solutions

$$m = m(z + z_1), \qquad \mu = \mu(z - z_1), \tag{1.216}$$

† In this case, we may have $(d^2v/dq_1{}^2) - (dv/dq_1) = \alpha$, or $v = Ae^{q_1} - \alpha q_1 + B$, α, A, and B are absolute constants.

and Eq. (1.215) now becomes

$$2 \sin z_1 + (\partial/\partial z_1)[m(z + z_1)\mu(z - z_1)] = 0, \tag{1.217}$$

a functional differential equation. Upon differentiation with respect to z, Eq. (1.217) is expressible as

$$\frac{m''(z + z_1)}{m(z + z_1)} = \frac{\mu''(z - z_1)}{\mu(z - z_1)} = -K^2,$$

where K is an absolute constant and the primes indicate differentiation with respect to the indicated argument. The classical solutions to this system are

$$m(z + z_1) = a \cos K(z + z_1) + b \sin K(z + z_1),$$

$$\mu(z - z_1) = \alpha \cos K(z - z_1) + \beta \sin K(z - z_1).$$

When these are substituted into Eq. (1.217), evaluated at $z = 0$, and odd and even functions of z_1 equated, one finds

$$\begin{aligned} \alpha b - \beta a &= 0, \\ \alpha a + \beta b &= 4, \\ K &= \tfrac{1}{2}. \end{aligned} \tag{1.218}$$

These are satisfied by setting $a = \alpha = 0$, $b\beta = 4$, or $b/2 = 2/\beta = A$. Equations (1.211) and (1.213) now yield one of the desired Bäcklund transformations

$$\begin{aligned} \frac{1}{2}(q + q_1) &= \frac{1}{A} \sin\left(\frac{z - z_1}{2}\right), \\ \frac{1}{2}(p - p_1) &= A \sin\left(\frac{z + z_1}{2}\right). \end{aligned} \tag{1.219}$$

We now discuss an application of this transformation to the optical pulse equation of Lamb [48] (see also his literature survey).

The basic equation for Lamb's problem is

$$\frac{\partial^2 \sigma}{\partial \xi\, \partial \tau} = \sin \sigma, \tag{1.220}$$

an equation which also arises in the theory of surfaces of constant negative curvature (see Eisenhart [41, p. 280]). Here $\tau = \Omega(t - x)/c$, $\xi = (\delta\Omega/c)x$, and Ω, δ, c, are physical constants. Also, σ is defined in terms of the pulse envelope E, that is $E = \partial\sigma/\partial\tau$. The general solution of

Eq. (1.220) is not known. Particular solutions are obtainable by employing the Bäcklund transformation, Eq. (1.219), which becomes, in the notation of Eq. (1.220),

$$\begin{aligned} \frac{1}{2}\frac{\partial}{\partial \xi}(\sigma_1 + \sigma_0) &= \frac{1}{a_1}\sin\left(\frac{\sigma_1 - \sigma_0}{2}\right), \\ \frac{1}{2}\frac{\partial}{\partial \tau}(\sigma_1 - \sigma_0) &= a_1 \sin\left(\frac{\sigma_1 + \sigma_0}{2}\right). \end{aligned} \tag{1.221}$$

This provides a vehicle for relating pairs of solutions to Eq. (1.220). Thus if σ_0 is a known solution of Eq. (1.220), another solution σ_1 is obtained by solving the pair of first-order equations (1.221).

By inspection, we see that $\sigma_0 = 0$ is a particular solution of Eq. (1.220). From Eqs. (1.221) it follows that

$$\sigma_1 = 4 \tan^{-1}[\exp(a_1\tau + \xi/a_1)] \tag{1.222}$$

is another particular solution. Since $E = \partial\sigma/\partial\tau$, the pulse envelope is

$$E = \operatorname{sech} \tfrac{1}{2}\Omega(t - x/v), \qquad v = c/(1 - 4\delta),$$

where a_1 has been chosen to normalize E.

Determination of σ_1 from σ_0 required integration of the two elementary first-order equations. (1.221). It is known from the theory of surfaces (Eisenhart [41, p. 286]) that there exists a relationship among four solutions of Eq. (1.220) *which does not involve quadratures.*† Let a transformation from σ_i to σ_j with a constant a_k be represented as shown in Fig. 1-3a. The aforementioned relationship involving four solutions is

$$\tan\left(\frac{\sigma_3 - \sigma_0}{4}\right) = \frac{a_1 + a_2}{a_1 - a_2} \tan\left(\frac{\sigma_1 - \sigma_2}{4}\right). \tag{1.223}$$

The sequence of transformations shown in Fig. 1-3b can be combined to generate the purely trigonometric relation of Eq. (1.223). This procedure when continued yields solutions of increasing complexity as in the work of Lamb [48].

Equation (1.120) is transformed into $4\phi_{\xi\eta} = G(\phi)$ by setting $\xi = x + t$, $\eta = x - t$. Thus the methods of this section are directly applicable to the transmission-line wave problems of Section 1.6.

† We can think of Eq. (1.222) as σ_0 in Eq. (1.221) and generate an additional solution by integration. A sequence σ_n of solutions is generated by continued repetition of the process.

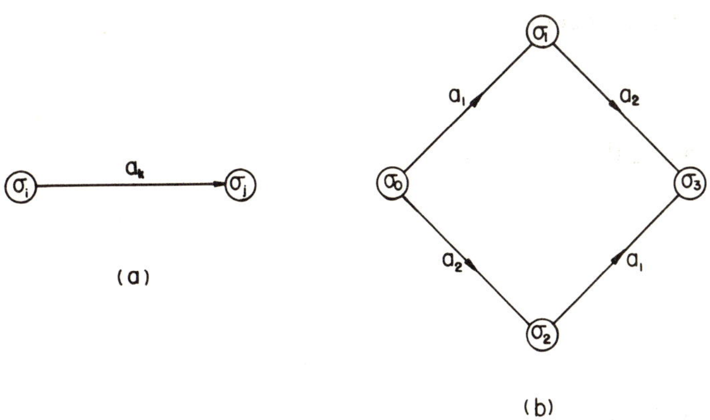

FIG. 1-3. Bäcklund transformation sequence.

1.14 FIRST INTEGRALS

A first (intermediate) integral of a second-order equation cannot be expected to contain more than one arbitrary function, for the elimination of two or more arbitrary functions from a first-order relation will, in general, lead to a differential equation of order higher than two. Indeed, a first integral may contain no arbitrary elements, or it may contain arbitrary constants. In Volume I (page 58ff.) the method of Mongé was discussed. There we showed that any second-order equation having a first integral

$$\phi(u, v) = 0, \tag{1.224}$$

where u and v are definite expressions depending upon x, y, z, $p = z_x$, $q = z_y$, and ϕ is arbitrary, must take the form†

$$Rz_{xx} + Sz_{xy} + Tz_{yy} + U(z_{xx}z_{yy} - z_{xy}^2) = V, \tag{1.225}$$

where R, S, T, U, and V are functions of x, y, z, p, and q.

Additional techniques for first integral calculation are found in the work of Forsyth [36].

An alternative method to those of the classicists was suggested by Jones and Ames [49]. It consists of selecting a first-order equation and

† We often refer to this as the Mongé–Amperé equation.

generating from it all possible second-order equations. Thus let us begin with the general nonlinear first-order equation

$$F(x, y, u, p, q) = 0. \tag{1.226}$$

Upon taking the x and y derivatives of Eq. (1.226), we find, respectively,

$$F_p u_{xx} + F_q u_{yx} + F_u u_x + F_x = 0, \tag{1.227}$$

and

$$F_q u_{yy} + F_p u_{xy} + F_u u_y + F_y = 0. \tag{1.228}$$

Additional equations are obtained by addition[†] (subtraction)

$$F_p u_{xx} + (F_q \pm F_p) u_{xy} \pm F_q u_{yy} + F_u(u_x \pm u_y) + F_x \pm F_y = 0, \tag{1.229}$$

and by elimination of u_{xy} between Eqs. (1.227) and (1.228),

$$u_{yy} - (F_p/F_q)^2 u_{xx} + (F_u/F_q^2)(F_q u_y - F_p u_x) + (1/F_q^2)(F_q F_y - F_p F_x) = 0. \tag{1.230}$$

If u is a twice-differentiable solution of Eq. (1.226), then u satisfies Eqs. (1.227)–(1.230). Nonlinear equations taking these forms occur in a number of areas. Before proceeding with the analysis, a few physical problems will be discussed.

1. THE FORM $u_{yy} - \{\phi^2(u)\, u_x\}_x = 0$ (1.231)

a. Transonic Gas Flow. The steady nearly uniform transonic flow of a gas, considered by Tomotika and Tamada [50] and Tamada [51], has the equation

$$W_{\psi\psi} = (2WW_\phi)_\phi\,, \tag{1.232}$$

where

$$W = \frac{w(\gamma + 1)}{2}, \qquad w = \int_1^q \left(\frac{\rho}{q}\right) dq.$$

Here γ is the ratio of specific heats, ρ is density, q is velocity magnitude, ψ is the stream function, and ϕ is the potential function.

† The integrability condition $u_{xy} = u_{yx}$ is assumed to be true.

b. One-Dimensional Gas Dynamics. The dynamics of a one-dimensional gas with pressure p, density ρ, and velocity u are governed by the well-known equations

$$\rho_t + (\rho u)_x = 0, \tag{1.233a}$$

$$u_t + uu_x + \rho^{-1}p_x = 0, \tag{1.233b}$$

$$p = p(\rho). \tag{1.233c}$$

These equations are transformed into the form of Eq. (1.231) by introducing the stream function ψ defined by $\rho = \psi_x$, $-\rho u = \psi_t$. When t and ψ are selected as the new independent variables, the basic transformation laws become

$$(\ )_x = (\ )_\psi \psi_x; \qquad (\ )_t|_x = (\ )_\psi \psi_t + (\ )_t|_\psi .$$

Upon application of these to Eqs. (1.233a) and (1.233b), we have

$$-\rho^{-2}\rho_t = u_\psi , \qquad u_t = -p_\psi ,$$

or

$$-(1/\rho)_{tt} = [p(\rho)]_{\psi\psi} . \tag{1.234}$$

Setting $\lambda = 1/\rho$, Eq. (1.234) becomes

$$\lambda_{tt} = -[p(\lambda)]_{\psi\psi} = [-(dp/d\lambda)\lambda_\psi]_\psi , \tag{1.235}$$

that is, a form completely analogous to Eq. (1.231) if we make the association $u \to \lambda$, $y \to t$, $x \to \psi$, and $\phi^2(\lambda) = -dp/d\lambda$.

c. Shallow Water Wave Theory. Shallow water waves (Stoker [31]) are approximately modeled by the equations

$$\bar{\rho}_t + (\bar{\rho} u)_x = 0, \qquad u_t + uu_x + \bar{p}_x/\bar{\rho} = 0,$$

where the "density" $\bar{\rho} = \alpha Z$ with α the *constant water density* and Z is the height above the reference plane. The "pressure" $\bar{p} = g\bar{\rho}^2/2\alpha$. Evidently, this situation is a special case of the previous example.

d. Wave Propagation. In Section 1.5 we derived Eq. (1.101):

$$\lambda_{tt} = [C^2(\lambda)\lambda_X]_X$$

for the stretch λ. With the proper identification of variables this equation is of the form Eq. (1.231). (See also Johnson [52].)

e. Longitudinal Wave Propagation on a Traveling Threadline. In a series of papers, Ames and his co-workers [7, 8, 26] have examined this topic in great detail. The governing equations (see e.g., [8, 26]) are

$$V_t + VV_x = (\lambda/\rho_0)\sigma_x, \tag{1.236a}$$

$$\lambda_t + V\lambda_x - \lambda V_x = 0, \tag{1.236b}$$

$$\sigma = \sigma(\lambda) = E_0 r(\lambda), \tag{1.236c}$$

where V, λ, σ, and ρ_0 are, respectively, longitudinal velocity, stretch, stress, and (constant) density. Upon setting $\lambda^{-1} = \psi_x$, $-\lambda^{-1}V = \psi_t$, the transformation of Eqs. (1.236) to the ψ, t-plane generates the relation

$$\lambda_{tt} = \mu^2[r'(\lambda)\lambda_\psi]_\psi, \qquad \mu^2 = E_0/\rho_0. \tag{1.237}$$

Equation (1.237) is analogous to Eq. (1.231) when the identification $y \to t$, $x \to \psi$, $u \to \lambda$, and $\phi^2 \to \mu^2 r'$ is made.

2. THE FORM $u_{yy} - \phi^2(u_x)\, u_{xx} - 0$ (1.238)

a. Vibrations of a Nonlinear String. Zabusky [53], motivated by the work of Fermi *et al.* [54] employed the equation

$$u_{tt} = (1 + \epsilon u_x)^\alpha u_{xx} \tag{1.239}$$

to describe the vibrations of a finite nonlinear string. The analogy between Eqs. (1.239) and (1.238) is direct if we set $y \to t$, $x \to x$, $\phi^2 \to (1 + \epsilon u_x)^\alpha$.

b. Propagation of Sound Waves in a Gas. The equation for the displacement, ξ, of a sound wave in a gas (see, e.g., Coulson [55]) is

$$\xi_{tt} = \mu^2\{1 + \xi_x\}^{-(\gamma+1)}(\partial^2\xi/\partial x^2), \tag{1.240}$$

where $p = k\rho^\gamma$ and $\mu^2 = \gamma p_0/\rho_0$, p_0 and ρ_0 being the normal average pressure and density, respectively. Equation (1.240) is exactly of the same form as Eq. (1.238).

c. Wave Propagation in Solids. In Section 1.5 we derived Eq. (1.100):

$$x_{tt} = C^2(x_X)x_{XX}$$

for the axial coordinate x as a function of its original abscissa X and time t. Here

$$C^2(x_X) = \rho_0^{-1}\, d\sigma/dx_X, \qquad \text{where} \quad \sigma = \sigma(\lambda) = \sigma(x_X).$$

Clearly, Eq. (1.100) is analogous to Eq. (1.238) after the obvious identifications are made.

1.15 DEVELOPMENT OF FIRST INTEGRALS

In this section, we shall demonstrate how some first integrals for Eqs. (1.231) and (1.238) may be developed and applied. This material is taken from Jones and Ames [49] and Ames [21, 56].

First, we assume that Eq. (1.226) has the form $F = F(u, p, q)$ and in Eq. (1.230) we set

$$(F_p/F_q)^2 = \phi^2(u). \tag{1.241}$$

This generates the two first-order equations

$$F_p + \phi(u)F_q = 0, \tag{1.242}$$

and

$$F_p - \phi(u)F_q = 0. \tag{1.243}$$

An elementary application of characteristics (Lagrange's method) demonstrates that $F = q - \phi(u)p$ is a solution† of Eq. (1.242), and $F = q + \phi(u)p$ is a solution of Eq. (1.243).

The final determination of Eq. (1.230) can now be completed. It is demonstrated by setting Eq. (1.243) into Eq. (1.230). This results in

$$u_{yy} - \phi^2(u)u_{xx} - \phi(u)\phi'(u)u_x^2 + \phi'(u)u_x u_y = 0,$$

which can be written

$$u_{yy} - [\phi^2(u)u_x]_x + \phi'(u)u_x[u_y + \phi(u)u_x] = 0. \tag{1.244}$$

Integrating $F = q + \phi(u)p = 0$, we obtain the general solution

$$G[u, x - \phi(u)y] = 0, \tag{1.245}$$

where G is arbitrary. Also, since $q + \phi(u)p = 0$, this means that Eq. (1.245) is a solution of

$$u_{yy} - [\phi^2(u)u_x]_x = 0,$$

that is, Eq. (1.231)!

† The general solution $\psi[F, q - \phi(u)p] = 0$, ψ arbitrary, is not used here because of complications introduced in subsequent steps of the analysis.

Approaching Eq. (1.242) in the same way, it is seen that

$$H[u, x + \phi(u)y] = 0 \tag{1.246}$$

is also a solution to Eq. (1.231) for arbitrary H.

The next order of equation complexity, within the framework of Eq. (1.230), arises from the assumption that $F = F(p, q)$ and $(F_p/F_q)^2$ is a function of p or q. For example, if

$$(F_p/F_q)^2 = \phi^2(p),$$

the two first-order equations

$$F_p + \phi(p)F_q = 0, \tag{1.247}$$

$$F_p - \phi(p)F_q = 0, \tag{1.248}$$

are generated. A solution of Eq. (1.247) is

$$F = q - h(p),$$

where $h' = \phi$. To solve $F = 0$, we need to solve

$$u_y - h(u_x) = 0,$$

which becomes

$$u_{yx} - h'(u_x)u_{xx} = 0 \tag{1.249}$$

upon differentiation with respect to x. With $v = u_x$, Eq. (1.249) becomes

$$v_y - \phi(v)v_x = 0,$$

which has a solution

$$H[v, x + \phi(v)y] = 0,$$

or

$$H[u_x, x + \phi(u_x)y] = 0. \tag{1.250}$$

Moreover, Eq. (1.230) becomes

$$u_{yy} - \phi^2(u_x)u_{xx} = 0,$$

that is, Eq. (1.238), under these assumptions. Consequently, Eq. (1.250) is a first integral of Eq. (1.238).

In an entirely, analogous fashion, we find that

$$G[u_x, x - \phi(u_x)y] = 0 \tag{1.251}$$

is also a first integral of Eq. (1.238) for arbitrary G.

These implicit solutions or first integrals can be employed to study the properties of the wave propagation. For example the evolution of discontinuities from smooth† initial data can be examined. In addition, one can generate series solutions with ease from the first integrals. The breakdown development will be given in this section and the series solution in the next section.

The finite "time" to the evolution of a discontinuity in u_x or higher order derivatives can be calculated from Eq. (1.245), (1.246), (1.250), or (1.251). For example, from the total x-derivative of Eq. (1.245), we obtain

$$u_x = -\frac{G_x + G_\omega}{G_u - yG_\omega\phi'(u)}, \qquad \omega = x - \phi(u)y, \tag{1.252}$$

which becomes arbitrarily large when

$$y = G_u/G_\omega\phi'(u).$$

Since the initial "time" of the occurrence is usually of most interest, we write

$$y_c = \min[G_u/G_\omega\phi'(u)], \tag{1.253}$$

where the minimum is evaluated over the appropriate range of the quantities in Eq. (1.253). For real problems, general interest is in positive values of y_c.

In some circumstances, the general solutions are employable in simpler, but still implicit, forms. Thus, instead of Eq. (1.246) we may find that

$$u = h[x + \phi(u)y] \tag{1.254}$$

is applicable. [Note the resemblance to a traveling wave if we think of $\phi(u)$ as a velocity.] In this case, the x-derivative becomes

$$u_x = h'(\omega)/[1 - h'(\omega)\phi'(u)y], \qquad \omega = x + \phi(u)y,$$

which is unbounded when

$$y = 1/h'(\omega)\phi'(u),$$

and the critical minimum value is

$$y_c = \min[h'(\omega)\phi'(u)]^{-1}. \tag{1.255}$$

† The term "smooth" is employed herein to mean no discontinuities in the function or its derivatives of all orders.

Similar calculations for Eq. (1.238) can be carried out from Eqs. (1.250) and (1.251). A discontinuity in the first derivative is discovered by inquiring when the second derivative becomes unbounded. Thus from Eq. (1.251) we find

$$u_{xx} = -\frac{G_x + G_\omega}{G_p - yG_\omega \phi'(p)}, \qquad p = u_x, \quad \omega = x - \phi(p)y, \tag{1.256}$$

and, consequently

$$y_c = \min \frac{G_p}{G_\omega \phi'(p)}. \tag{1.257}$$

As a first physical example where this theory can be employed, we consider the one-dimensional gas dynamics of Section 1.14, Eq. (1.235). According to the results for Eq. (1.231), that is Eqs. (1.245) and (1.246), solutions of Eq. (1.235) are

$$\lambda = H[\psi + t(-dp/d\lambda)^{1/2}], \qquad \text{or} \qquad 1/\rho = H[\psi + t(\rho^2\, dp/d\rho)^{1/2}],$$

and

$$1/\rho = G[\psi - t(\rho^2\, dp/d\rho)^{1/2}]. \tag{1.258}$$

These implicit displays of density immediately provide a vehicle for calculating the time to breakdown of ρ_x. In the general case, we find from Eq. (1.258)

$$\rho_x = \frac{2G'\rho^2(p')^{1/2}}{tG'(2\rho p' + \rho^2 p'') - 2\rho^{-1}(p')^{1/2}},$$

whereupon ρ_x is unbounded if

$$t = \frac{2(dp/d\rho)^{1/2}}{G'(2\rho^2\, dp/d\rho + \rho^3\, d^2p/d\rho^2)}.$$

The critical time is

$$t_c = \frac{2\min(p')^{1/2}}{\max[G'(2\rho^2 p' + \rho^3 p'')]}, \tag{1.259}$$

where the inclusion of G' introduces the effect of the initial state. For the special polytropic case, $p = A\rho^\gamma$, Eqs. (1.258) and (1.259) become

$$1/\rho = G[\psi - t(A\gamma\rho^{\gamma+1})^{1/2}],$$

and

$$t_c = \left[(A\gamma)^{1/2}\left(\frac{\gamma+1}{2}\right)\max\{G'\rho^{(\gamma+3)/2}\}\right]^{-1}. \tag{1.260}$$

Equation (1.260) clearly demonstrates that there is no breakdown if $G' = 0$ or if $\gamma = 0$ or -1.

A second example concerns the longitudinal wave propagation on a traveling threadline. Herein we shall use Eq. (1.100), which is of the same form as Eq. (1.238). Suppose that

$$\sigma = 2E_1(\lambda - 1) + E_1(\lambda - 1)^2,$$

so that $d\sigma/d\lambda = 2E_1\lambda = 2E_1 x_X$. Then Eq. (1.100) becomes

$$x_{tt} = C_1^2(\partial x/\partial X)(\partial^2 x/\partial X^2),$$

where $C_1^2 = 2E_1/\rho_0$. Using the simplified form $x_X = H[X + \phi(x_X)t]$ of Eq. (1.250), we find

$$x_X = H[X + C_1(x_X)^{1/2}t],$$

and from Eq. (1.257) the time to breakdown is

$$t_c = [2 \min(x_X)^{1/2}]/(C_1 \max H').$$

If the string is always in tension, $\lambda = x_X \geqslant 1$. For the special case $H(\omega) = \sin \omega$, the exact value of t_c is

$$t_c = 2/C_1 = \{2\rho_0/E_1\}^{1/2}.$$

Additional examples and discussion of the results are available in the work of Ames [21, 56].

1.16 LAGRANGE SERIES SOLUTIONS

While useful for the determination of such quantitative results as the breakdown times for waves, the implicit nature of the solutions and first integrals of Section 1.15 inhibits their use in the final display of the solutions. Alternatively, we can obtain a series solution to the equations $F = 0$, which are

$$u_y \pm \phi(u)u_x = 0 \tag{1.261}$$

in the first case, and

$$u_y \pm h(u_x) = 0 \tag{1.262}$$

in the second. Lagrange expansions are discussed in various contexts by Goursat [40], Bellman [57], Banta [58] (in connection with finite amplitude sound waves), and Ames and Jones [59] (for a Mongé–Amperé equation).

A Lagrange series is now constructed for the problem $u_y + \phi(u)u_x = 0$, $u(x, 0) = f(x)$ as a typical example of the methodology. Suppose u has a Taylor's series expansion about $y = 0$,

$$u(x, y) = \Gamma_0(x) + \sum_{n=1}^{\infty} \Gamma_n(x) y^n/n!$$

$$\Gamma_n(x) = \partial^n u(x, y)/\partial y^n|_{y=0} \,. \tag{1.263}$$

As is immediately obvious, this form is inconvenient, since the derivatives are with respect to y and not x, while the "initial" data are in the x variable. Replacement of the y-derivatives is carried out by using the differential equation and an inductive scheme due to Goursat [40, p. 405] (see also Banta [58]). If $u_y + \phi(u)u_x = 0$, then

$$\frac{\partial^n u}{\partial y^n} = (-1)^n \frac{\partial^{n-1}}{\partial x^{n-1}} \left[\phi^n \frac{\partial u}{\partial x}\right], \qquad n \geqslant 1, \tag{1.264}$$

as can easily be verified by induction. Consequently, for $n \geqslant 1$,

$$\Gamma_n(x) = (-1)^n \frac{\partial^{n-1}}{\partial x^{n-1}} \left\{\phi^n \frac{\partial u}{\partial x}\right\}\bigg|_{y=0},$$

a form that contains only x-derivatives! If the processes of x-differentiation and evaluation at $y = 0$ are interchangeable, the series takes the form

$$u(x, y) = f(x) + \sum_{n=1}^{\infty} \left(\frac{y^n}{n!}\right) \frac{d^{n-1}}{dx^{n-1}} \left\{[\phi(f(x))]^n \frac{df}{dx}\right\},$$

which is valid out to the first singularity—that is, out to the smallest breakdown "time" y_c.

To integrate $u_y + h(u_x) = 0$, which is no longer quasi-linear, note that it becomes

$$u_{yx} + h'(u_x)u_{xx} = 0 \tag{1.265}$$

upon differentiation with respect to x. With $v = u_x$, Eq. (1.265) becomes $v_y + h'(v)\ v_x = 0$, an equation of the same form as that previously analyzed. Upon solving for v, u is recovered by integration.

These Lagrange series can be employed to study the waveform in its transition from the smooth initial data $f(x)$ to the onset of breakdown.

1.17 BREAKDOWN THEORY OF JEFFREY–LAX

Discontinuity evolution in solutions of nonlinear hyperbolic equations possessing *discontinuous* initial data has been a topic of research interest for many years. Studies include those of Friedrichs [60], Courant and Friedrichs [61], Lax [62], Thomas [63], and Jeffrey [64–66]. The waves associated with these problems have a clearly defined wavefront, a characteristic along which the initial discontinuity propagates, until, in some cases, at some critical time it tends to a jump discontinuity. The condition for the nonoccurrence of such a jump, known as an *exceptional* case and the determination of the exact time of jump formation on the wavefront in the genuinely nonlinear case have been given in detail for hyperbolic systems of first and higher order by Thomas [63] and Jeffrey [64–66].

The evolution of discontinuities from *smooth* initial data was probably first examined in a simple problem by Riemann [67]. His conjecture concerned the conditions for a simple wave to develop a discontinuity. Ludford [68] has reexamined this conjecture in the context of the initial-value problem for unsteady isentropic perfect gas flow. He employed a variant of the hodograph method, which basically is an unfolding process for initial curves in the hodograph plane. An estimate of the time to breakdown of the solution is obtained. At this critical time, the derivative of the solution becomes unbounded. For the oscillations of a nonlinear string, Zabusky [53] (see also Volume I) used the Riemann invariants and Ludford's unfolding method to develop both a solution and an estimate of the time to breakdown. We shall discuss another example of Ludford's method in Section 1.20. Lastly, the work of Lax [69] and Jeffrey [70] also employ the Riemann invariants to develop comparison theorems that provide upper and lower bounds for the critical time of singularity occurrence. Jeffrey's work [70] is more general, and he has given several examples utilizing his theorems. Consequently, we present those details here and some applications in Section 1.18.

We shall consider a reducible† system of two homogeneous quasi-linear first-order partial-differential equations

$$U_t + AU_x = 0, \qquad U = \begin{bmatrix} u_1 \\ u_2 \end{bmatrix}, \qquad A = \begin{bmatrix} a_{11} & a_{12} \\ a_{21} & a_{22} \end{bmatrix}. \tag{1.266}$$

It is assumed that the eigenvalues $\lambda^{(i)}$ $(i = 1, 2)$ of $| A - \lambda I | = 0$ are real

† The coefficients of A, Eq. (1.266), are functions only of the dependent variables u_1 and u_2, and these coefficients are continuous, piecewise-differentiable functions of u_1 and u_2.

and distinct in the region of the (u_1, u_2) dependent variable space in question. Thus the characteristic curves are $C^{(1)}$: $dx/dt = \lambda^{(1)}$ and $C^{(2)}$: $dx/dt = \lambda^{(2)}$. (For the general quasi-linear theory see Volume I, page 72.) If, in addition, the *left eigenvectors* $l^{(i)}$—that is, $\lambda^{(i)} l^{(i)} = l^{(i)} A$—are linearly independent, the system so considered is said to be *totally hyperbolic*.

Premultiplying Eq. (1.266) by the row vector $l^{(i)}$ and using the relationship $l^{(i)}A = \lambda^{(i)}l^{(i)}$ generates the equations

$$l^{(i)}(U_t + \lambda^{(i)} U_x) = 0, \qquad i = 1, 2, \tag{1.267}$$

which are valid along the $C^{(1)}$ and $C^{(2)}$ characteristic curves, respectively. The expression (Volume I, pages 72–78) $(U_t + \lambda^{(i)} U_x)$ of Eq. (1.267) simply denotes differentiation of U with respect to time along the ith characteristic. Let us now parametrize the $C^{(1)}$ characteristic curves by the requirement that the $C^{(1)}$ characteristic through the point $(0, t_0)$ has associated with it the constant value $\beta(t_0)$, where $\beta(t)$ is *any* monotonic differentiable function of t. Thus β has a different constant value along each $C^{(1)}$ characteristic. A monotonic differentiable function $\alpha(t)$ is employed to parametrize the $C^{(2)}$ characteristics, so that α has a different constant value along each $C^{(2)}$ characteristic. Actually, any parametrization can be used, subject only to the monotonic differentiable condition. Consequently, we shall use the pair $\alpha(t) = t$, $\beta(t) = t$, but, to avoid confusion, we shall retain the symbols α and β as new independent variables.

In terms of α and β the two equations represented by Eq. (1.267) can be written

$$\begin{aligned} l^{(1)} U_\alpha &= 0, \qquad \text{along} \quad C^{(1)} \text{ characteristics,} \\ l^{(2)} U_\beta &= 0, \qquad \text{along} \quad C^{(2)} \text{ characteristics.} \end{aligned} \tag{1.268}$$

Writing the row vector $l^{(i)} = [l_1^{(i)}, l_2^{(i)}]$, these become

$$\begin{aligned} l_1^{(1)} u_{1\alpha} + l_2^{(1)} u_{2\alpha} &= 0, \qquad \text{along} \quad C^{(1)} \text{ characteristics,} \\ l_1^{(2)} u_{1\beta} + l_2^{(2)} u_{2\beta} &= 0, \qquad \text{along} \quad C^{(2)} \text{ characteristics.} \end{aligned} \tag{1.269}$$

The left eigenvector $l^{(i)}$, which is arbitrary up to a nonzero scale factor, might not be in the appropriate form to make Eqs. (1.269) exact differentials. Consequently Eqs. (1.269) are multiplied by the integrating factors $\bar{q}_1$ and $\bar{q}_2$, if necessary, so that they become exact. Then, upon integrating with respect to α and β, respectively, we find the relations

$$\begin{aligned}
&\int \bar{q}_1 l_1^{(1)}\, du_1 + \int \bar{q}_1 l_2^{(1)}\, du_2 = r(\beta), \qquad \text{along} \quad C^{(1)} \text{ characteristics,} \\
&\int \bar{q}_2 l_1^{(2)}\, du_1 + \int \bar{q}_2 l_2^{(2)}\, du_2 = s(\alpha), \qquad \text{along} \quad C^{(2)} \text{ characteristics,}
\end{aligned} \tag{1.270}$$

where the integration constants are known as the *Riemann invariants*.

As is usual in the discussion of reducible equations (see Courant and Friedrichs [61, Sect. 21]) the relationship $(u_1, u_2) \to (r, s)$ is assumed to be one to one. The discontinuities that can occur in the resulting simple wave flows when this condition is not satisfied have been discussed by Courant and Friedrichs [61] and Jeffrey and Taniuti [65].

From Eqs. (1.270) and the definition of differentiation along a characteristic, it follows that alternative forms of Eqs. (1.269), in terms of $r(\beta)$ and $s(\alpha)$, are

$$\frac{dr}{d\alpha} = \frac{\partial r}{\partial t} + \lambda^{(1)} \frac{\partial r}{\partial x} = 0, \qquad \text{along} \quad C^{(1)} \text{ characteristics,} \tag{1.271a}$$

$$\frac{ds}{d\beta} = \frac{\partial s}{\partial t} + \lambda^{(2)} \frac{\partial s}{\partial x} = 0, \qquad \text{along} \quad C^{(2)} \text{ characteristics.} \tag{1.271b}$$

Equations (1.271) are fundamental in that they are used to derive important differential equations that apply along the characteristics. The singularity in the solutions of those equations determines the time and position of singularity occurrence in solutions of Eq. (1.266) when suitable auxiliary conditions are given.

Since $dr/d\alpha$ in Eq. (1.271a) involves the behavior of r_t and r_x along $C^{(1)}$ characteristics, it will be used to determine the differential equations satisfied by these functions along $C^{(1)}$ characteristics. Differentiating Eq. (1.271a) with respect to α and β, we obtain

$$\left(\frac{\partial}{\partial t} + \lambda^{(1)} \frac{\partial}{\partial x}\right)\left(\frac{\partial r}{\partial t} + \lambda^{(1)} \frac{\partial r}{\partial x}\right) = 0,$$

and

$$\left(\frac{\partial}{\partial t} + \lambda^{(2)} \frac{\partial}{\partial x}\right)\left(\frac{\partial r}{\partial t} + \lambda^{(1)} \frac{\partial r}{\partial x}\right) = 0, \tag{1.272}$$

which are respectively valid along $C^{(1)}$ characteristics. Since $\lambda^{(1)} \neq \lambda^{(2)}$, it follows immediately that

$$\frac{\partial}{\partial t}\left(\frac{\partial r}{\partial t} + \lambda^{(1)} \frac{\partial r}{\partial x}\right) = 0, \qquad \text{and} \qquad \frac{\partial}{\partial x}\left(\frac{\partial r}{\partial t} + \lambda^{(1)} \frac{\partial r}{\partial x}\right) = 0, \tag{1.273}$$

along $C^{(1)}$ characteristics. Since $\lambda^{(1)}$ and $\lambda^{(2)}$ are finite and r_t and r_x are related by Eq. (1.271a), it will suffice to determine the time to solution breakdown, due to the Riemann invariant r, by finding that time (if any) at which r_x becomes unbounded.

Performing the operation indicated in

$$\frac{\partial}{\partial x}\left(\frac{\partial r}{\partial t} + \lambda^{(1)} \frac{\partial r}{\partial x}\right) = 0,$$

we find that

$$r_{tx} + \lambda_r^{(1)}(r_x)^2 + \lambda_s^{(1)} s_x r_x + \lambda^{(1)} r_{xx} = 0 \tag{1.274}$$

must hold along $C^{(1)}$ characteristics. By a similar analysis, the equation describing the variation of s_x along $C^{(2)}$ characteristics is found to be

$$s_{tx} + \lambda_s^{(2)}(s_x)^2 + \lambda_r^{(2)} r_x s_x + \lambda^{(2)} s_{xx} = 0. \tag{1.275}$$

This equation determines the time of solution breakdown due to the Riemann invariant s.

Equation (1.274) can be simplified by noting that the first and last terms combine to form $d(\partial r/\partial x)/d\alpha$. Then, applying the operator identity $d/d\alpha = d/d\beta + (\lambda^{(1)} - \lambda^{(2)})\partial/\partial x$ to $s(\alpha)$, we see that $ds/d\alpha = (\lambda^{(1)} - \lambda^{(2)})$ $(\partial s/\partial x)$, whereupon Eq. (1.274) reduces to

$$\frac{d}{d\alpha}\left(\frac{\partial r}{\partial x}\right) + \frac{\partial \lambda^{(1)}}{\partial r}\left(\frac{\partial r}{\partial x}\right)^2 + [\lambda^{(1)} - \lambda^{(2)}]^{-1} \frac{ds}{d\alpha} \frac{\partial \lambda^{1)}}{\partial s} \frac{\partial r}{\partial x} = 0. \tag{1.276}$$

Equation (1.276) is a homogeneous nonlinear ordinary-differential equation along the $C^{(1)}$ characteristics. To achieve a further simplification, we set $v_1 = (\partial r/\partial x)f_1$, thereby converting Eq. (1.276) into

$$\frac{dv_1}{d\alpha} + \frac{1}{f_1}\left\{\frac{\partial \lambda^{(1)}}{\partial r}\right\} v_1{}^2 = 0, \tag{1.277}$$

provided f_1 is chosen to satisfy

$$f_1^{-1}(df_1/ds) = [\lambda^{(1)} - \lambda^{(2)}]^{-1}(\partial \lambda^{(1)}/\partial s). \tag{1.278}$$

Upon setting the integration constant equal to unity, f_1 is found to be

$$f_1 = \exp\left\{\int (\lambda^{(1)} - \lambda^{(2)})^{-1}(\partial \lambda^{(1)}/\partial s)\, ds\right\}.$$

If we set

$$q_1 = \int (\lambda^{(1)} - \lambda^{(2)})^{-1}(\partial \lambda^{(1)}/\partial s)\, ds,$$

then Eq. (1.277) for the variable v_1 along the $C^{(1)}$ characteristics becomes

$$(dv_1/d\alpha) + e^{-q_1}(\partial\lambda^{(1)}/\partial r)v_1^2 = 0. \tag{1.279}$$

Similar reasoning with $v_2 = (\partial s/\partial x) f_2$, $f_2 = \exp q_2$,

$$q_2 = \int (\lambda^{(2)} - \lambda^{(1)})^{-1}(\partial\lambda^{(2)}/\partial r)\, dr$$

shows that the equation for v_2 along the $C^{(2)}$ characteristics is

$$(dv_2/d\beta) + e^{-q_2}(\partial\lambda^{(2)}/\partial s)v_2^2 = 0. \tag{1.280}$$

Equations (1.279) and (1.280) will be taken as the starting point for the determination of asymptotic estimates of the critical time t_c at which the solution to Eq. (1.266), together with the Lipschitz continuous (LC)† initial condition

$$U(x, 0) = \Phi(x), \tag{1.281}$$

first develops a singularity due to v_1 or v_2 becoming unbounded along their respective characteristics. By virtue of Eqs. (1.270), $\Phi(x)$ determines the values of r and s associated with the $C^{(1)}$ and $C^{(2)}$ characteristics passing through each point of the x-axis.

Equations (1.279) and (1.280) are both of the form

$$dw/dx = G(x, w),$$

in which $G(x, w)$ is LC with respect to w, since the coefficients of $A(U)$ are LC, which implies that the coefficients of v_1^2 and v_2^2 are at least LC. This implies (see Coddington and Levinson [71, Theorem 2.2, p. 10]) that the solution w of this equation is unique. Consequently, the solutions v_1 and v_2 of Eqs. (1.279) and (1.280) are unique.

The basic nonlinear Eq. (1.279) and (1.280) cannot be integrated in the form displayed, so we shall instead resort to a comparison theorem. Let us begin by considering Eq. (1.279) which is appropriate to a $C^{(1)}$ characteristic through the point $x = \xi$ on $t = 0$. The coefficient $e^{-q_1}\, \partial\lambda^{(1)}/\partial r$ of that equation is a function of α and $v_1(\alpha)$ when both the point $x = \xi$ and the initial vector $\Phi(x)$ have been specified. Denoting

† The function $f(x)$ is Lipschitz continuous on $a \leqslant x \leqslant b$ if $|f(x_1) - f(x_2)| \leqslant M\,|x_1 - x_2|$ for a constant M and for all x_1, x_2 in $a \leqslant x \leqslant b$. An LC function is continuous and the modulus of the derivative is bounded by M.

this coefficient by $-A(\alpha, \xi)$, we compare the solutions v_1 and $\bar{v}_1$ of the two equations

$$dv_1/d\alpha = A(\alpha, \xi)v_1^2, \tag{1.282}$$

and

$$d\bar{v}_1/d\alpha = A_0(\xi)\bar{v}_1^2, \tag{1.283}$$

where $A_0(\xi) = A(\alpha_0, \xi)$ and $\alpha = \alpha_0$ at $x = \xi$, on the initial line. The *comparison solution* $\bar{v}_1$ is the solution of a *constant coefficient equation.*

We now write $B(\alpha, v_1(\alpha), \xi) = A(\alpha, \xi)\, v_1^2$, which is LC with constant $K(\xi)$ and $B_0(\bar{v}_1(\alpha), \xi) = A_0(\xi)\, \bar{v}_1^2$, and assume

$$|\, B(\alpha, w, \xi) - B_0(w, \xi)| \leqslant M(\xi), \tag{1.284}$$

for some range of α, say $\alpha_0 \leqslant \alpha \leqslant \alpha_1$, where α_1 is chosen such that w remains uniformly bounded for all values of $A(\alpha, \xi)$ and $A_0(\xi)$, ξ on the initial line. That it is possible to choose that value α_1 will be established subsequently. In this notation the final forms of the equations to be compared are

$$dv_1/d\alpha = B(\alpha, v_1(\alpha), \xi), \tag{1.285}$$

$$d\bar{v}_1/d\alpha = B_0(\bar{v}_1(\alpha), \xi). \tag{1.286}$$

Using an obvious extension of a standard comparison theorem given by Coddington and Levinson [71, Theorem 2.1, p. 8] it follows that

$$|\, v_1(\alpha) - \bar{v}_1(\alpha)| \leqslant |\, v_1(\alpha_0) - \bar{v}_1(\alpha_0)| \exp\{K(\xi)|\, \alpha - \alpha_0\,|\} + \frac{M(\xi)}{K(\xi)} [\exp\{K(\xi)|\, \alpha - \alpha_0\,|\} - 1]. \tag{1.287}$$

It will be convenient to adopt the same initial conditions for the comparison solution $\bar{v}_1$ as for the genuine solution v_1, we may set $v_1(\alpha_0) = \bar{v}_1(\alpha_0)$ at all points of the initial line to obtain

$$|\, v_1(\alpha) - \bar{v}_1(\alpha)| \leqslant [M(\xi)/K(\xi)][\exp\{K(\xi)|\, \alpha - \alpha_0\,|\} - 1]. \tag{1.288}$$

This provides a bound for the modulus of the error between the required solution v_1 and the simple comparison solution $\bar{v}_1$ at a point with coordinate α on the determining $C^{(1)}$ characteristic through $x = \xi$ on the initial line. A similar result is also valid for the solution v_2 of Eq. (1.280) and its related comparison solution $\bar{v}_2$. Later we shall integrate the comparison Eq. (1.283), so that when the form of $e^{-q_1}(\partial\lambda^{(1)}/\partial r)$ is specified and $K(\xi)$ and $M(\xi)$ are known, inequality (1.288) may be used to estimate the modulus of its difference from v_1 for $\alpha_0 \leqslant \alpha \leqslant \alpha_1$.

The discussion of the estimate for v_2 is strictly analogous to that for v_1. So it now remains to show the existence of the number α_1 introduced in connection with condition (1.284). This is established prior to determining the asymptotic estimate of the critical time.

To determine the behavior of the comparison solution $\bar{v}_1$, say, as a function of the time t, we need only recall that the parametrization $\alpha(t) = t$ has been used along the $C^{(1)}$ characteristics, so that direct integration of Eq. (1.283) yields

$$\bar{v}_1(t) = \frac{\bar{v}_{01}(\xi)}{1 - tA_0(\xi)\bar{v}_{01}(\xi)}, \tag{1.289}$$

where $\bar{v}_{01}(\xi)$ denotes the initial value of $\bar{v}_1$ at the point $x = \xi$ on the initial line $t = 0$ through which the defining characteristic passes. This expression shows that when $A_0(\xi)\bar{v}_{01}(\xi)$ is positive, the comparison solution $\bar{v}_1$ becomes unbounded on the defining $C^{(1)}$ characteristics at elapsed critical times $T_c^{(1)}(A_0(\xi), \xi)$ determined by the expression

$$T_c^{(1)}(A_0(\xi), \xi) = [A_0(\xi)\bar{v}_{01}(\xi)]^{-1}. \tag{1.290}$$

The superscript 1 and the argument ξ are used to signify that $T_c^{(1)}(A_0(\xi), \xi)$ is the comparison solution critical time on the $C^{(1)}$ characteristic that passes through the point with coordinate $x = \xi$ on the initial line $t = 0$.

Similar reasoning demonstrates that the comparison solution $\bar{v}_2$ becomes unbounded on the defining $C^{(2)}$ characteristics at a critical time determined by the expression

$$T_c^{(2)}(\bar{A}_0(\eta), \eta) = [\bar{A}_0(\eta)\bar{v}_{02}(\eta)]^{-1}. \tag{1.291}$$

The superscript 2 signifies that $T_c^{(2)}(\bar{A}_0(\eta), \eta)$ is the comparison solution critical time on the $C^{(2)}$ characteristic passing through the point with coordinate $x = \eta$ on the initial line $t = 0$. In this equation, $\bar{A}_0(\eta)$ has been used to denote the constant coefficient in the comparison equation for $\bar{v}_2$ that corresponds to the coefficient $A_0(\xi)$ in Eq. (1.283).

Provided the derivatives $(\partial\lambda^{(1)}/\partial r)$ and $(\partial\lambda^{(2)}/\partial s)$ do not change sign along the initial line, by always choosing the signs of the Riemann invariants so that these derivatives are negative, the coefficients $A_0(\xi)$ and $\bar{A}_0(\eta)$ may always be taken as positive, thereby ensuring that the elapsed critical times given by Eqs. (1.290) and (1.291) are positive whenever $\bar{v}_{01}$, $\bar{v}_{02}$ are positive.

Let us now establish the existence of the number α_1 used in Eq. (1.284) and examine the relationship of the comparison solution critical time

$T_c^{(1)}(A(\xi), \xi)$ to the critical time $t_c^{(1)}(\xi)$ that is appropriate to v_1. Again the result for v_2 is strictly analogous, so only the details of the argument for the variable v_1 will be presented.

We begin by considering Eq. (1.283) defined along the $C^{(1)}$ characteristic through the point $x = \xi$ on the initial line (hereafter replacing α by t) together with the new comparison equation

$$d\tilde{v}_1/dt = N\tilde{v}_1^2, \tag{1.292}$$

in which N is some positive constant. We shall assume that $\tilde{v}_1 = v_1$ when $t = 0$, so that Eqs. (1.282) and (1.292) have identical initial conditions. If Eq. (1.282) is now subtracted from this equation, it is obvious that we can write the result in the form

$$(d/dt)[\tilde{v}_1 - v_1] = [N - A(t, \xi)]\tilde{v}_1^2 + A(t, \xi)[\tilde{v}_1^2 - v_1^2].$$

Using the initial condition $v_1 = \tilde{v}_1$ when $t = 0$, it then follows that the rate and direction of change of the difference $(\tilde{v}_1 - v_1)$ is determined by the magnitude and sign of $N - A(t, \xi)$. When, for example, $A(t, \xi) > 0$ and it is possible to choose a positive N dependent on ξ, which we shall write as $N(\xi)$, such that $N(\xi) - A(t, \xi) > 0$ throughout the range of integration with respect to time t, we can immediately conclude that $v_1 < \tilde{v}_1$ and so $\tilde{v}_1$ then provides an upper bound for $\tilde{v}_1$ on the defining characteristic. Since we have seen from Eq. (1.290) that for a comparison equation of this kind $\tilde{v}_1$ will cease to be defined beyond the time

$$T_c^{(1)}(N(\xi), \xi) = [N(\xi)\tilde{v}_{01}(\xi)]^{-1},$$

it immediately follows that along the characteristic that is involved $T_c^{(1)}(N(\xi), \xi)$ provides a lower bound for the time for which the exact solution v_1 exists.

Provided $A(t, \xi)$ does not change sign along a characteristic (i.e., $\partial\lambda^{(1)}/\partial r$ and $\partial\lambda^{(2)}/\partial s$ remain negative everywhere), by taking $N(\xi)$ to be the least upper bound $n(\xi)$ of $A(t, \xi)$ along the characteristic, we obtain the expression

$$T_{\inf}^{(1)}(\xi) = 1/n(\xi)\tilde{v}_{01}(\xi)$$

for the greatest lower bound of the times $T_c^{(1)}(N(\xi), \xi)$. This then is the best estimate that can be obtained of the lower bound for the critical time (for the genuine solution v_1) associated with the $C^{(1)}$ characteristic through $x = \xi$ on the initial line.

To obtain the analytic form of $T_{\inf}^{(1)}(\xi)$, we must now utilize the results

$$N(\xi) = -(\partial\lambda^{(1)}/\partial r)\exp\{-q_1(r, s)\}, \qquad \tilde{v}_{01}(\xi) = [(\partial r/\partial x)\exp\{q_1(r, s)\}]_{t=0}\,.$$

Recalling that $x = \xi$ on the initial line determines a particular $C^{(1)}$ characteristic, and so corresponds to some definite value of r which we shall denote by $r_0(\xi)$ while s, which varies along the $C^{(1)}$ characteristic, is some function of time with the initial value $s = s_0(\xi)$ at $t = 0$, since on the initial line $r(x, t)$ and $s(x, t)$ are of the form $r_0(x) = r(x, 0)$ and $s_0(x) = s(x, 0)$; the expression for $T_{\inf}^{(1)}(\xi)$ then becomes

$$T_{\inf}^{(1)}(\xi) = \frac{-1}{\left(\dfrac{\partial r}{\partial x}\right)_{\substack{t=0\\x=\xi}} \exp\{q_1(r_0(\xi), s_0(\xi))\} \sup\left[\left(\dfrac{\partial\lambda^{(1)}}{\partial r}\right)\exp\{-q_1(r_0(\xi), s)\}\right]_s}, \tag{1.293}$$

where $\sup[\cdot]_s$ denotes the supremum with respect to s, the variable r_0 being held constant. Each $C^{(1)}$ characteristic has such a time $T_{\inf}^{(1)}(\xi)$ associated with it, and if we denote the greatest lower bound of $T_{\inf}^{(1)}(\xi)$ by ${}_*T^{(1)}$, then ${}_*T^{(1)}$ is the desired lower bound of the times for which the comparison solution exists.

The critical time ${}_*T^{(1)}$ associated with the $C^{(1)}$ characteristics thus has the form

$${}_*T^{(1)} = \inf\left[\frac{-1}{\left(\dfrac{\partial r}{\partial x}\right)_{\substack{t=0\\x=\xi}} \exp\{q_1(r_0(\xi), s_0(\xi))\} \sup\left[\left(\dfrac{\partial\lambda^{(1)}}{\partial r}\right)\exp\{-q_1(r_0(\xi), s)\}\right]_s}\right]_\xi, \tag{1.294}$$

and is the least upper bound of the numbers α_1 involved in Eq. (1.284) as ξ ranges over the initial line. That this bound exists follows immediately from the fact that for a properly posed problem, $\Phi(x)$ and the initial Riemann invariant distributions $r_0(x) = r(x, 0)$ and $s_0(x) = s(x, 0)$ are continuous and finitely bounded functions, so that $T_{\inf}^{(1)}(\xi)$, which is a continuous function of them, must itself be finitely bounded. The time ${}_*T^{(1)}$ must be positive, so that a breakdown in the solution due to the $C^{(1)}$ characteristics will occur, whenever

$$\left(\frac{\partial r}{\partial x}\right)_{\substack{t=0\\x=\xi}} > 0.$$

This establishes the existence of a number $\alpha_1 < {}_*T^{(1)}$ with the property that the function w in Eq. (1.284) remains uniformly bounded for all points ξ on the initial line provided $\alpha_0 < \alpha < \alpha_1$.

A similar argument when applied to the $C^{(2)}$ characteristics gives rise to a critical time ${}_*T^{(2)}$ associated with the $C^{(2)}$ characteristics of the form

$$ {}_*T^{(2)} = \inf\left[\frac{-1}{\left(\frac{\partial s}{\partial x}\right)_{\substack{t=0\\x=\eta}} \exp\{q_2(r_0(\eta), s_0(\eta))\} \sup\left[\left(\frac{\partial\lambda^{(1)}}{\partial s}\right)\exp\{-q_2(r, s_0(\eta))\}\right]_r}\right]_\eta . \tag{1.295} $$

Since both the $C^{(1)}$ and $C^{(2)}$ characteristics can give rise to a breakdown of the solution, it follows that the number $t_{\inf}$, defined to be the least positive number of ${}_*T^{(1)}$ and ${}_*T^{(2)}$, provides the lower bound for the time of existence of a solution of the comparison equations. Since $\tilde{v}_i$ majorizes v_i, $t_{\inf}$ is the best estimate obtainable by this method of the lower bound for the time of existence of a solution of the original system.

Returning to Eq. (1.292) and this time identifying the number N dependent on ξ with the number $N'(\xi)$ which now satisfies the inequality $N'(\xi) - A(t, \xi) < 0$, the previous argument then establishes that $\tilde{v}_1 < v_1$. Thus on the defining $C^{(1)}$ characteristic, the corresponding time $T_c^{(1)}(N'(\xi), \xi)$ provides an upper bound to the time of existence of a solution for $\tilde{v}_1$ and thus also for v_1. Continuing the argument in an analogous fashion then leads to the determination of a number $t_{\sup}$ which is the least positive number of the two expressions

$$ {}^*T^{(1)} = \inf\left[\frac{-1}{\left(\frac{\partial r}{\partial x}\right)_{\substack{t=0\\x=\xi}} \exp\{q_1(r_0(\xi), s_0(\xi))\} \inf\left[\left(\frac{\partial\lambda^{(1)}}{\partial r}\right)\exp\{-q_1(r_0(\xi), s)\}\right]_s}\right]_\xi , \tag{1.296} $$

and

$$ {}^*T^{(2)} = \inf\left[\frac{-1}{\left(\frac{\partial s}{\partial x}\right)_{\substack{t=0\\x=\eta}} \exp\{q_2(r_0(\eta), s_0(\eta))\} \inf\left[\left(\frac{\partial\lambda^{(2)}}{\partial s}\right)\exp\{-q_2(r, s_0(\eta))\}\right]_r}\right]_\eta . \tag{1.297} $$

The actual value t_c of the time of existence of a solution of the original system of equations thus satisfies the inequality

$$ t_{\inf} < t_c < t_{\sup} . $$

The numbers $t_{\inf}$ and $t_{\sup}$ are to be interpreted in the sense that the solution is certainly bounded for $t < t_{\inf}$, while the solution is certainly unbounded for $t > t_{\sup}$.

When, as assumed by Ludford [68], the Riemann invariant distributions $r_0(x)$ and $s_0(x)$ differ only slightly from the constant values $\hat{r}_0$ and $\hat{s}_0$, respectively, then $(t_{\text{sup}} - t_{\text{inf}})$ is small and the bounds provide a good estimate of t_c. Under these conditions, since r and s are constant along respective characteristics, they also can differ only slightly from $\hat{r}_0$ and $\hat{s}_0$ at all subsequent times along the characteristics until breakdown of the solution, so that continuous functions of r and s will vary only slightly from constant values. Consequently, the expressions occurring in Eqs. (1.293)–(1.297), which define t_{sup} and t_{inf}, may then be approximated by the much simpler expressions: t_{inf} is the least positive number of the two quantities

$$\frac{-1}{\max\limits_{r,s}\left[\left(\dfrac{\partial\lambda^{(1)}}{\partial r}\right)\exp\{q_1(\hat{r}_0, \hat{s}_0) - q_1(\hat{r}_0, s)\}\right]\max\left(\dfrac{\partial r}{\partial x}\right)_{t=0}}, \tag{1.298}$$

$$\frac{-1}{\max\limits_{r,s}\left[\left(\dfrac{\partial\lambda^{(2)}}{\partial s}\right)\exp\{q_2(\hat{r}_0, \hat{s}_0) - q_2(r, \hat{s}_0)\}\right]\max\left(\dfrac{\partial s}{\partial x}\right)_{t=0}}, \tag{1.299}$$

and t_{sup} is the least positive number of the two quantities

$$\frac{-1}{\min\limits_{r,s}\left[\left(\dfrac{\partial\lambda^{(1)}}{\partial r}\right)\exp\{q_1(\hat{r}_0, \hat{s}_0) - q_1(\hat{r}_0, s)\}\right]\max\left(\dfrac{\partial r}{\partial x}\right)_{t=0}}, \tag{1.300}$$

$$\frac{-1}{\min\limits_{r,s}\left[\left(\dfrac{\partial\lambda^{(2)}}{\partial s}\right)\exp\{q_2(\hat{r}_0, \hat{s}_0) - q_2(r, \hat{s}_0)\}\right]\max\left(\dfrac{\partial s}{\partial x}\right)_{t=0}}. \tag{1.301}$$

A further simplification is possible if the functions $q_1(r, s)$ and $q_2(r, s)$ are represented approximately by the first two terms of their Taylor series expansions, and the defining relation for $q_1(r, s)$ and the corresponding relation for $q_2(r, s)$ are used. To see this we write

$$q_1(\hat{r}_0, s) = q_1(\hat{r}_0, \hat{s}_0) + (s - \hat{s}_0)(\partial q_1/\partial s)_0 + O((s - \hat{s}_0)^2),$$

which becomes

$$q_1(\hat{r}_0, \hat{s}_0) - q_1(\hat{r}_0, s) = \left\{\frac{\hat{s}_0 - s}{\lambda_0^{(1)} - \lambda_0^{(2)}}\right\}\left(\frac{\partial\lambda^{(1)}}{\partial s}\right)_0 + O((s - \hat{s}_0)^2),$$

where the suffix 0 refers to the initial values. A similar result may be obtained for $q_2(r, s)$. Consequently, when r and s differ only slightly from constant values $\hat{r}_0$ and $\hat{s}_0$, we may use as our estimates of t_{inf} and t_{sup}

the alternative expressions: $t_{\inf}$ is the least positive number of the two quantities

$$\frac{-1}{\max_{r,s}\left[\left(\frac{\partial\lambda^{(1)}}{\partial r}\right)\exp\left\{\left(\frac{\hat{s}_0 - s}{\lambda_0^{(1)} - \lambda_0^{(2)}}\right)\left(\frac{\partial\lambda^{(1)}}{\partial s}\right)_0\right\}\right]\cdot\max\left(\frac{\partial r}{\partial x}\right)_{t=0}}, \tag{1.302}$$

$$\frac{-1}{\max_{r,s}\left[\left(\frac{\partial\lambda^{(2)}}{\partial s}\right)\exp\left\{\left(\frac{\hat{r}_0 - r}{\lambda_0^{(2)} - \lambda_0^{(1)}}\right)\left(\frac{\partial\lambda^{(2)}}{\partial r}\right)_0\right\}\right]\cdot\max\left(\frac{\partial s}{\partial x}\right)_{t=0}}, \tag{1.303}$$

and $t_{\sup}$ is the least positive number of the two quantities

$$\frac{-1}{\min_{r,s}\left[\left(\frac{\partial\lambda^{(1)}}{\partial r}\right)\exp\left\{\left(\frac{\hat{s}_0 - s}{\lambda_0^{(1)} - \lambda_0^{(2)}}\right)\left(\frac{\partial\lambda^{(1)}}{\partial s}\right)_0\right\}\right]\cdot\max\left(\frac{\partial r}{\partial x}\right)_{t=0}}, \tag{1.304}$$

$$\frac{-1}{\min_{r,s}\left[\left(\frac{\partial\lambda^{(2)}}{\partial s}\right)\exp\left\{\left(\frac{\hat{r}_0 - r}{\lambda_0^{(2)} - \lambda_0^{(1)}}\right)\left(\frac{\partial\lambda^{(2)}}{\partial r}\right)_0\right\}\right]\cdot\max\left(\frac{\partial s}{\partial x}\right)_{t=0}}. \tag{1.305}$$

It is at once obvious that the solution will only break down due to the $C^{(1)}$ characteristics when max $(\partial r/\partial x)_{t=0} > 0$, and, similarly, it will only break down due to the $C^{(2)}$ characteristics when $\max(\partial s/\partial x)_{t=0} > 0$.

A much simplified version of these results was obtained by Lax [69] using different comparison theorems in which he assumed that $r = \hat{r}_0$, $s = \hat{s}_0$ in order to study the existence of solutions of a certain nonlinear string equation.

1.18 APPLICATION OF THE JEFFREY–LAX METHOD

The mixed initial-boundary value problem for the one-dimensional motion of a polytropic gas in a closed tube extending from $x = 0$ to $x = l$ has been examined for wave breakdown by Ludford [68], Jeffrey [70], and Ames [56]. In this section, we shall apply the Jeffrey–Lax theory to the preceding problem, and as in Ludford's paper convert the original problem to a pure, initial-value problem by a suitable periodic extension outside the interval $0 \leqslant x \leqslant l$. The breakdown time t_c so obtained will be compared with that developed by the technique of Section 1.15.

Governing equations for this gas dynamics problem [see Eqs. (1.233)] are

$$U_t + AU_x = 0,$$

in which

$$U = \begin{bmatrix} \rho \\ u \end{bmatrix}, \qquad A = \begin{bmatrix} u & \rho \\ c^2/\rho & u \end{bmatrix},$$

where ρ, u, $c^2 = \partial p/\partial \rho$ are density, velocity, and the square of the speed of sound, respectively. We shall assume the gas is polytropic, i.e., $p = A\rho^\gamma$, where A and γ are constants.

From the quasi-linear theory (see Volume I) we have the following results:

$$\lambda^{(1)} = u + c, \qquad \lambda^{(2)} = u - c, \tag{1.306}$$

$$l^{(1)} = [c/\rho, 1], \qquad l^{(2)} = [c/\rho, -1].$$

Along $C^{(1)}$ characteristics we have the Riemann invariant,

$$u + 2c/(\gamma - 1) = -r. \tag{1.307}$$

Along $C^{(2)}$ charactristics we have the Riemann invariant,

$$u - 2c/(\gamma - 1) = -s, \tag{1.308}$$

where the minus signs are introduced to make $\partial\lambda^{(1)}/\partial r$ and $\partial\lambda^{(2)}/\partial s$ negative as required.

In Eqs. (1.298)–(1.301) we need $\partial\lambda^{(1)}/\partial r$ and $\partial\lambda^{(2)}/\partial s$. From Eq. (1.306), $\lambda^{(1)} = u + c$, so that

$$\frac{\partial\lambda^{(1)}}{\partial r} = \frac{\partial u}{\partial r} + \frac{dc}{d\rho}\frac{\partial\rho}{\partial r}. \tag{1.309}$$

Since the gas is polytropic, $c^2 = A\gamma\rho^{\gamma-1}$ and $dc/d\rho = (c/2\rho)(\gamma - 1)$.

Adding and subtracting the invariants, we have

$$r + s = -2u, \qquad r - s = -4c/(\gamma - 1).$$

Consequently,

$$\partial u/\partial r = -\tfrac{1}{2}, \tag{1.310}$$

and

$$1 = \left(\frac{-4}{\gamma - 1}\right)\frac{dc}{d\rho}\frac{\partial\rho}{\partial r}. \tag{1.311}$$

Combining Eqs. (1.309)–(1.311), we have

$$\partial\lambda^{(1)}/\partial r = -[(\gamma + 1)/4],$$

and by similar reasoning

$$\partial\lambda^{(1)}/\partial s = (\gamma - 3)/4.$$

Both are independent of r and s!

To apply the estimates, Eqs. (1.298)–(1.301) or the simplified forms, Eqs. (1.302)–(1.305), we must convert the initial-boundary value problem to a *pure initial-value problem.* The initial values of u and ρ, specified on $0 < x < l$, determine the initial values

$$r_0(x) = r(x, 0) = -u(x, 0) - \frac{2c(x, 0)}{\gamma - 1},$$

and

$$s_0(x) = s(x, 0) = -u(x, 0) + \frac{2c(x, 0)}{\gamma - 1}$$

on $0 < x < l$. The boundary conditions are $u(0, t) = u(l, t) = 0$ for all $t > 0$. Thus, by $r + s = -2u$, we have

$$r_0(0) + s_0(0) = r_0(l) + s_0(l) = 0.$$

We then extend $r_0(x) + s_0(x)$ to the interval $-l \leqslant x \leqslant l$ and thence to the entire initial line, $-\infty < x < \infty$, by defining it to be an *even* function in $-l \leqslant x \leqslant l$ which is periodic of period $2l$. $r_0(x) - s_0(x)$ is similarly extended as an *odd* function in $-l \leqslant x \leqslant l$ which is periodic of period $2l$. Since the extension converts the problem to a pure initial-value problem, the boundary conditions are disregarded.

When $\max(\partial r/\partial x)_{t=0}$ and $\max(\partial s/\partial x)_{t=0}$ are both positive, t_{inf}, from Eqs. (1.302)–(1.305), is the lesser of the two numbers

$$\frac{4}{(\gamma + 1) \max\limits_{s} \left[\exp\left\{\dfrac{(\gamma - 3)(\hat{s}_0 - s)}{8c_0}\right\}\right] \max\left(\dfrac{\partial r_0}{\partial x}\right)},$$

$$\frac{4}{(\gamma + 1) \max\limits_{r} \left[\exp\left\{\dfrac{(3 - \gamma)(\hat{r}_0 - r)}{8c_0}\right\}\right] \max\left(\dfrac{\partial s_0}{\partial x}\right)}.$$

Similarly t_{sup} is the lesser of the two numbers

$$\frac{4}{(\gamma + 1) \min\limits_{s} \left[\exp\left\{\dfrac{(\gamma - 3)(\hat{s}_0 - s)}{8c_0}\right\}\right] \max\left(\dfrac{\partial r_0}{\partial x}\right)},$$

$$\frac{4}{(\gamma + 1) \min\limits_{r} \left[\exp\left\{\dfrac{(3 - \gamma)(\hat{r}_0 - r)}{8c_0}\right\}\right] \max\left(\dfrac{\partial s_0}{\partial x}\right)}.$$

If r and s in these results are replaced by their constant values r_0 and s_0, then t_{inf} and t_{sup} coincide, and we obtain

$$t_c = 4/(\gamma + 1)\beta, \qquad \text{where} \quad \beta = \max\{\max(\partial r_0/\partial x), \max(\partial s_0/\partial x)\}. \tag{1.312}$$

Equation (1.312) has a similar form to that of Eq. (1.260), although the notation is somewhat different. Since the methods of Section 1.15 are applied to solitary waves (that is, there is no energy partition between waves moving in two directions) the critical time calculated from Eq. (1.260) must be multiplied by 2 to exactly compare the results. Stated alternatively, the initial data of the method of Section 1.15 must be taken as one-half that of the Jeffrey–Lax method.

1.19 DYNAMICS OF A MOVING THREADLINE

Equations modeling the wave propagation and vibration of a traveling threadline have been formulated by Ames *et al.* [8] and presented in Section 1.2 [Eqs.(1.14)–(1.17)]. In this section we shall give the two- and three-dimensional forms of those equations and discuss the nonlinear dynamics of the system in two dimensions. In Section 1.20 the three-dimensional problem will be examined.

Let u, v; V, T, and m be the components of transverse displacement, axial velocity, tension, and mass per unit length of the string, respectively. In these dimensionless variables the dimensionless (Eulerian) equations for the three-dimensional dynamics are those of:

transverse momentum:

$$V^2 u_{xx} + 2V u_{xt} + u_{tt} = T u_{xx}/m(1 + u_x^2 + v_x^2), \tag{1.313}$$

$$V^2 v_{xx} + 2V v_{xt} + v_{tt} = T v_{xx}/m(1 + u_x^2 + v_x^2), \tag{1.314}$$

longitudinal momentum:

$$V_t + V V_x = [1/m(1 + u_x^2 + v_x^2)^{1/2}][T/(1 + u_x^2 + v_x^2)^{1/2}]_x, \tag{1.315}$$

conservation of mass:

$$[m(1 + u_x^2 + v_x^2)^{1/2}]_t + [mV(1 + u_x^2 + v_x^2)^{1/2}]_x = 0, \tag{1.316}$$

constitutive law:

$$T = T(m, m_t). \tag{1.317}$$

As was experimentally verified by Ames *et al.* [8], this system may respond by plane motion or by a three-dimensional vibration, which has been called *ballooning* (Ames *et al.* [8]). Breakdown (jump) phenomena are observed in both responses. Data showing the plane vibration jump are given in Fig. 1-4 and the balloon jump is displayed in Fig. 1-5. We

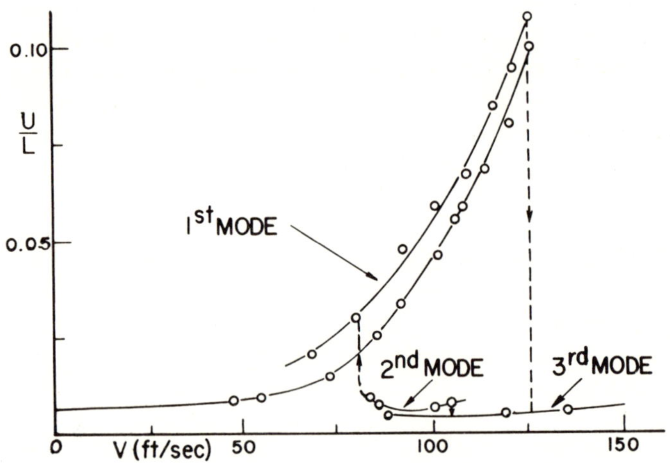

FIG. 1-4. Amplitude response for plane motion.

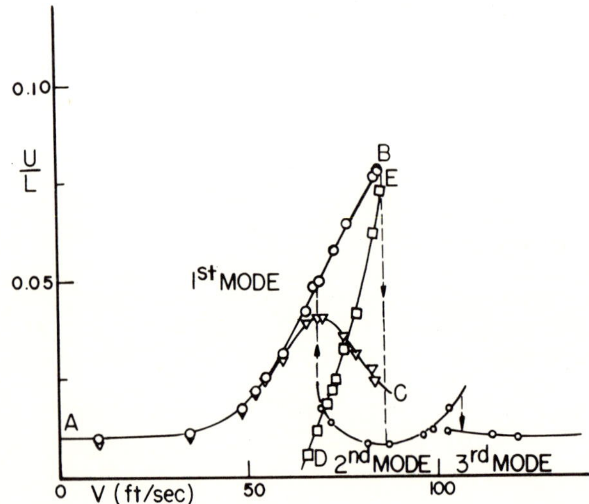

FIG. 1-5. Amplitude response for ballooning motion: A–B, A–C—up- and down-stream horizontal component; D–E—up-stream vertical component.

shall first discuss some of the peculiarities of the plane motion and then turn our attention to ballooning.

In the case of plane oscillation an alternative formulation is sometimes to be preferred. Thus Lee [72] chooses u and V as before but prefers λ, σ, and ρ which represent stretch per unit length, engineering uniaxial stress, and mass density per unit volume. In this notation the equations are those of:

transverse momentum:

$$V^2 u_{xx} + 2Vu_{xt} + u_{tt} = \lambda\sigma u_{xx}/\rho(1 + u_x^2), \tag{1.318}$$

longitudinal momentum:

$$V_t + VV_x = [\lambda/\rho(1 + u_x^2)^{1/2}][\sigma/(1 + u_x^2)^{1/2}]_x, \tag{1.319}$$

conservation of mass:

$$[\lambda^{-1}(1 + u_x^2)^{1/2}]_t + [V\lambda^{-1}(1 + u_x^2)^{1/2}]_x = 0, \tag{1.320}$$

constitutive equation:

$$\sigma = \sigma(\lambda, \lambda_t). \tag{1.321}$$

The following analysis is based upon the assumption that the material is strain rate independent ($\sigma = \sigma(\lambda)$) and we specify that $\lambda > 1$, $d\sigma/d\lambda > 0$.

Linearized and simplified nonlinear forms of Eqs. (1.318)–(1.321) have been extensively analyzed by Ames *et al.* [7, 8, 26]. Instead of discussing those here, we shall directly examine the complete Eqs. (1.318)–(1.321). With $p = u_x$ and $q = u_t$, they become the following set of quasi-linear first-order equations:

$$U_t + AU_x = 0, \tag{1.322}$$

where

$$U^{\mathrm{T}} = [p, q, V, \lambda],$$

and

$$A = \begin{bmatrix} 0, & -1, & 0, & 0 \\ V^2 - \sigma\lambda/\rho(1 + p^2), & 2V, & 0, & 0 \\ p\lambda\sigma/(1 + p^2)^2\rho, & 0, & V, & (\lambda\, d\sigma/d\lambda)/(1 + p^2)\rho \\ -V\lambda p/(1 + p^2), & -\lambda p/(1 + p^2), & -\lambda, & V \end{bmatrix}.$$

The characteristic directions are obtained by calculating the eigenvalues of A (see Volume I, Chapter 7), with the result that

$$\Gamma_{1,2} = \left.\frac{dx}{dt}\right|_{1,2} = V \pm \frac{\lambda c}{(1 + p^2)^{1/2}}, \tag{1.323}$$

and

$$\Gamma_{3,4} = \left.\frac{dx}{dt}\right|_{3,4} = V \pm \frac{\lambda\bar{c}}{(1 + p^2)^{1/2}}, \qquad (1.324)$$

where

$$c^2 = \rho^{-1}\frac{d\sigma}{d\lambda}, \qquad \bar{c}^2 = \frac{\sigma}{\rho\lambda}. \qquad (1.325)$$

From these we recognize c and $\bar{c}$ as the (Lagrangian) elastic and transverse wave propagation speeds, that is, wave speeds with respect to the material particles. In this Eulerian frame, those wave speeds are modified, as shown in Eqs. (1.323) and (1.324), by the local particle velocity, the configuration form, and the stretch of the element. The assumptions $\lambda > 1$, $d\sigma/d\lambda > 0$, assures us of the genuine hyperbolic character of Eqs. (1.318)–(1.320). Across a characteristic manifold, we have† for $\Gamma_{1,2}$,

$$\frac{dp}{0} = \frac{dq}{0} = \frac{dV}{\pm c} = \frac{-d\lambda}{(1 + p^2)^{1/2}}, \qquad (1.326)$$

and for $\Gamma_{3,4}$

$$\frac{dp}{1} = \frac{dq}{-V \pm \lambda\bar{c}/(1 + p^2)^{1/2}} = \frac{-dV}{\pm\lambda p\bar{c}/(1 + p^2)^{3/2}} = \frac{d\lambda}{0}. \qquad (1.327)$$

Equations (1.326) demonstrate that across an *elastic* wave front the transverse velocity ($u_t = q$) and slope of the string ($u_x = p$) undergo no change, while the changes of stretch and axial velocity are mutually related. Equations (1.327) explain that across a *transverse* wave front only the stretch remains constant and all other quantities p, q, and V undergo changes. Stated alternatively, Eqs. (1.326) imply that the longitudinal (elastic) waves which propagate along the characteristics $\Gamma_{1,2}$ have no effect on the transverse motion. Similarly, from Eqs. (1.327) we infer that the transverse oscillations have no effect on the stretch (λ) and therefore the strain, but do affect the axial velocity. As we shall see, in Section 1.20, entirely analogous results are obtained in the ballooning motion situation.

The two-dimensional Eulerian equations (1.318)–(1.321) possess an equivalent Lagrangean form. In many situations the use of the Lagrange system is preferable (see for example Rakhmatulin [73], Schultz *et al.* [74], and Schultz [75]). However, in this case the Eulerian form is preferred, since the string passes through eyelets with an unprescribed

† From the eigenvectors of A (Volume I, Chapter 7, or Jeffrey and Taniuti [65]).

history. If the Lagrange equations are used, the boundary conditions must be assigned at a series of mass points. Since this cannot generally be done, an Eulerian form is chosen. In that development such difficulties do not arise, since assigning boundary conditions in the Eulerian reference frame is the same as installing physical constraints in a fixed laboratory frame and allowing the material to pass through them. Further discussion and comparison of the two frames is found in the work of Lee [72].

1.20 BALLOONING VIBRATION OF A MOVING THREADLINE

The three-dimensional problem has been treated by Ames *et al.* [8] and Shih [76]. In this section we shall discuss the application of characteristics, the occurrence of singularities, and the breakdown of the solution during the ballooning oscillation.

Directly from Eqs. (1.313)–(1.316) we find by standard methods (see Volume I or Jeffrey and Taniuti [65]) that the relations across a characteristic manifold are

$$(2V - \lambda)\, du_t + (V^2 - A^2)\, du_x = 0, \tag{1.328}$$

$$(2V - \lambda)\, dv_t + (V^2 - A^2)\, dv_x = 0, \tag{1.329}$$

$$(V - \lambda)\, dV + B^2 m^{-1}\, dm + A^2(u_x\, du_x + v_x\, dv_x)/(1 + u_x^2 + v_x^2) = 0, \tag{1.330}$$

$$dV + (V - \lambda) m^{-1}\, dm + (V - \lambda)(u_x\, du_x + v_x\, dv_x)/(1 + u_x^2 + v_x^2) = 0, \tag{1.331}$$

where λ is an eigenvalue and A and B are defined as

$$A = \left\{\frac{T}{m(1 + u_x^2 + v_x^2)}\right\}^{1/2}, \qquad B = \left\{\frac{-\partial T/\partial m}{(1 + u_x^2 + v_x^2)}\right\}^{1/2}. \tag{1.332}$$

Since $[(\ )_t]_x - [(\ )_x]_t = 0$, we may write

$$du_t + \lambda\, du_x = 0, \tag{1.333}$$

and

$$dv_t + \lambda\, dv_x = 0. \tag{1.334}$$

From Eqs. (1.328) and (1.333) or from Eqs. (1.329) and (1.334) the same characteristic condition $(\lambda - V)^2 - A^2 = 0$ is obtained which yields the eigenvalues (characteristics) for this subsystem

$$\lambda_A{}^{\pm} = V \pm A. \tag{1.335}$$

Setting Eq. (1.335) into Eqs. (1.330) and (1.331), we find

$$\frac{dV}{A} = \pm \frac{u_x\,du_x + v_x\,dv_x}{1 + u_x^2 + v_x^2} = \frac{dm}{0}. \tag{1.336}$$

Consequently, perturbations of both transverse displacements, u and v, propagate along the same characteristics

$$\Gamma_A^{\pm} : \frac{dx}{dt} = V \pm A,$$

and along these

$$du_t + (V \pm A)\,du_x = 0, \qquad dv_t + (V \pm A)\,dv_x = 0. \tag{1.337}$$

Further, we observe that these transverse oscillations have no effect on the strain of the string, but do affect the axial velocity (compare Section 1.19).

For the entire system, Eqs. (1.328)–(1.331), (1.333), and (1.334), the characteristic condition is

$$\{(\lambda - V)^2 - A^2\}\{(\lambda - V)^2 - B^2\} = 0,$$

thereby generating the additional eigenvalues

$$\lambda_B^{\pm} = V \pm B. \tag{1.338}$$

Setting these into Eqs. (1.328), (1.333), (1.329), (1.334), and (1.331) we find

$$du_t/0 = du_x/0 = dv_t/0 = dv_x/0 = dV/B = \pm(dm/m). \tag{1.339}$$

Consequently, the longitudinal waves propagate along the characteristics

$$\Gamma_B^{\pm} : \frac{dx}{dt} = V \pm B,$$

and along these

$$dV \pm B\frac{dm}{m} + \frac{u_x(du_t + V\,du_x) + v_x(dv_t + V\,dv_x)}{1 + u_x^2 + v_x^2} = 0. \tag{1.340}$$

Further, we note that the longitudinal waves have no effect on the transverse motion (compare Section 1.19).

The analysis for the singularities and breakdown of the oscillation is simplified by introduction of two variables p and q defined as

$$p = (u_x^2 + v_x^2)^{1/2}, \tag{1.341}$$

and

$$q = \int (u_x \, du_t + v_x \, dv_t)/p, \tag{1.342}$$

which satisfy the relation

$$p_t - q_x = 0. \tag{1.343}$$

Upon multiplying Eq. (1.313) by u_x/p and Eq. (1.314) by v_x/p and adding the results, we obtain

$$q_t + 2Vp_t + (V^2 - A^2)p_x = 0. \tag{1.344}$$

Employing Eqs. (1.341) and (1.342), Eqs. (1.315) and (1.316) may be written as

$$V_t + VV_x + B^2 m^{-1} m_x = -A^2 pp_x/(1 + p^2), \tag{1.345}$$

$$m^{-1} m_t + V m^{-1} m_x = -p(p_t + vp_x)/(1 + p^2) - V_x. \tag{1.346}$$

If we now multiply Eq. (1.346) by B and add to (or subtract from) Eq. (1.345), we obtain, with the aid of Eq. (1.344),

$$\xi_t + \lambda_B{}^+ \xi_x = 0, \tag{1.347}$$

$$\eta_t + \lambda_B{}^- \eta_x = 0, \tag{1.348}$$

where ξ and η are the Riemann invariants defined by

$$\begin{aligned} \xi &= V + \int B \frac{dm}{m} + \int p(dq + V\,dp)/(1 + p^2), \\ \eta &= V - \int B \frac{dm}{m} + \int p(dq + V\,dp)/(1 + p^2). \end{aligned} \tag{1.349}$$

If V and B are expressible as functions of ξ and η, the quasi-linear equations (1.347) and (1.348) are reducible and can be linearized by interchange of the dependent and independent variables. Thus we have

$$\begin{gathered} \xi_x = Jt_\eta, \qquad \xi_t = -Jx_\eta, \qquad \eta_x = -Jt_\xi, \qquad \eta_t = Jx_\xi, \\ J = \xi_x \eta_t - \xi_t \eta_x = 2B\xi_x \eta_x. \end{gathered} \tag{1.350}$$

The resulting equations in the ξ, η-plane are

$$x_\eta - \lambda_B{}^+ t_\eta = 0, \qquad x_\xi - \lambda_B{}^- t_\xi = 0. \tag{1.351}$$

From Eqs. (1.350) we can write

$$t_\xi = -(2B\xi_x)^{-1}, \qquad t_\eta = (2B\eta_x)^{-1}. \tag{1.352}$$

Consequently if t_ξ or t_η vanishes at some point in the ξ, η-plane, Eqs. (1.352) indicate that, for finite B, ξ_x or η_x has become infinite. From Eqs. (1.349) it is apparent that a strong discontinuity is formed. At the corresponding point in the ξ, η-plane, J has become infinite, indicating termination of the utility of the transformation. When V approaches a critical value numerically equal to B, λ_B^- goes to zero. If η_t is to possess a nonzero value, then η_x must become arbitrarily large and breakdown of the solution occurs.

The three-dimensional response, which we have called *ballooning*, has an elliptic cross section whose major axis lies in the direction of the excitation (see Fig. 1-5). With increasing velocity the cross section is observed to approach a circle with a subsequent jump. In order to analyze this motion, it is convenient to express the governing equations in terms of r and θ as dependent variables. Thus, we set

$$u = r \cos \theta, \qquad v = r \sin \theta, \tag{1.353}$$

where r and θ are both functions of x and t. If we multiply the first (resulting) equation by sin θ, the second by cos θ, and subtract one from the other, there results

$$\begin{aligned} &\theta_{tt} + 2V\theta_{xt} + (V^2 - A^2)\theta_{xx} \\ &\qquad = -(2/r)\{r_t\theta_t + V(r_t\theta_x + r_x\theta_t) + (V^2 - A^2)r_x\theta_x\}. \end{aligned} \tag{1.354}$$

Alternatively, if we multiply the first equation by cos θ and the second by sin θ, we obtain by addition

$$(1/r)\{r_{tt} + 2Vr_{xt} + (V^2 - A^2)r_{xx}\} = \theta_t^2 + 2V\theta_x\theta_t + (V^2 - A^2)\theta_x^2. \tag{1.355}$$

Let us now consider ballooning motion with similar cross sections of the form

$$r = f(x)\,\rho(\theta). \tag{1.356}$$

Using this expression, Eqs. (1.354) and (1.355) become

$$\begin{aligned} &\theta_{tt} + 2V\theta_{xt} + (V^2 - A^2)\theta_{xx} + 2\rho'\rho^{-1}\{\theta_t^2 + 2V\theta_x\theta_t + (V^2 - A^2)\theta_x^2\} \\ &\qquad + 2f'f^{-1}\{V\theta_t + (V^2 - A^2)\theta_x\} = 0, \end{aligned} \tag{1.357}$$

and

$$\begin{aligned} &\rho'\rho^{-1}\{\theta_{tt} + 2V\theta_{xt} + (V^2 - A^2)\theta_{xx} + 2f'f^{-1}[V\theta_t + (V^2 - A^2)\theta_x]\} \\ &\qquad + (V^2 - A^2)f''f^{-1} = [1 - (\rho''/\rho)]\{\theta_t^2 + 2V\theta_x\theta_t + (V^2 - A^2)\theta_x^2\}, \end{aligned} \tag{1.358}$$

where, as usual, primes denote total differentiation. Subtracting Eq. (1.357) multiplied by ρ'/ρ from Eq. (1.358), we find

$$\{1 + 2(\rho'\rho^{-1})^2 - \rho''\rho^{-1}\}\{\theta_t^2 + 2V\theta_x\theta_t + (V^2 - A^2)\theta_x^2\} = (V^2 - A^2)f''f^{-1}. \tag{1.359}$$

Equations (1.357) and (1.359) constitute the pair of equations of transverse motion used to solve the problem for any cross section.

For elliptic cross sections of semimajor axis $a(x)$ and semiminor axis $b(x)$, we set

$$f(x) = [a(x)b(x)]^{1/2}, \tag{1.360}$$

that is, the geometric mean of semimajor and semiminor axes. The function $\rho(\theta)$ is then expressible as

$$\rho(\theta) = (b/a)^{1/2}/(1 - e^2 \cos^2 \theta)^{1/2},$$

where $e = (a^2 - b^2)^{1/2}/a$ is the eccentricity of the ellipse. For elliptic ballooning $1 + 2(\rho'\rho^{-1})^2 - \rho''\rho^{-1} = \rho^4$,so that Eq. (1.359) becomes

$$\rho^4\{\theta_t^2 + 2V\theta_x\theta_t + (V^2 - A^2)\theta_x^2\} = (V^2 - A^2)f''f^{-1}. \tag{1.361}$$

Additionally Eq. (1.358) becomes

$$\begin{aligned}\theta_{tt} + 2V\theta_{xt} + (V^2 - A^2)\theta_{xx} + 2f'f^{-1}\{V\theta_t + (V^2 - A^2)\theta_x\}\\ + 2\rho'\rho^{-5}(V^2 - A^2)f''f^{-1} = 0.\end{aligned} \tag{1.362}$$

At this point we must recall that V and A are also unknown. Consequently two additional equations must be considered to solve the system.

Circular ballooning is a special case which may be examined by letting $a \to b$ in Eqs. (1.361) and (1.362). In this case $e = 0$, $\rho = 1$, and these equations are somewhat simplified. This has been completely solved by Ames *et al.* [8].

After these preliminaries we wish to examine the existence of an equilibrium state. In such a state the functions p, V, m, and T will vary essentially with x. For elliptic ballooning they may also vary cyclically with respect to t within each revolution ($\theta = 2\pi$) or period ($t = \tau$). For circular ballooning, they are completely independent of t. In order to maintain a steady configuration of the balloon at equilibrium, we assume that the angular velocity, $\omega = \theta_t$, is a function of θ only and that a condition of constancy of the period τ along the longitudinal axis is satisfied. That is to say that

$$\tau = \int_0^{2\pi} \frac{d\theta}{\omega(\theta)} = \frac{2\pi}{K}, \tag{1.363}$$

where K is a constant numerically equal to the average angular velocity over one revolution.

If $t_0 = t_0(x)$ designates the time when $\theta = 0$, then the angular displacement may be expressed as $\theta = \theta(t - t_0)$, from which we find

$$\theta_{xx} = -\omega t_0'' + \omega\omega'(t_0')^2, \qquad \theta_{tt} = \omega\omega', \qquad \theta_{xt} = \omega'\theta_x = -\omega\omega' t_0'.$$

Setting these into Eqs. (1.361) and (1.362), we obtain

$$(V^2 - A^2)\{(t_0')^2 - (\rho^4\omega^2)^{-1}f''f^{-1}\} - 2Vt_0' + 1 = 0 \tag{1.364}$$

and

$$(V^2 - A^2)\{t_0'' + 2f'f^{-1}t_0' + f''f^{-1}\rho^{-2}(\rho^{-2}\omega^{-1})'\} - 2Vf'f^{-1} = 0. \tag{1.365}$$

The continued treatment of the general case is complicated. As an alternative various models are proposed in searching for analytic solutions and properties. In some cases the solution does not exist.

Case 1. A necessary condition for the existence of a solution is that

$$t_0't_0'' + f'f^{-1}(t_0')^2 + f''f^{-1}\rho^{-2}(\rho^{-2}\omega^{-1})'t_0' + f'f''f^{-2}(\rho^2\omega)^{-2} \neq 0. \tag{1.366}$$

This follows from Eqs. (1.364) and (1.365) and the observation that V and A must be finite.

Case 2. If V and A are time varying, as in the elliptic ballooning case, then $\rho^2\omega$ must be a function of θ for a solution to exist. Thus a model possessing conserved angular momentum cannot have a solution.

Let us assume that† $(\rho^2\omega)' = 0$. Recalling that t_0 and f are functions only of x, the time (t)-derivatives of Eqs. (1.364) and (1.365) are

$$(VV_t - AA_t)\{(t_0')^2 - (\rho^4\omega^2)^{-1}f''f^{-1}\} - V_t t_0' = 0,$$

and

$$(VV_t - AA_t)\{(t_0'' + 2f'f^{-1}t_0'\} - V_t f'f^{-1} = 0.$$

Combination of these two equations yields a condition which contradicts Eq. (1.366). Therefore $(\rho^2\omega)' \neq 0$.

† Beginning with Eq. (1.361), we have continued to use ρ', ρ'' to designate derivatives with respect to θ even though ρ is also a function of x.

Case 3. If $\rho = 1$, as in circular ballooning, then

$$t_0' = C + Ef^{-2}, \tag{1.367}$$

where C and E are constants.

Since p, v, m, and T vary only with x, Eq. (1.347) becomes

$$Vm^{-1}m_x + V_x = -pVp_x/(1 + p^2),$$

which integrates to

$$mV(1 + p^2)^{1/2} = \text{constant}.$$

Also Eq. (1.315) becomes

$$mV(1 + p^2)^{1/2}V_x = \{T/(1 + p^2)^{1/2}\}_x ,$$

which integrates to

$$V^2 - A^2 = C_1V. \tag{1.368}$$

When Eq. (1.368) is substituted into Eq. (1.365) and $\omega = \theta_t$ is assumed constant as a result of symmetry, we have

$$C_1\{t_0'' + 2f'f^{-1}t_0'\} = 2f'f^{-1},$$

which integrates to Eq. (1.367) with $C = C_1^{-1}$.

Case 4. If T and therefore m are constant, no equilibrium solution exists for circular ballooning ($\rho = 1$).

From Eqs. (1.346) and (1.347), by integration we find $V^2 - A^2 =$ constant and $V/A =$ constant. The combination of these implies that either both V and A are constant or $V^2 - A^2 = 0$. The latter implies that $C_1 = 0$ and Eq. (1.357) reduces to

$$Vf'f^{-1}\theta_t = 0,$$

which is impossible since none of the variables can be zero for this problem. Consequently no solution exists for a stable circular ballooning.

Case 4, due to Shih [76], verifies the experimental results observed by Ames *et al.* [8]. In those tests it was observed that the ballooning vibration attempts to maintain a constant tension in the string. The motion is stable during the time it has an elliptic cross section. When the envelope achieves a circular cross section, the motion breaks down by jumping to second-mode plane motion. Then $\theta_t = \omega = 0$.

References

1. Jones, S. E., and Ames, W. F., *J. Math. Anal. Appl.* **17**, 484 (1967).
2. Burgers, J. M., *Advan. Appl. Mech.* **1**, 171 (1948).
3. Forsythe, G. E., and Wasow, W. R., "Finite Difference Methods for Partial Differential Equations," p. 141. Wiley, New York, 1960.

4a. Ames, W. F., "Nonlinear Partial Differential Equations in Engineering," Vol. I. Academic Press, New York, 1965.

4b. Levin, S. A., *J. Math. Anal. Appl.* **30**, 197 (1970).

4c. Kečkić, J. D., Elektrotehnic Fac., Univ. of Belgrade, Yugoslavia. Personal communication (1971).

5. Chu, C-w., *Quart. Appl. Math.* **23**, 275 (1965).
6. Montroll, E. W., "Lectures on Nonlinear Rate Equations, Especially Those with Quadratic Nonlinearities." Theoretical Physics Institute, Univ. of Colorado, 1967.
7. Ames, W. F., and Vicario, A. A., Jr., *Develop. Mech.* **5**, 733, Proc. 11th Midwest. Mech. Conf. (1969).
8. Ames, W. F., Lee, S. Y., and Zaiser, J. N., *Internat. J. Nonlinear Mech.* **3**, 449 (1968).
9. Fisher, R. A., *Ann. Eugen.* **7**, 355 (1937).
10. Kolmogorov, A., Petrovsky, I., and Piscounov, N., *Bull. Univ. Etat Moscou* (Ser. Intermat.), A. I, 1 (1937).
11. Benedict, M. (ed.), "Encyclopedia of Chemical Technology," Vol. V, p. 76. Wiley (Interscience), New York, 1950.
12. Cohen, K., "Science and Engineering of Nuclear Power," Vol. 2, p. 19. Addison-Wesley, Reading, Massachusetts, 1949.
13. Montroll, E., and Newell, G., *J. Appl. Phys.* **23**, 184 (1952).
14. Gray, E. P., and Kerr, D. E., *Ann. Physics* **17**, 276 (1959).
15. Montroll, E., *J. Appl. Probability* **4**, 281 (1967).
16. Cole, J. D., *Quart. Appl. Math.* **9**, 225 (1951).
17. Vein, P. R., Personal communication (1969).
18. Mei, C. C., Proc. 12th Internat. Congr. Appl. Mech., p. 321. Stanford Univ., Stanford, California, 1968.
19. Earnshaw, S., *Philos. Trans. Roy. Soc. London Ser. A* **150**, 133 (1858).
20. Nowinski, J. L., *Trans. ASME Ser. B* **87B**, 523 (1965).
21. Ames, W. F., *J. Math. Anal. Appl.* **34**, 214 (1971).
22. James, H. M., and Guth, E., *J. Appl. Phys.* **16**, 643 (1945).
23. Taylor, G. I., "Scientific Papers," Vol. 1, p. 467. Cambridge Univ. Press, London and New York, 1958.
24. Truesdell, C., *J. Rational Mech. Anal.* **1**, 125 (1952).
25. Seth, B. R., *Proc. Indian. Acad. Sci. Sect. A* **A25**, 151 (1948).
26. Ames, W. F., Lee, S. Y., and Vicario, A. A., Jr., *Internat. J. Non-linear Mech.* **5**, 413 (1970).
27. Scott, A., "Active and Nonlinear Wave Propagation in Electronics." Wiley, New York, 1970.
28. Scott, A., *Amer. J. Phys.* **37**, 52 (1969).
29. Abramowitz, M., and Stegun, L. A., "Handbook of Mathematical Functions." U.S. Department of Commerce, Washington, D.C., 1964.
30. Korteweg, D. J., and deVries, G., *Phil. Mag.* **39**, 422 (1895).
31. Stoker, J. J., "Water Waves." Wiley (Interscience), New York, 1957.
32. Zabusky, N. J., A Synergetic Approach to Problems of Non-linear Dispersive Wave

Propagation and Interaction, Chapter *in* "Nonlinear Partial Differential Equations" (W. F. Ames, ed.), p. 223. Academic Press, New York, 1967.
33. Miura, R. M., *J. Math. Phys.* **9**, 1202 (1968).
34. Miura, R. M., Gardner, C. S., and Kruskal, M. D., *J. Math. Phys.* **9**, 1204 (1968).
35. Lax, P. D., *Commun. Pure Appl. Math.* **21**, 467 (1968).
36. Forsyth, A. R., "Theory of Differential Equations," Vols. 5 and 6. Dover, New York, 1959.
37. Ames, W. F., Ad-hoc Exact Techniques for Nonlinear Partial Differential Equations, Chapter *in* "Non-linear Partial Differential Equations" (W. F. Ames, ed.), p. 55. Academic Press, New York, 1967.
38. Dasarathy, B. V., Personal communication (1970).
39. Varley, E., *Commun. Pure Appl. Math.* **15**, 91 (1962).
40. Goursat, E., "A Course in Mathematical Analysis," Vol. I. Dover, New York, 1959.
41. Eisenhart, L. P., "A Treatise on the Differential Geometry of Curves and Surfaces," Chapter 8, p. 271ff. Dover, New York, 1960.
42. Bäcklund, J. O., *Math. Ann.* **17**, 285 (1880).
43. Bäcklund, J. O., *Lunds Univ. Arsskr. Afd. 2 KFLH* **19** (1883).
44. Clairin, M. J., *Ann. École Norm. 3e Ser.* **19**, 15 (1902).
45. Clairin, M. J., *Ann. Toulouse, 2e Ser.* **5**, 437 (1903).
46. Lamb, G. L., Jr., *Phys. Lett.* **25A**, 181 (1967).
47. Lamb, G. L., Jr., *Phys. Lett.* **28A**, 548 (1969).
48. Lamb, G.L., Jr., "Propagation of Ultrashort Optical Pulses," *in* "Festschrift for P. M. Morse." MIT Press, Cambridge, Massachusetts, to be published.
49. Jones, S. E., and Ames, W. F., *Quart. Appl. Math.* **25**, 302 (1967).
50. Tomotika, S., and Tamada, K., *Quart. Appl. Math.* **7**, 381 (1949).
51. Tamada, K., Studies on the Two-dimensional Flow of a Gas, with Special Reference to the Flow Through Various Nozzles. Ph.D. Thesis, Univ. of Kyoto, Japan, 1950.
52. Johnson, G. D., On a Nonlinear Vibrating String, Ph.D. Dissertation, Univ. of California, Los Angeles, California, 1967.
53. Zabusky, N. J., *J. Math. Phys.* **3**, 1028 (1962).
54. Fermi, E., Pasta, J. R., and Ulam, S., "Studies of Nonlinear Problems," Los Alamos Rept. #1940, May, 1955. *See also* S. Ulam, "A Collection of Mathematical Problems," Chapter 7, p. 8. Wiley (Interscience), New York, 1960.
55. Coulson, C. A., "Waves," p. 88. Oliver and Boyd, London; Wiley (Interscience), New York, 1955.
56. Ames, W. F., *Internat. J. Non-linear Mech.* **5**, 605 (1970).
57. Bellman, R. E., "Perturbation Techniques in Mathematics, Physics and Engineering" (Athena Series). Holt, New York, 1964.
58. Banta, E. D., *J. Math. Anal. Appl.* **10**, 166 (1965).
59. Ames, W. F., and Jones, S. E., *J. Math. Anal. Appl.* **21**, 479 (1968).
60. Friedrichs, K. O., *Amer. J. Math.* **70**, 555 (1948).
61. Courant, R., and Friedrichs, K. O., "Supersonic Flow and Shock Waves," Sections 48, 49. Wiley (Interscience), New York, 1948.
62. Lax, P. D., *Ann. Math. Studies (Princeton)* **33**, 211 (1954).
63. Thomas, T. Y., *J. Math. Mech.* **6**, 455 (1957).
64. Jeffrey, A., *Arch. Rat. Mech. Anal.* **14**, 27 (1963).
65. Jeffrey, A., and Taniuti, T., "Nonlinear Wave Propagation," Chapter 2. Academic Press, New York, 1964.
66. Jeffrey, A., *J. Math. Mech.* **15**, 585 (1966).
67. Riemann, B., *Abh. Ges. Wiss. Göttingen* **8**, 43 (1860).

68. Ludford, G. S. S., *Proc. Cambridge Philos. Soc.* **48**, 499 (1952).
69. Lax, P. D., *J. Math. Phys.* **5**, 611 (1964).
70. Jeffrey, A., *J. Math. Mech.* **17**, 331 (1967).
71. Coddington, E. A., and Levinson, N., "Theory of Ordinary Differential Equations." McGraw-Hill, New York, 1955.
72. Lee, S. Y., *Develop. Mech.* **5**, 543, Proc. 11th Midwest Mech. Conf. (1969).
73. Rakhmatulin, K. A., *Prikl. Mat. Meh.* **16**, 23 (1952).
74. Schultz, A. B., Tuschak, P. A., and Vicario, A. A., Jr., *J. Appl. Mech.* **34**, 392 (1967).
75. Schultz, A. B., *Int. J. Solids Struct.* **4**, 799 (1968).
76. Shih, L. Y., *Internat. J. Non-linear Mech.* **6**, 427 (1971).

CHAPTER 2

Applications of Modern Algebra

2.0 INTRODUCTION

Tobias Dantzig in "Number, The Language of Science" (1930) observed:

> The mathematician may be compared to a designer of garments, who is utterly oblivious of the creatures whom his garments may fit. To be sure, his art originated in the necessity for clothing such creatures, but this was long ago; to this day a shape will occasionally appear which will fit into the garment as if the garment had been made for it. Then there is no end of surprise and of delight!

The structures of modern algebra, as applied to the analyses of nonlinear problems, have generated a number of surprises. It is these that concern us here.

In recent years, we have witnessed a revival of interest in applying the techniques of modern algebra to the equations of analysis. Of course, this concept is not original with this era, for as early as 1881, Lie [1] had begun his coordination of the apparently disconnected methods of integration of ordinary differential equations, both linear and nonlinear. The unifying concept was found to be the algebraic structure known as continuous transformation groups. The classical work on continuous groups is that of Eisenhart [2]. An elementary treatment of the applications to ordinary differential equation is found in the work of Ames [3] together with additional bibliography. In an article published in 1924, Dickson [4] showed how some differential equations could be integrated with the aid of group theory thereby continuing to build an important link between the techniques of algebra and those of differential equations.

Birkhoff [5] in 1950 suggested that the reduction of the number of independent variables in partial differential equations could also be attacked by algebraic methods. An additional motivation for employing abstract algebraic structures stems from the realization that these are not based upon linear operators, linear superposition, or any other linear assuption and therefore may be of assistance in circumventing the "curse of linearity," which has too long limited our capacity to solve the nonlinear problems of science and technology.

In this chapter, our principal attention will be focused on the utilization of the group concept, although other abstract algebraic entities such as rings (Mikusinski [6]), semigroups (Feller [7], Yosida [8]), and non-associative algebras (Aris [9]) are playing an increasing role.

2.1 THE SIMILARITY METHOD OF MORGAN

The general theory of Morgan [10] and Michal [11] for developing similarity solutions of partial differential equations was succinctly discussed in Volume I. This significant contribution to applied mathematics employed one-parameter continuous groups of transformations. Later extensions by Manohar [12] to special forms of n-parameter groups are also found in Volume I. In this section we shall present the basic definitions and theorems of Morgan [10] and discuss their applications. These will form the foundation for discussion of the subsequent deductive theory.

Consider the one-parameter groups S, G, and E_k of the form

$$E_k: \begin{cases} G: \begin{cases} S: \{x^{i*} = f^i(x^1,\ldots, x^m; a), & i = 1, 2,\ldots, m, \quad m \geqslant 2, \\ y_j{}^* = f_j(y_j; a), & j = 1, 2,\ldots, n, \quad n \geqslant 1, \end{cases} \\ \left[\dfrac{\partial^l y_j}{\partial(x^1)^{l_1} \cdots \partial(x^m)^{l_m}}\right]^* \\ \qquad = [f^{(l)}_{(j;l_1,\ldots,l_m)}] \left(\dfrac{\partial^l y_j}{\partial(x^1)^l}, \dfrac{\partial^l y_j}{\partial(x^1)^{l-1}\,\partial(x^2)^1}, \ldots, \dfrac{\partial^l y_j}{\partial(x^m)^l},\right. \\ \qquad\qquad \left.\dfrac{\partial^{l-1} y_j}{\partial(x^1)^{l-1}}, \ldots, y_j, x^1,\ldots, x^m; a\right), \\ \sum\limits_i l_i = l \leqslant k, \end{cases} \tag{2.1}$$

where the functions f are continuous in the parameter a. The identity element is denoted by a_0, thus, e.g., $x^i = f^i(x^1,\ldots, x^m; a_0)$. The value of the parameter for the transformation inverse to that given by a is denoted by a^*; thus, e.g., if $y_j^* = f_j(y_j; a)$, then $y_j = f_j(y_j^*; a^*)$. The transformations $x^i \to x^{i*}$ form a subgroup S of G.† Subsequently the x^i and y_j will be identified with the independent and dependent variables, respectively, of a system of partial differential equations.

Let the set of functions $\{y_j\}$, $y_j = y_j(x^1,..., x^m)$ be differentiable in x^i up to order k, and append to the transformations of G the transformations of the partial derivatives of the y_j with respect to the x^i. That is to say, consider the set of functions $\{y_j^*\}$ defined by

$$y_j^*(x^{1*},..., x^{m*}) = f_j\{y_j[f^1(x^{1*},..., x^{m*}; a^*),..., f^m(x^{1*},..., x^{m*}; a^*)]; a\}.$$

E_k is a continuous group called the *kth enlargement of* G for $\{y_j\}$, where the functions $[f^{(l)}_{(j; l_1,..., l_m)}]$ are defined so that

$$\left\{\frac{\partial^l y_j}{\partial(x^1)^{l_1} \cdots \partial(x^m)^{l_m}}\right\}^* = \frac{\partial^l y_j^*}{\partial(x^{1*})^{l_1} \cdots \partial(x^{m*})^{l_p}}.$$

By elementary group theory (Eisenhart [2]), G has $m + n - 1$ functionally independent (when considered as functions of the $m + n$ independent variables x^i, $i = 1,..., m$; y_j, $j = 1, 2,..., n$) absolute invariants,‡ designated by

$$\eta_1(x^1,..., x^m),..., \eta_{m-1}(x^1,..., x^m);$$

$$g_j(y_1,..., y_n; x^1,..., x^m), \qquad j = 1, 2,..., n.$$

The g_j can be so chosen that the Jacobian $\partial(g_1,..., g_n)/\partial(y_1,..., y_n) \neq 0$ and the rank of the Jacobian matrix $\partial(\eta_1,..., \eta_{m-1})/\partial(x^1,..., x^m)$ is equal to $m - 1$.

A *differential form of the kth order in m independent variables* is a function, usually in class $C^{(1)}$ or greater, of the type

$$\Phi(x^1,..., x^m; y_1,..., y_n,..., \partial^k y_1/\partial(x^1)^k,..., \partial^k y_n/\partial(x^m)^k), \tag{2.2}$$

whose arguments $\xi^1,..., \xi^p$ are the variables $x^1,..., x^m$, functions $y_1,..., y_n$ dependent on them, and the partial derivatives of the y_j with respect to the x^i up to the kth order. The function Φ is said to be *conformally invariant* under E_k if

$$\Phi(\xi^{1*},..., \xi^{p*}) = F(\xi^1,..., \xi^p; a)\Phi(\xi^1,..., \xi^p), \tag{2.3}$$

† For a discussion of the basic group theory aspects, the reader is referred to Eisenhart [2], Cohen [13], or Birkhoff and MacLane [14].

‡ Here η is an absolute invariant if $\eta(x^1,..., x^m) = \eta(x^{1*},..., x^{m*})$. g is an absolute invariant if

$$g(y_1,..., y_n; x^1,..., x^m) = g(y_1^*,..., y_n^*; x^{1*},..., x^{m*}).$$

where Φ is exactly the same function of the ξ's as it is of the ξ^*'s, and F is some function of the ξ's and the parameter a. If F is a function of a only, Φ is said to be *constant conformally invariant* under E_k. In the event that $F \equiv 1$, Φ is said to be *absolutely constant conformally invariant* under E_k. A system *of partial differential equations*

$$\phi_\gamma[x^1,..., x^m; y_1,..., y_n; \partial y_1/\partial x^1,..., \partial^k y_n/\partial(x^m)^k] = 0, \quad \gamma = 1,..., N \leqslant n, \tag{2.4}$$

is said to be *invariant* under G, of Eq. (2.1), if each of the differential forms $\phi_1,..., \phi_N$ is conformally invariant under E_k. By *invariant solutions* of a system of partial differential equations is meant that class of solutions of the system which have the property that the y_j are exactly the same functions of the x^i as the y_j^* are of the x^{i*}.

The principal results obtained by Morgan [10] are contained in the following theorems:

Theorem 2.1-1. *If a differential form ϕ_γ is conformally invariant under E_k and if $\{I_j\}$ is any set of functions such that, when $y_j = I_j(x^1,..., x^m)$, then $y_j^* = I_j(x^{1*},..., x^{m*})$, then under a transformation from $\{y_1,..., y_n; x^1,..., x^m\}$ to the functionally independent set $\{g_1,..., g_n; \eta_1,..., \eta_{m-1}; \hat{x}\}$, consisting of one of the sets of absolute invariants and one of the x^i, there exist differential forms σ_γ and A_γ such that*

$$\begin{aligned} &\phi_\gamma[x^1,..., x^m; I_1,..., I_n; \partial I_1/\partial x^1,..., \partial^k I/\partial(x^m)^k] \\ &\quad = [A_\gamma(\eta_1,..., \eta_{m-1}, F_1,..., F_n; \partial F_1/\partial\eta_1,..., \partial^k F_n/\partial(\eta_{m-1})^k)] \\ &\qquad \cdot [\exp \sigma_\gamma(\hat{x}, \eta_1,..., \partial^k F_n/\partial(\eta_{m-1})^k)], \end{aligned} \tag{2.5}$$

where the functions F_j are defined by

$$\begin{aligned} &F_j[\eta_1(x^1,..., x^m),..., \eta_{m-1}(x^1,..., x^m)] \\ &\quad \equiv g_j[I_1(x^1,..., x^m),..., I_n(x^1,..., x^m); x^1,..., x^m]. \end{aligned} \tag{2.6}$$

Theorem 2.1-2. *If each of the ϕ_γ of Eq.* (2.4) *is conformally invariant and if and only if the set of functions $\{I_j\}$ is a solution to the system, Eq.* (2.4), *then from Eq.* (2.5), *with the arguments of A_γ given there,*

$$A_\gamma = 0, \qquad \gamma = 1, 2,..., N \leqslant n. \tag{2.7}$$

Theorem 2.1-3. *If the set of functions $\{F_j\}$ is any solution to the system, Eq.* (2.7), *then the functions $\{I_j\}$, given by the inverse transformation of Eq.* (2.6) *is a solution to Eqs.* (2.4). *The resultant set $\{I_j\}$ is an invariant solution for each value of the group parameter a. Conversely, any invariant solution $\{I_j\}$ of Eq.* (2.4) *yields a solution $\{F_j\}$ of Eq.* (2.7) *upon transforming variables to a set of functionally independent invariants of G.*

For the purposes of solving the system of Eqs. (2.4), we can summarize the impact of the foregoing theorems as follows: Sufficient conditions for reducing Eqs. (2.4) to Eqs. (2.7), which has *one less* independent variable, are that Eqs. (2.3) be invariant under a group of the form Eq. (2.1) and that invariant solutions exist. Solutions to Eq. (2.7) yield invariant solutions to Eqs. (2.4).

Equations (2.7) constitute a *similarity representation* of the system of differential Eqs. (2.4); the underlying transformation from $\{y_1, ..., y_n; x^1, ..., x^m\}$ to $\{g_1, ..., g_n; \eta_1, ..., \eta_{m-1}\}$ is called a *similarity transformation*, and the variables $\{g_1, ..., g_n; \eta_1, ..., \eta_{m-1}\}$ are called *similarity variables*.

Theorems 2.1-2 and 2.1-3 consider systems consisting of partial differential equations alone, without regard for auxiliary conditions. The usual manner of application to systems which possess auxiliary conditions is to examine the equations by themselves. If a set of similarity variables is found, we test to see if the auxiliary conditions are also expressible, without inconsistency, in terms of these similarity variables. If so, these variables are termed the *similarity variables for the composite system* of equations and auxiliary conditions. The resultant composite system, expressed in terms of these similarity variables, is called a similarity *representation of the composite*. All too often, however, the set of similarity variables for the equations alone are found to be inappropriate for the auxiliary conditions, and the cycle must be repeated.

A second weakness of the classical method is the lack of a systematic procedure for establishing the required set of $m + n - 1$ functionally independent absolute invariants. The invariants have been determined by trial or inspection, which has been possible because of the simple groups used. Furthermore, rather than deducing groups under which the hypotheses of Theorems 2.1-2 and 2.1-3 are satisfied, applications of the theorems have been based on particular assumed transformation groups. Some of these studies have been very successful. In the next section, we shall sketch the classical method, give an example and additional references, and prepare the foundation for developing a *deductive* theory which, while complicated, removes the above objections to the classical method.

2.2 APPLICATION OF THE MORGAN METHOD

The essence of this theory is that the determination of similarity solutions for a system of partial differential equations is equivalent to the determination of the invariant solutions of those equations under an appropriate one (or more)-parameter group of transformations. In the elementary applications of this theory, a group is assumed and consequently the general form of the invariants is prescribed. The requirement of invariance of the equations under the assumed group generates a set of simple simultaneous algebraic equations whose solution determines the specific form† of the invariants.

A number of recent studies have employed the Morgan theory and its extensions to n-parameter groups in the development of similarity solutions. Among these we find a general treatment of the laminar boundary layer equations by Manohar [12], the nonlinear diffusion equation $r^{1-m}\{r^{m-1}c^n c_r\}_r = c_t$ by Ames [15], the power-law Ostwald–de Waele model of laminar quasi-two-dimensional boundary layer flow by Na and Hansen [16], and the two-dimensional laminar incompressible boundary layer equations of non-Newtonian fluids by Lee and Ames [17], for power-law fluids, and Hansen and Na [18] in the general case for a variety of relations between shearing stress and rate of strain. The ready availability of Volume I and Hansen [19] suggests that this simple procedure, despite its limitations, will become a standard tool of the applied mathematician and engineer.

Two groups of one-parameter transformations very often suffice for treatment of the boundary-layer equations. Na *et al.* [20], employing Lie's theory of infinitesimal transformation groups (see Section 2.3), have shown that the only two possible groups for the two-dimensional laminar boundary layer equations are the linear and spiral groups.‡ In the notation of Eq. (2.1), these are $f(x; a) = a^k x$ and $f(x; a) = x + \ln a$, respectively.

In illustration of the elementary application of the Morgan theory, we present a portion of the Lee and Ames [17] analysis for forced convective heat transfer in power-law fluids. Let the dimensionless quantities u, v, U_e, and θ be velocity components, free-stream velocity, and temperature, respectively. With N_{PR} as the Prandtl number and n as the power-law exponent, the basic steady-state equations become

$$u_x + v_y = 0, \tag{2.8}$$

$$uu_x + vu_y = U_e(dU_e/dx) + [|u_y|^{n-1}u_y]_y, \tag{2.9}$$

$$u\theta_x + v\theta_y = (1/N_{\mathrm{PR}})[\theta^{r-1}\theta_y]_y. \tag{2.10}$$

† Some arbitrary parameters may still be present. These may be employed in deciding whether the auxiliary conditions can be expressed in terms of these equation similarity variables.

‡ These may be supplemented by certain translations, e.g., $x \to x + b$. This was for the classical flat plate theory with the usual boundary conditions.

It is to be noted that the heat conductivity k has been assumed to depend upon the temperature, thus $k = k'\theta^{r-1}$, and the Prandtl number is defined using k'.

A stream function ψ, defined by the relations

$$u = \psi_y , \qquad v = -\psi_x ,$$

is now introduced so that Eq. (2.8) is satisfied identically. In ψ Eqs. (2.9) and (2.10) become

$$\psi_y\psi_{xy} - \psi_x\psi_{yy} = U_e(dU_e/dx) + [| \psi_{yy} |^{n-1}\psi_{yy}]_y , \tag{2.11}$$

and

$$\psi_y\theta_x - \psi_x\theta_y = (1/N_{\text{PR}})[\theta^{r-1}\theta_y]_y , \tag{2.12}$$

where we do not specify U_e *a priori*, but permit the analysis to describe free-stream velocities permitting similarity solutions. Since Eq. (2.11) does not contain θ, we shall first examine its similarity representation, if any, and then introduce Eq. (2.12).

If the *linear group*, with real parameter $a > 0$,

$$G_1: \{S: \{x^* = a^{\alpha_1}x, y^* = a^{\alpha_2}y, \psi^* = a^{\alpha_3}\psi, U_e^* = a^{\alpha_4}U_e\}, \tag{2.13}$$

is selected and applied to Eq. (2.11), we find

$$a^{-2\alpha_3+2\alpha_2+\alpha_1}\left\{\frac{\partial\psi^*}{\partial y^*}\frac{\partial^2\psi^*}{\partial x^*\,\partial y^*} - \frac{\partial\psi^*}{\partial x^*}\frac{\partial^2\psi^*}{(\partial y^*)^2}\right\}$$

$$= a^{\alpha_1-2\alpha_4}U_e^*\frac{dU_e^*}{dx^*} + a^{-n\alpha_3+(2n+1)\alpha_2}\frac{\partial}{\partial y^*}\left\{\left|\frac{\partial^2\psi^*}{(\partial y^*)^2}\right|^{n-1}\right\}.$$

If this equation is to be constant conformally invariant under the enlargement of G_1, then the simple simultaneous equations

$$\alpha_1 + 2\alpha_2 - 2\alpha_3 = \alpha_1 - 2\alpha_4 = (2n+1)\alpha_2 - n\alpha_3 \tag{2.14}$$

must hold.

As described in Volume I, and as is easily verified, the invariants of G_1 [Eq. (2.13)] are

$$\eta = yx^{-\alpha_2/\alpha_1}, \qquad f(\eta) = \psi x^{-\alpha_3/\alpha_1}, \qquad h_e(\eta) = U_e x^{-\alpha_4/\alpha_1}. \tag{2.15}$$

Thus the values of the ratios α_2/α_1, α_3/α_1, and α_4/α_1 are essential to continuation of the analysis. Dividing by α_1, the set of Eqs. (2.14) consists only of

$$1 + (1-2n)\alpha_2/\alpha_1 + (n-2)\alpha_3/\alpha_1 = 0, \tag{2.16}$$

and

$$\alpha_2/\alpha_1 - \alpha_3/\alpha_1 + \alpha_4/\alpha_1 = 0. \tag{2.17}$$

These two equations in three unknowns do not have a unique solution thereby providing a degree of freedom in fitting boundary conditions or other restrictions.

For example, flat-plate flow has U_e = constant, hence $dU_e/dx = 0$. From Eqs. (2.15) we find that $\alpha_4 = 0$. Consequently, it follows directly from Eqs. (2.16) and (2.17) that $\alpha_2/\alpha_1 = \alpha_3/\alpha_1 = 1/(n+1)$. The invariants of G_1 are therefore uniquely determined to be

$$\eta = yx^{-1/(n+1)}, \qquad f(\eta) = \psi x^{-1/(n+1)}, \qquad h_e = U_e = \text{const.}$$

a form used by Acrivos *et al.* [21] but obtained there by an alternative method.

When conservation of momentum is required, as in two-dimensional jet problems, then

$$\int_{-\infty}^{\infty} u^2\, dy = \int_{-\infty}^{\infty} \left(\frac{\partial \psi}{\partial y}\right)^2 dy = \text{const.} \tag{2.18}$$

Upon application of G_1 to Eq. (2.18), we find absolute invariance is required so that $\alpha_2 - 2\alpha_3 = 0$. Setting this into Eq. (2.16) results in $\alpha_2/\alpha_1 = 2\alpha_3/\alpha_1 = 2/3n$. Hence the invariants (similarity variables) are

$$\eta = yx^{-2/3n}, \qquad f = \psi x^{-1/3n}, \qquad h_e = U_e x^{+1/3n}, \tag{2.19}$$

a result obtained by Gutfinger and Shinnar [22] and Kapur [23] by alternative methods. Since $U_e = U_e(x)$, h_e can only be a function of the similarity variable η if $U_e = Cx^{-1/3n}$, where C is constant.

Next, if the general form of (Falkner–Skan) wedge flow is considered, that is $U_e = Cx^m$, then the requirement that h_e only depends upon η generates the requirement that $\alpha_4/\alpha_1 = m$. Using this relation, the solutions of Eqs. (2.16) and (2.17) are

$$\alpha_2/\alpha_1 = [(n-2)m+1]/(n+1), \qquad \alpha_3/\alpha_1 = [(2n-1)m+1]/(n+1), \tag{2.20}$$

with the corresponding similarity variables $\eta = yx^{-\alpha_2/\alpha_1}$, $f = \psi x^{-\alpha_3/\alpha_1}$. Thus it is possible to transform the power-law momentum equation [Eq. (2.11)] to an ordinary differential equation (the similarity representation) for wedge flows with an arbitrary wedge angle. An interesting case is that when $m = \frac{1}{3}$, for in this case $\alpha_2/\alpha_1 = \frac{1}{3}$, $\alpha_3/\alpha_1 = \frac{2}{3}$, and the invariants are independent of n.

Turning now to the energy Eq. (2.12), the constant conformal invariance requirement under the enlargement of G_1, supplemented with $\theta^* = a^{\alpha_5}\theta$, generates the algebraic equation

$$\alpha_1 + \alpha_2 - \alpha_3 - \alpha_5 = 2\alpha_2 - r\alpha_5\,. \tag{2.21}$$

Using the general wedge flow, $\alpha_4/\alpha_1 = m$, Eqs. (2.16), (2.17), and (2.21) yield the solution given by Eq. (2.20) together with

$$\alpha_5/\alpha_1 = [(n-1)(1-3m)]/(n+1)(r-1). \tag{2.22}$$

Here it is seen, if $r \neq 1$ (i.e., constant conductivity), that m can be any real value including zero. Thus any kind of mainstream flow velocity is applicable provided a unique boundary condition for θ is prescribed. Since the θ invariant is $g = \theta x^{-\alpha_5/\alpha_1}$, the boundary conditions on θ must be a power function, x^{α_5/α_1}, when $y = 0$. If $r = 1$, there is no solution for α_5/α_1 except as $m \to \frac{1}{3}$, in which case α_5/α_1 is arbitrary, but the main stream flow is restricted to $U_e = x^{1/3}$.

Additional cases and physical discussion are found in the cited literature.

Lastly the similarity representation for our original problem, Eqs. (2.11) and (2.12) consists of ordinary differential equations for f and g as functions of η *together with boundary conditions inferred from those of the original system.* As previously observed, the original system boundary conditions have not been involved in the analysis. At this point, we must ascertain whether these can be written in the "trial" similarity variables without inconsistencies. In illustration of this point we select forced convective flow under right-angle wedge geometry, described by Eqs. (2.11) and (2.12) with $r = 1$, together with the boundary conditions in the quarter plane $x \geqslant 0$, $y \geqslant 0$,

$$\begin{gathered} u(x>0, y=0)=0, \quad v(x>0, y=0)=0, \quad \theta(x>0, y=0)=x^l \ (l>0), \\ u(x, y\to\infty) = U_e = x^{1/3}, \quad u(x=0, y>0)=0, \\ \theta(x=0, y>0)=0, \quad \theta(x>0, y\to\infty)=0. \end{gathered} \tag{2.23}$$

These physical conditions transform into the conditions

$$\begin{gathered} \psi_y(x>0, y=0)=0, \quad \psi_x(x>0, y=0)=0, \quad \theta(x>0, y=0)=x^l, \\ \psi_y(x>0, y\to\infty) = x^{1/3}, \quad \psi_y(x=0, y>0)=0, \\ \theta(x=0, y>0)=0, \quad \theta(x>0, y\to\infty)=0 \end{gathered} \tag{2.24}$$

in the stream function–temperature function notation.

From Eqs. (2.20) and (2.22) we have, with $r = 1$, the similarity variables

$$\eta = yx^{-1/3}, \qquad f = \psi x^{-2/3}, \qquad g = \theta x^{-t}, \tag{2.25}$$

where t is an arbitrary real number. Using these we see that $(x > 0, y = 0)$ becomes $\eta = 0$, and $(x = 0, y > 0)$, $(x > 0, y \to \infty)$ both become $\eta \to \infty$. With this information at hand we find, from Eq. (2.25), that $\psi_y = x^{1/3}\, df/d\eta$, so that $\psi_y(x > 0, y = 0) = 0$ becomes $f'(0) = 0$. Now $\psi_x = \frac{2}{3}x^{-1/3}f(\eta) - \frac{1}{3}\eta x^{-1}\, df/d\eta$, so that $\psi_x(x > 0, y = 0) = 0$ requires that $f(0) = 0$. The third condition $\theta(x > 0, y = 0) = x^l$ forces $t = l$, whereupon $g(0) = \theta(x > 0,\ y = 0)\, x^{-t} = x^l \cdot x^{-l} = 1$. The fourth and fifth conditions are satisfied if $f' = 1$ as $n \to \infty$. Furthermore, since $\theta(x, y) = g(\eta)x^l$, the last two conditions are satisfied if $g = 0$ as $\eta \to \infty$.

The complete similarity representation is obtained by transforming Eqs. (2.11) and (2.12), with $r = 1$, into ordinary differential equations by employing Eqs. (2.25). Thus we find that complete representation to be[†]

$$\begin{aligned} &(f')^2 - 2ff'' = 1 + 3(d/d\eta)[(f'')^n], \\ &g'' + \tfrac{2}{3}N_{\mathrm{PR}}fg' = 0, \\ f = 0, \quad &f' = 0, \quad g = 1, \quad \text{at} \quad \eta = 0, \\ &f' = 1, \quad g = 0, \quad \text{as} \quad \eta \to \infty. \end{aligned} \tag{2.26}$$

Additional examples and detailed calculations are available in the cited literature.

We should also mention here that wave mechanics has benefited recently from the construction of similarity solutions. Among these we find the work of Schultz [24] on the large dynamic deformations caused by a force traveling on an extensible string, similarity solutions for spherical shock waves in a polytropic gas by Latter [25], the equations for anisentropic gases by Ames [3], shocks in plasmas by Friedhoffer [27], solutions of the nonlinear wave equation $\phi_{tt} + 2\phi\phi_t - \phi_{xx} = 0$ by Rosen [28], and longitudinal waves on a moving threadline by Vicario [29].

2.3 DETERMINATION OF GROUPS BY FINITE TRANSFORMATIONS

During the time of the evolution of the Birkhoff–Morgan–Michal theory, a considerable number of studies with similar goals (i.e., the reduction of the number of independent variables in systems of partial

[†] With the velocity gradient in the y-direction always positive, we can, with care, ignore the absolute value sign.

differential equation) have been carried out. Among the more successful ones are those of Strumpf [30], who restricted his attention to the equation

$$\psi_y(\nabla^2\psi)_x - \psi_x(\nabla^2\psi)_y = \nu\nabla^4\psi, \tag{2.27}$$

which stems from the Navier–Stokes equations for steady, incompressible two-dimensional viscous flow. The class of solutions for the stream function ψ is of the form $\psi = x^n\psi_1(z)$, where $z = \eta(x, y)$ and n is a real number. A set of transformations is defined which reduces the original equation to an ordinary differential with ψ_1 and η as dependent and independent variables, respectively. This procedure is related to that of "separation of variables" discussed in Volume I.

In a series of three basic papers v. Krzywoblocki and Roth [31–33] undertook a study of the Morgan–Michal method with the goal of *developing a method for obtaining the proper groups for a given system of equations*. This effort, successful as it was, did not include the auxiliary (initial and boundary) conditions. In addition to this limitation, these early articles (done prior to 1962) were published in a relatively obscure place and hence did not reach the proper audience. Had their work been better advertised, much of it would not have had to be redone.

We shall sketch the method using Laplace's equation

$$u_{xx} + u_{yy} = 0 \tag{2.28}$$

as a vehicle. Our goal is to discover how groups G, with one parameter (a), may be found such that a particular system of equations [Eq. (2.28) in this case] is conformally invariant under G. Let G have the general form

$$G\colon \begin{cases} \bar{x} = f_1(x, y; a), \qquad \bar{y} = f_2(x, y; a), \\ \bar{u} = f_4(x, y; a)u.^\dagger \end{cases} \tag{2.29}$$

Under this transformation we find

$$(\partial^2\bar{u}/\partial\bar{x}^2) + (\partial^2\bar{u}/\partial\bar{y}^2) = P(u, f_4, u_x, u_y, f_{4x}, f_{4y}, \ldots), (\partial^2 u/\partial x^2)$$
$$+ Q(u, f_4, u_x, \ldots)(\partial^2 u/\partial y^2) + \sum_{i=1}^{12} R_i(u, f_4, \ldots), \tag{2.30}$$

where R_i, $i = 1, \ldots, 12$, P, and Q are defined subsequently. For conformal

† It is not difficult to extend this transformation to $\bar{u} = f_3(x, y, u; a)$ as we shall do presently.

invariance (see Section 2.1 for the definition) the right-hand side of Eq. (2.30) must be equal to

$$S(u, x, y, u_x, u_y, u_{xx}, u_{yy}; a)[u_{xx} + u_{yy}]. \tag{2.31}$$

This occurs if $P = Q$ and if $\sum_{i=1}^{12} R_i = 0$. If we choose

$$R_1 = R_2 = \cdots = 0, \tag{2.32}$$

then it follows that $\sum_{i=1}^{12} R_i = 0$, but alternative choices are available.† Consequently, our basic equations become

$$R_1 = u \frac{\partial^2 f_4}{\partial x^2} \left[\left(\frac{\partial x}{\partial \bar{x}}\right)^2 + \left(\frac{\partial x}{\partial \bar{y}}\right)^2 \right] = 0, \tag{2.33a}$$

$$R_2 = u \frac{\partial^2 f_4}{\partial y^2} \left[\left(\frac{\partial y}{\partial \bar{x}}\right)^2 + \left(\frac{\partial y}{\partial \bar{y}}\right)^2 \right] = 0, \tag{2.33b}$$

$$R_3 = u \frac{\partial^2 f_4}{\partial x\, \partial y} \left[\frac{\partial x}{\partial \bar{x}} \frac{\partial y}{\partial \bar{x}} + \frac{\partial x}{\partial \bar{y}} \frac{\partial y}{\partial \bar{y}} \right] = 0, \tag{2.33c}$$

$$R_4 = u \frac{\partial f_4}{\partial x} \left[\frac{\partial^2 x}{\partial \bar{x}^2} + \frac{\partial^2 x}{\partial \bar{y}^2} \right] = 0, \tag{2.33d}$$

$$R_5 = u \frac{\partial f_4}{\partial y} \left[\frac{\partial^2 y}{\partial \bar{x}^2} + \frac{\partial^2 y}{\partial \bar{y}^2} \right] = 0, \tag{2.33e}$$

$$J: \qquad R_6 = f_4 \frac{\partial^2 u}{\partial x\, \partial y} \left[\frac{\partial x}{\partial \bar{x}} \frac{\partial y}{\partial \bar{x}} + \frac{\partial x}{\partial \bar{y}} \frac{\partial y}{\partial \bar{y}} \right] = 0, \tag{2.33f}$$

$$R_7 = f_4 \frac{\partial u}{\partial x} \left[\frac{\partial^2 x}{\partial \bar{x}^2} + \frac{\partial^2 x}{\partial \bar{y}^2} \right] = 0, \tag{2.33g}$$

$$R_8 = f_4 \frac{\partial u}{\partial y} \left[\frac{\partial^2 y}{\partial \bar{x}^2} + \frac{\partial^2 y}{\partial \bar{y}^2} \right] = 0, \tag{2.33h}$$

$$R_9 = \frac{\partial u}{\partial x} \frac{\partial f_4}{\partial x} \left[\left(\frac{\partial x}{\partial \bar{x}}\right)^2 + \left(\frac{\partial x}{\partial \bar{y}}\right)^2 \right] = 0, \tag{2.33i}$$

$$R_{10} = \frac{\partial u}{\partial y} \frac{\partial f_4}{\partial y} \left[\left(\frac{\partial y}{\partial \bar{x}}\right)^2 + \left(\frac{\partial y}{\partial \bar{y}}\right)^2 \right] = 0, \tag{2.33j}$$

$$R_{11} = \frac{\partial u}{\partial x} \frac{\partial f_4}{\partial y} \left[\frac{\partial x}{\partial \bar{x}} \frac{\partial y}{\partial \bar{x}} + \frac{\partial x}{\partial \bar{y}} \frac{\partial y}{\partial \bar{y}} \right] = 0, \tag{2.33k}$$

$$R_{12} = \frac{\partial u}{\partial y} \frac{\partial f_4}{\partial x} \left[\frac{\partial y}{\partial \bar{x}} \frac{\partial x}{\partial \bar{x}} + \frac{\partial y}{\partial \bar{y}} \frac{\partial x}{\partial \bar{y}} \right] = 0, \tag{2.33l}$$

$$P = Q, \qquad \left(\frac{\partial x}{\partial \bar{x}}\right)^2 + \left(\frac{\partial x}{\partial \bar{y}}\right)^2 = \left(\frac{\partial y}{\partial \bar{y}}\right)^2 + \left(\frac{\partial y}{\partial \bar{x}}\right)^2 \neq 0. \tag{2.33m}$$

† Another choice consists of $R_1' = R_2' = \cdots = 0$, where $R_1' = R_1 + R_2$, $R_2' = R_3 + R_4, \ldots$.

The function S of Eq. (2.31) must satisfy some conditions resulting from the group properties, but it is otherwise arbitrary.

Equations (2.33a–l) are satisfied if $u \equiv 0$, but we shall seek nontrivial solutions. The definition of a group requires each element to have an inverse. Therefore, the Jacobian associated with G cannot be zero, i.e.

$$\frac{\partial(\bar{x}, \bar{y}, \bar{u})}{\partial(x, y, u)} = \begin{vmatrix} \dfrac{\partial \bar{x}}{\partial x} & \dfrac{\partial \bar{x}}{\partial y} & \dfrac{\partial \bar{x}}{\partial u} \\ \dfrac{\partial \bar{y}}{\partial x} & \dfrac{\partial \bar{y}}{\partial y} & \dfrac{\partial \bar{y}}{\partial u} \\ \dfrac{\partial \bar{u}}{\partial x} & \dfrac{\partial \bar{u}}{\partial y} & \dfrac{\partial \bar{u}}{\partial u} \end{vmatrix} \neq 0. \tag{2.34}$$

If in Eq. (2.29) $f_4 = 0$, then $\partial\bar{u}/\partial x = \partial\bar{u}/\partial y = \partial\bar{u}/\partial u = 0$, in which case the Jacobian is equal to zero. Consequently, it is necessary that $f_4 \neq 0$.

Similarily each element of the subgroup

$$\bar{G}\colon \bar{x} = f_1(x, y; a), \quad \bar{y} = f_2(x, y; a) \tag{2.35}$$

must have an inverse, and therefore neither

$$\frac{\partial(\bar{x}, \bar{y})}{\partial(x, y)} = \frac{\partial \bar{x}}{\partial x}\frac{\partial \bar{y}}{\partial y} - \frac{\partial \bar{x}}{\partial y}\frac{\partial \bar{y}}{\partial x},$$

nor

$$\frac{\partial(x, y)}{\partial(\bar{x}, \bar{y})} = \frac{\partial x}{\partial \bar{x}}\frac{\partial y}{\partial \bar{y}} - \frac{\partial x}{\partial \bar{y}}\frac{\partial y}{\partial \bar{x}},$$

may be equal to zero.

Of course, we wish to avoid imposing restrictions upon u and its derivatives. Therefore, in examining Eqs. (2.33a–l) we will never allow u or its derivatives to vanish. For example in Eq. (2.33i), $\partial u/\partial x \neq 0$ and $(\partial x/\partial\bar{x})^2 + (\partial x/\partial\bar{y})^2 \neq 0$; consequently $\partial f_4/\partial x = 0$. In this way we generate the equations J' for which J is satisfied. These are

$$J'\colon \begin{aligned} &\frac{\partial^2 x}{\partial \bar{x}^2} + \frac{\partial^2 x}{\partial \bar{y}^2} \equiv 0, && (2.36\text{a}) \\ &\frac{\partial^2 y}{\partial \bar{x}^2} + \frac{\partial^2 y}{\partial \bar{y}^2} \equiv 0, && (2.36\text{b}) \\ &\frac{\partial x}{\partial \bar{x}}\frac{\partial y}{\partial \bar{x}} + \frac{\partial x}{\partial \bar{y}}\frac{\partial y}{\partial \bar{y}} \equiv 0, && (2.36\text{c}) \\ &\frac{\partial f_4}{\partial x} \equiv 0, && (2.36\text{d}) \\ &\frac{\partial f_4}{\partial y} \equiv 0, && (2.36\text{e}) \\ &\left(\frac{\partial x}{\partial \bar{x}}\right)^2 + \left(\frac{\partial x}{\partial \bar{y}}\right)^2 = \left(\frac{\partial y}{\partial \bar{y}}\right)^2 + \left(\frac{\partial y}{\partial \bar{x}}\right)^2 \not\equiv 0. && (2.36\text{f}) \end{aligned}$$

We note here that the conditions of Eqs. (2.36) do not completely specify a group† or class of groups. On the other hand groups are easily found which satisfy Eqs. (2.36). Among these are the following groups, together with their invariants (a is taken as a real parameter):

A_1: $\bar{x} = a^{\alpha}x$,
$\bar{y} = a^{\beta}y$, α, β, γ constants,
$\bar{u} = a^{\gamma}u$,
Invariants: Specified by Eq. (2.15).

B_1: $\bar{x} = x + \gamma_1 a$,
$\bar{y} = y + \gamma_2 a$, γ_1, γ_2 constants,
$\bar{u} = (\exp a)u$,
Invariants: $\eta = \gamma_2 x - \gamma_1 y$; $F(\eta) = \gamma_1 \ln u - x$, or $\gamma_2 \ln u - y$.

C_1: $\bar{x} = x \cos a - y \sin a$,
$\bar{y} = x \sin a + y \cos a$,
$\bar{u} = u$,
Invariants: $\eta = x^2 + y^2$; $F(\eta) = u$.

The unsteady three-dimensional motion of a viscous perfect fluid ($p = \rho RT$) is considered in the second paper of v. Krzywoblocki and Roth [32] using a three-dimensional generalization of the group B_1.

Several modifications of the finite transformation, Eq. (2.29), are of considerable use. The first of these is

$$G: \begin{cases} \bar{x} = f_1(x, y; a), \\ \bar{y} = f_2(x, y; a), \\ \bar{u} = f_4(x, y; a)u + f_5(a). \end{cases} \tag{2.37}$$

Determination of the proper groups for the Burgers' equation

$$u_y + uu_x = u_{xx} \tag{2.38}$$

will serve to discuss the procedure. If we require that Eq. (2.38) be conformally invariant under Eq. (2.37), then

$$\frac{\partial \bar{u}}{\partial \bar{y}} + \bar{u}\frac{\partial \bar{u}}{\partial \bar{x}} - \frac{\partial^2 \bar{u}}{\partial \bar{x}^2} = 0. \tag{2.39}$$

† Thus there is sufficient freedom in this procedure to add additional restrictions, such as those determined by the boundary and initial conditions. This was not done by v. Krzywoblocki and Roth, although in principle all the mechanism for so doing is present.

The quantities $\partial\bar{u}/\partial\bar{x}$, $\partial^2\bar{u}/\partial\bar{x}^2$, $\partial\bar{u}/\partial\bar{y}$ are determined by elementary calculation and substituted in Eq. (2.39). Since we do not wish to place restrictions on u or its derivative, we collect like terms in u and its derivative and list these below:

$$\frac{\partial u}{\partial y}\left\{f_4\frac{\partial y}{\partial\bar{y}} + f_4 f_5\frac{\partial y}{\partial\bar{x}} - f_4\frac{\partial^2 y}{\partial\bar{x}^2} - 2\frac{\partial x}{\partial\bar{x}}\frac{\partial y}{\partial\bar{x}}\frac{\partial f_4}{\partial x} - 2\left(\frac{\partial y}{\partial\bar{x}}\right)^2\frac{\partial f_4}{\partial y}\right\} \tag{2.40a}$$

$$\frac{\partial u}{\partial x}\left\{f_4\frac{\partial x}{\partial\bar{y}} + f_4 f_5\frac{\partial x}{\partial\bar{x}} - f_4\frac{\partial^2 x}{\partial\bar{x}^2} - 2\left(\frac{\partial x}{\partial\bar{x}}\right)^2\frac{\partial f_4}{\partial x} - 2\frac{\partial x}{\partial\bar{x}}\frac{\partial y}{\partial\bar{x}}\frac{\partial f_4}{\partial y}\right\} \tag{2.40b}$$

$$u\left\{\frac{\partial f_4}{\partial y}\frac{\partial y}{\partial\bar{y}} + \frac{\partial f_4}{\partial x}\frac{\partial x}{\partial\bar{y}} + f_5\left(\frac{\partial f_4}{\partial x}\frac{\partial x}{\partial\bar{x}} + \frac{\partial f_4}{\partial y}\frac{\partial y}{\partial\bar{x}}\right) - \frac{\partial^2 x}{\partial\bar{x}^2}\frac{\partial f_4}{\partial x}\right.$$
$$\left. - \frac{\partial^2 y}{\partial\bar{x}^2}\frac{\partial f_4}{\partial y} - \left(\frac{\partial x}{\partial\bar{x}}\right)^2\frac{\partial^2 f_4}{\partial x^2} - 2\frac{\partial x}{\partial\bar{x}}\frac{\partial y}{\partial\bar{x}}\frac{\partial^2 f_4}{\partial x\,\partial y} - \left(\frac{\partial y}{\partial\bar{x}}\right)^2\frac{\partial^2 f_4}{\partial y^2}\right\} \tag{2.40c}$$

$$u^2\left\{f_4\frac{\partial f_4}{\partial x}\frac{\partial x}{\partial\bar{x}} + \frac{\partial f_4}{\partial y}\frac{\partial y}{\partial\bar{x}}\right\} \tag{2.40d}$$

$$\frac{\partial^2 u}{\partial y^2}\left\{-f_4\left(\frac{\partial y}{\partial\bar{x}}\right)^2\right\} \tag{2.40e}$$

$$u\frac{\partial u}{\partial y}\left\{f_4^2\frac{\partial y}{\partial\bar{x}}\right\} \tag{2.40f}$$

$$\frac{\partial^2 u}{\partial x\,\partial y}\left\{-2f_4\frac{\partial y}{\partial\bar{x}}\frac{\partial x}{\partial\bar{x}}\right\} \tag{2.40g}$$

$$u\frac{\partial u}{\partial x}\left\{f_4^2\frac{\partial x}{\partial\bar{x}}\right\} \tag{2.40h}$$

$$\frac{\partial^2 u}{\partial x^2}\left\{-f_4\left(\frac{\partial x}{\partial\bar{x}}\right)^2\right\}. \tag{2.40i}$$

Since u_x, u, u^2, u_{yy}, uu_y, and u_{xy} do not appear as separate terms in Eq. (2.38), their coefficients must vanish identically. Consequently, Eqs. (2.40b–g) may be equated to zero immediately. Since $u_{xx} = u_y + uu_x$, Eq. (2.40i) becomes $(u_y + uu_x)\{-f_4(\partial x/\partial\bar{x})^2\}$. Consequently the coefficient of u_{xx} combines with those of u_y and uu_x, and the resulting coefficients are required to vanish. Accordingly, we find the following set of equations describe the desired finite transformation:

$$\frac{\partial y}{\partial \bar{x}} = 0, \tag{2.41}$$

$$\frac{\partial f_4}{\partial x} = 0, \tag{2.42}$$

$$\frac{\partial f_4}{\partial y} = 0, \tag{2.43}$$

$$\frac{\partial x}{\partial \bar{y}} - \frac{\partial^2 x}{\partial \bar{x}^2} + f_5 \frac{\partial x}{\partial \bar{x}} = 0, \tag{2.44}$$

$$f_4 \frac{\partial x}{\partial \bar{x}} - \left(\frac{\partial x}{\partial \bar{x}}\right)^2 = 0, \tag{2.45}$$

$$-\left(\frac{\partial x}{\partial \bar{x}}\right)^2 + \frac{\partial y}{\partial \bar{y}} = 0, \tag{2.46}$$

Equations (2.41)–(2.46) are satisfied by[†]

$$\bar{y} = [f(a)]^2 y, \tag{2.47}$$

$$\bar{x} = f(a)x + H(a)y, \tag{2.48}$$

where $f(a) = [f_4(a)]^{-1}$ and $H(a) = f_5(a)[f(a)]^2$. With these definitions and Eq. (2.37), we have

$$\bar{u} = [f(a)]^{-1}u + H(a)[f(a)]^{-2}. \tag{2.49}$$

Next we require that the set of transformations given by Eqs. (2.47)–(2.49) form a group, i.e., the system is *closed* under the group operation of composition, has a *unique identity* and a *unique inverse*, and satisfies the *associative law*. These requirements place restrictions on the parameter functions $f(a)$ and $H(a)$. Denoting members of the transformation by

$$\begin{aligned} \bar{x}_i &= f(a_i)x + H(a_i)y, \\ \bar{y}_i &= [f(a_i)]^2 y, \\ \bar{u}_i &= [f(a_i)]^{-1}u + H(a_i)[f(a_i)]^{-2}, \end{aligned}$$

[†] Functions of a can be added to Eqs. (2.47) and (2.48) to achieve a modest generalization.

we note that if $\bar{x}_1$, $\bar{y}_1$, $\bar{u}_1$ and $\bar{x}_2$, $\bar{y}_2$, $\bar{u}_2$ are transformations, then $\bar{x}_2(\bar{x}_1, \bar{y}_1, a_2)$, $\bar{y}_2(\bar{x}_1, \bar{y}_1, a_2)$, $\bar{u}_2(\bar{u}_1, a_2)$ must be a transformation. Thus

$$\bar{x}_2(\bar{x}_1, \bar{y}_1, a_2) = \bar{x}_3(x, y, a_3), \tag{2.50}$$

$$\bar{y}_2(\bar{x}_1, \bar{y}_1, a_2) = \bar{y}_3(x, y, a_3), \tag{2.51}$$

$$\bar{u}_2(\bar{u}_1, a_2) = \bar{u}_3(u, a_3). \tag{2.52}$$

In turn we employ Eqs. (2.47)–(2.49) to express these as

$$\begin{aligned}\bar{x}_3 &= f(a_2)\bar{x}_1 + H(a_2)\bar{y}_1\\ &= f(a_2)[f(a_1)x + H(a_1)y] + H(a_2)[f(a_1)]^2y\\ &= f(a_3)x + H(a_3)y,\end{aligned}$$

$$\begin{aligned}\bar{y}_3 &= [f(a_2)]^2\bar{y}_1\\ &= [f(a_2)]^2[f(a_1)]^2y\\ &= [f(a_3)]^2y,\end{aligned}$$

$$\begin{aligned}\bar{u}_3 &= [f(a_2)]^{-1}\bar{u}_1 + H(a_2)[f(a_2)]^{-2}\\ &= [f(a_2)]^{-1}\{[f(a_1)]^{-1}u + H(a_1)[f(a_1)]^{-2}\} + H(a_2)[f(a_2)]^{-2}\\ &= [f(a_3)]^{-1}u + H(a_3)[f(a_3)]^{-2}.\end{aligned}$$

The system satisfies the closure property if

$$f(a_3) = f(a_2)f(a_1), \tag{2.53}$$

$$H(a_3) = f(a_2)H(a_1) + H(a_2)[f(a_1)]^2. \tag{2.54}$$

From Eq. (2.53) we observe that interchange of a_1 and a_2 leaves Eq. (2.53) unaltered. Consequently, for closure, Eq. (2.54) must be unaltered under that same interchange. Thus

$$f(a_2)H(a_1) + H(a_2)[f(a_1)]^2 = f(a_1)H(a_2) + H(a_1)[f(a_2)]^2,$$

which upon rearrangement becomes

$$H(a_1)f(a_2)[1 - f(a_2)] = H(a_2)f(a_1)[1 - f(a_1)],$$

a result implying that

$$H(a) = f(a)[1 - f(a)]. \tag{2.55}$$

As a consequence of the preceding analysis, the finite transformation now takes the form

$$\begin{aligned}\bar{x} &= f(a)x + f(a)[1 - f(a)]y,\\ \bar{y} &= [f(a)]^2 y,\\ \bar{u} &= [f(a)]^{-1}u + [f(a)]^{-1}[1 - f(a)],\end{aligned} \tag{2.56}$$

where f is, for the moment, arbitrary.† It is immediately evident that the establishment of closure has produced a system for which the other properties are easily established. Thus we need one and only one a_0 such that $f(a_0) = 1$—this establishes the unique identity transformation. Since the system is linear, a unique inverse for each element follows immediately as does the associative law. Incorporation of boundary conditions and development of the absolute invariants is left for the next two sections.

Lastly we note the work of Hellums and Churchill [34, 35]. Their method of analysis consists of the following steps:

(a) The variables, parameters, boundary conditions, and initial conditions are placed in dimensionless form by the introduction of arbitrary reference variables. Each arbitrary function is also placed into dimensionless form by the introduction of the function in terms of the reference variables as a reference quantity.

(b) Each dimensionless parameter is equated to a constant. This procedure yields a system of algebraic equations in the reference quantities.

(c) The set of equations in (b) is solved to yield expressions of the reference quantities in terms of the parameters of the original problem. If the system is overdetermined, it is not possible to eliminate all parameters by choice of the reference quantities, and one parameter will appear in the problem for each algebraic equation which cannot be satisfied.

(d) If the system is underdetermined, that is, if all of the independent algebraic equations can be satisfied without specifying all of the reference quantities, this degree of freedom may be used to reduce the number of independent variables. The dimensionless variables are therefore combined in such a way as to eliminate the remaining arbitrary reference quantities.

(e) In problems involving arbitrary functions it is often important

† The arbitrary nature of f suggests using further conditions, such as boundary and initial conditions, to fix its form. We do this subsequently.

to determine what class of functions will admit a reduction in the number of independent variables. This can be resolved by finding those functions which leave one or more reference quantities arbitrary.

The Hellums–Churchill method is discussed, in the light of his general procedure, by Moran [36]. It is subsumed by the method of Section 2.2 but nevertheless has been a contribution because of its emphasis on the unity that exists in dimensional and similarity analysis. Its principle disadvantage is that considerable manipulation as well as insight and experience are often required to apply the procedure. An additional limitation, like that of the method of Section 2.2, is that not every similarity representation can be deduced under a restricted class of groups. On the other hand, it may be advantageously employed for motivating certain presentations where explicit invocation of the group concept is inappropriate.

2.4 INCORPORATION OF THE AUXILIARY CONDITIONS

In the general theory of Section 2.1, all possible transformation groups given by Eq. (2.1) are considered at the outset. Then those under which the system of differential equations does not transform conformally are eliminated from further consideration. That is, restrictions on the functions f of G are found to satisfy the conditions of Theorem 2.1-1 including group properties. There may exist many different groups satisfying all of these restrictions, and each predicts a similarity representation of the problem consisting of the *differential equations alone*. Generally not all of these, and perhaps none, will generate a similarity representation of the problem with auxiliary conditions.

Since a solution of a similarity representation of the equations alone is invariant under the group, any such solution can yield an invariant solution to the complete problem (equations with auxiliary conditions) only if the auxiliary conditions when transformed by the group can be satisfied by the invariant solution. Thus further restrictions on the functions f can be determined by the requirements that the auxiliary conditions be compatible with invariant solutions.

In Summary. If all the conditions placed on the f's to satisfy Theorem 2.1-1 (Morgan theorem) are met, a similarity representation is predicted for the equations. This can yield a similarity representation for the problem provided the conditions placed on the f's by the requirement of auxiliary condition compatibility with invariant solutions is also satisfied.

Inclusion of auxiliary conditions and development of the resulting deductive similarity theory was pioneered by Gaggioli and Moran [37, 38], Moran [36], Moran and Gaggioli [39–41]. Employing the notation of Section 2.1, their basic result is embodied in the following:

Theorem 2.4-1. *Let the auxiliary conditions be*

$$\beta_r\{\partial^s y_1/\partial(x^1)^s, \ldots, \partial y_n/\partial x^m, y_1, \ldots, y_n, x^1, \ldots, x^m\} = B_r(\sigma^1, \ldots, \sigma^t)$$

on Σ_r where

$$\Sigma_r\colon \{x^i = b_r{}^i(\sigma^1, \ldots, \sigma^t),\ t \leqslant m, \text{ for } \sigma^q \in [S_r{}^q, S_q{}^r]\}. \tag{2.57}$$

If $y^* = I_j(x^{1*}, \ldots, x^{m*})$ for all a, is an invariant solution, then

$$\begin{aligned}
&\beta_r{}^a\{\partial^s y_1/\partial(x^1)^s, \ldots, y_n, x^1, \ldots, x^m; a^*\}\\
&\quad \equiv \beta_r\{[f^{(s)}_{(1;s,0,\ldots)}]\\
&\qquad \times (\partial^s y_1/\partial(x^1)^s, \ldots, y_n, x^1, \ldots, x^m; a^*), \ldots, f_n(y_n; a^*), \ldots, f^m(x^1, \ldots, x^m; a^*)\}\\
&\quad = B_r(\sigma^1, \ldots, \sigma^t),
\end{aligned} \tag{2.58}$$

when

$$x^i = f^i[b_r{}^1(\sigma^1, \ldots, \sigma^t), \ldots, b_r{}^m(\sigma^1, \ldots, \sigma^t); a].^\dagger$$

(Alternate forms and a proof of this result are found in Moran and Gaggioli [39].)

Thus, with Eq. (2.58), the single auxiliary condition $\beta_r(\cdots) = B_r(\cdots)$ on Σ_r of Eq. (2.57) leads to a family of auxiliary conditions $\{\beta_r{}^a \cdots) = B_r(\cdots)\}$ on the family

$$\{\Sigma_r{}^a\colon [x^i = f^i(b_r{}^1(\sigma^1, \ldots, \sigma^t), \ldots, b_r{}^m(\sigma^1, \ldots, \sigma^t); a)]\}.$$

With the requirement that Eq. (2.58) or (2.59) be satisfied for each auxiliary condition $\beta_r(\cdots) = B_r(\cdots)$ on Σ_r, further restrictions on the f's will be imposed.

We shall discuss applications of this result in Section 2.6.

† Equivalently,

$$\begin{aligned}
&\beta_r\left\{[f^{(s)}_{(1;s,0,0\ldots)}]\left(\frac{\partial^s I_1}{\partial(z^1)^s}, \ldots, I_n, z^1, \ldots, z^m; a^*\right), \ldots, f_n(I_n; a^*), \ldots, f^m(z^1, \ldots, z^m; a^*)\right\}\\
&\quad = B_r(\sigma^1, \ldots, \sigma^t),
\end{aligned} \tag{2.59}$$

for

$$z^i = f^i[b_r{}^1, \ldots, b_r{}^m; a],$$

since

$$y_j{}^*(z^1, \ldots, z^m) = y_j(z^1, \ldots, z^m) = I_j(z^1, \ldots, z^m).$$

2.5 DETERMINATION OF ABSOLUTE INVARIANTS

Transformation groups (if any) have now been determined whose f's [Eq. (2.1)] are consistent with the twin requirements of equation invariance and auxiliary condition compatibility. It now remains to establish a set of functionally independent invariants for each group in order to complete the construction of the similarity representation.

Determination of the absolute invariants proceeds in a manner exactly analogous to that of the classical Lie theory (see, e.g., Eisenhart [2], Cohen [13], Ames [3]). It will be convenient to use the symbol of a group in our subsequent discussion. The *symbol* Q of the one-parameter group $\{\bar{z}^i = f^i(z^1, z^2,\dots, z^p; a); i = 1, 2,\dots, p\}$ is given by

$$Q \equiv \sum_{i=1}^{p} \xi^i(z^1,\dots, z^p) \frac{\partial}{\partial z^i}, \tag{2.60}$$

where

$$\xi^i(z^1,\dots, z^p) = (\partial f^i/\partial a)(z^1,\dots, z^p; a_0),$$

and a_0 is the value of a generating the identity element.

In terms of the symbol, the invariants are determined from the following result:

Theorem 2.5-1. *The function $I(z^1,\dots, z^p)$ is an absolute invariant of the transformation group with symbol Q if and only if $QI \equiv 0$. Furthermore, if $I_1 ,\dots, I_{p-1}$ are functionally independent solutions of $QI_j \equiv 0$, $j = 1,\dots, p - 1$, then any solution of $QI = 0$ can be expressed as*

$$I(z^1,\dots, z^p) = R[I_1 ,\dots, I_{p-1}],^{\dagger}$$

where R is a differentiable function.

Establishment of the invariants requires the solution of the linear partial differential equation $QI = 0$. We shall illustrate the computation for the group given by Eq. (2.56):

$$\begin{aligned}
\bar{x} &= f(a)x + f(a)[1 - f(a)]y = f^1(x, y; a),\\
\bar{y} &= [f(a)]^2 y = f^2(x, y; a),\\
\bar{u} &= [f(a)]^{-1}u + [f(a)]^{-1}[1 - f(a)] = f^3(u; a).
\end{aligned}$$

† This is the general solution.

Two absolute invariants are required. One must be an absolute invariant of the subgroup S defined by

$$S: \begin{cases} \bar{x} = f(a)x + f(a)[1 - f(a)]y, \\ \bar{y} = [f(a)]^2 y, \end{cases} \tag{2.61}$$

whose symbol is

$$(\partial f^1/\partial a)(x, y; a_0)(\partial/\partial x) + (\partial f^2/\partial a)(x, y; a_0)(\partial/\partial y),$$

or

$$f'(a_0)\{(x - y)(\partial/\partial x) + 2y(\partial/\partial y)\}.$$

The function $I_1\,(x, y)$ is an absolute invariant of S if and only if

$$f'(a_0)\{(x - y)(\partial I_1/\partial x) + 2y(\partial I_1/\partial y)\} = 0.$$

Thus when $f'(a_0) \neq 0$,† I_1 must satisfy

$$(x - y)(\partial I_1/\partial x) + 2y(\partial I_1/\partial y) = 0. \tag{2.62}$$

The general solution of Eq. (2.62) is given by $I_1 = g(\eta)$, where η is any nontrivial solution of Eq. (2.62) and g is arbitrary. A solution of Eq. (2.62) is easily obtained by the method of Lagrange (characteristics; see Volume I) as

$$\eta(x, y) = (x + y)y^{-1/2}. \tag{2.63}$$

Thus the general solution is

$$I_1 = g[(x + y)y^{-1/2}], \tag{2.64}$$

wherein g is arbitrary. That this is an absolute invariant can easily be verified. From Section 2.1, $I_1(x, y)$ is an absolute invariant if $I_1(\bar{x}, \bar{y}) = I_1(x, y)$. With Eq. (2.56), we find that

$$I_1(\bar{x}, \bar{y}) = g[(\bar{x} + \bar{y})\,\bar{y}^{-1/2}] = g[(x + y)y^{-1/2}],$$

which is what was required.

The remaining absolute invariant $I_2(u, x, y)$ must satisfy $QI_2 = 0$,

† The case $f'(a_0) = 0$ is not admissible if S is a one-parameter continuous transformation group, for this would imply $(\partial f^1/\partial a)(x, y; a_0) = (\partial f^2/\partial a)(x, y; a_0) = 0$. This condition is not allowable (see, e.g., Cohen [13, p. 12]).

where Q is the symbol of the entire group given by Eq. (2.56). Therefore I_2 is the solution of the equation

$$\left[\frac{\partial \bar{x}}{\partial a}(x, y; a_0)\right]\frac{\partial I_2}{\partial x} + \left[\frac{\partial \bar{y}}{\partial a}(x, y; a_0)\right]\frac{\partial I_2}{\partial y} + \left[\frac{\partial \bar{u}}{\partial a}(x, y, u; a_0)\right]\frac{\partial I_2}{\partial u} = 0, \tag{2.65}$$

which, for Eq. (2.56), becomes

$$f'(a_0)\{(x - y)(\partial I_2/\partial x) + 2y(\partial I_2/\partial y) - (u + 1)(\partial I_2/\partial u)\} = 0. \tag{2.66}$$

A solution is

$$I_2 = h[\gamma_1, \gamma_2],$$

wherein h is arbitrary and γ_1 and γ_2 are independent solutions of Eq. (2.66). It is readily verified that

$$\gamma_1 = y^{1/2}(u + 1), \qquad \gamma_2 = y^{-1/2}(x + y),$$

are solutions of Eq. (2.66). Hence

$$I_2 = h[y^{1/2}(u + 1), y^{-1/2}(x + y)], \tag{2.67}$$

is the absolute invariant we seek.

As a specific example, let us specialize the functions g and h of Eqs. (2.64) and (2.67) so that

$$I_1 = \eta = (x + y)y^{-1/2}, \qquad \text{and} \qquad (u + 1) = y^{-1/2}f(\eta).$$

With this transformation, Burgers' equation (2.38), is readily shown to transform into the similarity equation

$$f'' - ff' + \tfrac{1}{2}\eta f' + \tfrac{1}{2}f = 0,$$

or

$$(d/d\eta)\{f' - \tfrac{1}{2}f^2 + \tfrac{1}{2}\eta f\} = 0,$$

which integrates to a form of the Riccati equation

$$f' - \tfrac{1}{2}f^2 + \tfrac{1}{2}\eta f = \text{const.}$$

2.6 EXAMPLE OF DEDUCTIVE SIMILARITY METHOD

The example of this section is the classical one of steady two-dimensional laminar incompressible boundary layer flow over an infinite flat plate. It was first treated in the deductive format by Gaggioli and Moran [37, 39]. Our main interest herein is to demonstrate how the auxiliary conditions are introduced.

With $u(x, y)$, $v(x, y)$, $U(x)$ representing velocity components parallel and normal to the plate and the limit of u as $y \to \infty$, respectively, the governing equations are

$$\begin{aligned} uu_x + vu_y - UU_x - \nu u_{yy} &= 0, \qquad \nu \text{ const}, \\ u_x + v_y &= 0, \end{aligned} \tag{2.68}$$

for $x > 0$, $y \geqslant 0$, together with the auxiliary conditions

$$u(x, 0) = 0, \qquad v(x, 0) = 0, \qquad u(x, y \to \infty) = U(x), \tag{2.69}$$

and $U(x)$, $u(x, y)$, $v(x, y)$ analytic on $x > 0$, $y > 0$.

Our objective is to determine similarity representations predicted by the Morgan theory whose invariant solutions are compatible with the specified auxiliary conditions. The analysis of the auxiliary conditions, to determine necessary conditions on the transformations of the group, is enhanced by means of a change of variables. This is sometimes useful when unspecified functions [$U(x)$ here] appear in the auxiliary conditions. Upon setting

$$w(x, y) = u(x, y)/U(x),$$

Eqs. (2.68) and (2.69) become

$$\begin{aligned} Uww_x + vw_y - (1 - w^2)U_x - \nu w_{yy} &= 0, \\ U_x w + Uw_x + v_y &= 0, \end{aligned} \tag{2.70}$$

for $x > 0$, $y \geqslant 0$, together with the new auxiliary conditions

$$w(x, 0) = 0, \qquad v(x, 0) = 0, \qquad w(x, y \to \infty) = 1, \tag{2.71}$$

and w, v, U are analytic on $x > 0$, $y > 0$.

One could now introduce a general, initially unspecified group of transformations for all of the variables, but for this discussion† we shall assume

$$G: \bar{w} = f_w(w; a), \qquad \bar{U} = f_U(U; a), \qquad \bar{v} = f_v(v; a),$$
$$\bar{y} = f^y(y; a), \qquad \bar{x} = f^x(x; a). \tag{2.72}$$

The requirement of conformal invariance of the differential equations (2.70) (that is the application of the Morgan theory) generates the class

$$G: \begin{cases} \bar{w} = w, \\ \bar{U} = C_3(a)U, \\ \bar{v} = [C_1(a)]^{-1}v, \\ \bar{y} = C_1(a)y + C_2(a), \\ \bar{x} = [C_1(a)]^2 C_3(a)x + C_4(a), \end{cases} \tag{2.73}$$

where, in accord with the continuity of the f's, the C's are assumed to be at least continuous.

Next we must find the additional restrictions placed upon the C's of Eq. (2.73) by the requirement that each of the auxiliary conditions, Eq. (2.71), be satisfied by functions invariant under G. Using the terminology of Theorem 2.4-1, let us denote the invariant solution for w by $I_w(z^1, z^2)$.

For the first auxiliary condition of Eq. (2.71),

$$\beta_1(w, x, y) \equiv w = 0, \qquad \text{if} \quad y = 0, \quad \text{and} \quad x = \sigma > 0.$$

Thus, by Eq. (2.59),

$$f_w[I_w(z^1, z^2); a^*] = 0 \qquad \text{for} \quad z^1 = f^x(\sigma; a), \quad z^2 = f^y(0; a), \tag{2.74}$$

where $\bar{w} = I_w(\bar{x}, \bar{y}) = I_w[f^x(x; a), f^y(y; a)]$ for all values of a. With Eq. (2.73), Eq. (2.74) becomes

$$I_w[z^1; C_2(a)] = 0 \qquad \text{for} \quad z^1 = [C_1(a)]^2 C_3(a)\sigma + C_4(a),$$

since $C_3(a)$ must not vanish for any a, lest the corresponding inverse fail to exist. Moreover, since $C_3(a) \neq 0$, $C_3(a_0) = 1$, and $C_3(a)$ is continuous, it follows that $C_3(a) > 0$ for all a. Thus

$$I_w[z^1; C_2(a)] = 0 \qquad \text{for all} \quad z^1 > C_4(a). \tag{2.75}$$

† At the end of this section, we shall briefly describe the results obtained by lifting this restriction.

Suppose there exists a value of a, say a', such that $C_2(a) > 0$ and not constant in some neighborhood N of a'. Then with $I_w(x, y)$ analytic on $x > 0, y > 0$, it follows that $I_w(z^1, C_2(a')) = 0$, for all $z^1 > 0$. Repeating this argument for all other a in N wherein $C_2(a)$ is not constant, the analyticity gives $I_w \equiv 0$, for all $z^1 > 0$ and all z^2 on the open interval which is the image of N under C_2. Thus $I_w \equiv 0$, for $z^1 > 0$ and $z^2 > 0$. To avoid this, it must be required that

$$C_2(a) = \text{const}, \qquad \text{and/or} \qquad C_2(a) \leqslant 0.$$

If constant, $C_2(a)$ must be zero, since the identity requirement of the group is

$$y = C_1(a_0)y + C_2(a_0).$$

Therefore $C_2(a) \leqslant 0$. With this we write

$$\bar{y} = C_1(a)y + C_2(a) = C_1(a)y - | C_2(a)|,$$

or

$$y = [1/C_1(a)]\,\bar{y} + [| C_2(a)|/C_1(a)]. \tag{2.76}$$

For $a = \bar{a}$, the group inverse implies

$$y = C_1(\bar{a})\,\bar{y} + C_2(\bar{a}) = C_1(\bar{a})\,\bar{y} - | C_2(\bar{a})|. \tag{2.77}$$

Upon equating Eqs. (2.76) and (2.77), it follows that

$$-| C_2(\bar{a})| = | C_2(a)|/C_1(a). \tag{2.78}$$

Thus, whenever $C_1(a) > 0$, Eq. (2.78) requires $C_2(a) \equiv 0$. Now $C_1(a) > 0$ for all a. $C_1(a)$ may not vanish for any a, for if it did the corresponding inverse would not exist [Eq. (2.76)]. Further, the identity transformation requires $C_1(a_0) = 1$. Our conclusion follows from the continuity of C_1. Consequently for all a

$$C_2(a) \equiv 0.$$

By analogous procedures the other boundary conditions, $v(x, 0) = 0$, $w(x, y \to \infty) = 1$, are readily analyzed, and no further restrictions need be placed on the C's. That is, it is assured that the differential equations transform with conformal invariance and the auxiliary conditions are compatible with invariant solutions, under classes G of the form

$$G: \begin{cases} \bar{w} = w, \quad \bar{U} = C_3(a)U, \quad \bar{v} = [C_1(a)]^{-1}v, \\ \bar{y} = C_1(a)y, \quad \bar{x} = [C_1(a)]^2 C_3(a)x + C_4(a). \end{cases} \tag{2.79}$$

A number of distinct transformation groups may assume this required form. Among these, many may lead to the same solution of the problem. A solution which is invariant under one particular group of the required form may be invariant under others.

To determine absolute invariants of groups with the form of Eq. (2.79) we employ the results of Section 2.5. Four functionally independent absolute invariants of any group satisfying Eq. (2.79) are required. One of these, say η, must be an invariant of the subgroup

$$S: \quad \bar{y} = C_1(a)y, \quad \bar{x} = [C_1(a)]^2 C_3(a)x + C_4(a). \tag{2.80}$$

Then by Theorem 2.5-1, $\eta(x, y)$ is an absolute invariant if and only if $Q\eta \equiv 0$, where Q is the symbol of S. Thus

$$Q\eta = (px + r)(\partial\eta/\partial x) + y(\partial\eta/\partial y) \equiv 0, \tag{2.81}$$

where

$$p = [(2C_1C_1'C_3 + C_1^2C_3')/C_1']_{a=a_0}, \tag{2.82}$$

and

$$r = [C_4'/C_1']_{a=a_0}. \tag{2.83}$$

The general solution of Eq. (2.81) is given by $\eta = g(\lambda)$, where g is arbitrary and $\lambda(x, y)$ is a nontrivial solution of

$$(px + r)(\partial\lambda/\partial x) + y(\partial\lambda/\partial y) = 0. \tag{2.84}$$

The cases $p \neq 0$ and $p = 0$ generate two independent solutions.

For the case $p \neq 0$, one has by separation of variables (or by Lagrange's method)

$$\tilde{\lambda} = y/(px + r)^{1/p}, \tag{2.85}$$

so that

$$\tilde{\eta} = \tilde{g}[y/(px + r)^{1/p}] \tag{2.86}$$

is the general solution of Eq. (2.81) for $p \neq 0$. Since $Q\tilde{\lambda} = 0$, $\tilde{\lambda}$ is an absolute invariant of any group satisfying Eq. (2.80). For $\tilde{\lambda}$ to transform invariantly, i.e., $\tilde{\lambda}(x, y) = \tilde{\lambda}(\bar{x}, \bar{y})$, certain limitations must be placed upon the C's. With Eq. (2.80)†

$$\frac{\bar{y}}{(p\bar{x} + r)^{1/p}} = \frac{\tilde{C}_1(a)y}{\{p[\tilde{C}_1(a)]^2\tilde{C}_3(a)x + p\tilde{C}_4(a) + r]\}^{1/p}}, \tag{2.87}$$

† We use $\tilde{C}$ to indicate the parameter functions corresponding to the group whose invariant is $\tilde{\lambda}$.

which equals $y/(px + r)^{1/p}$ if and only if

$$\tilde{C}_3(a) = [\tilde{C}_1(a)]^{p-2} \qquad \tilde{C}_4(a) = (r/p)\{[\tilde{C}_1(a)]^p - 1\}. \tag{2.88}$$

At this point we note that satisfaction of Eqs. (2.80) alone, by a set of functions, does not assure that the set satisfies the group definition. If Eqs. (2.88) are not satisfied, $y/(px + r)^{1/p}$ does not transform invariantly and S is not a group (really a subgroup of G). Therefore when $p \neq 0$, the group G must satisfy Eqs. (2.88) as well as Eqs. (2.79), i.e.,

$$\tilde{G}: \begin{cases} \bar{w} = w, \quad \bar{U} = [\tilde{C}_1(a)]^{p-2}U, \quad \bar{v} = [\tilde{C}_1(a)]^{-1}v, \\ \bar{y} = [\tilde{C}_1(a)]y, \quad \bar{x} = [\tilde{C}_1(a)]^p[x + r/p] - r/p. \end{cases} \tag{2.89}$$

For the case $p = 0$, a solution of Eq. (2.84) is

$$\hat{\lambda} = y/\exp[x/r + k], \tag{2.90}$$

where k is a constant. Thus the general solution is

$$\hat{\eta} = \hat{g}(\hat{\lambda}),$$

in the case $p = 0$. Furthermore, since $\hat{\lambda}(x, y)$ is an absolute invariant, $\hat{\lambda}(x, y) = \hat{\lambda}(\bar{x}, \bar{y})$. With Eq. (2.80)

$$\frac{\bar{y}}{\exp[\bar{x}/r + k]} = \frac{\hat{C}_1(a)y}{\exp[\hat{C}_4(a)/r]\exp\{[\hat{C}_1(a)]^2\hat{C}_3(a)x/r\}},$$

which equals $y/\exp[x/r + k]$ if and only if

$$\hat{C}_4(a) = r\ln[\hat{C}_1(a)], \qquad [\hat{C}_1(a)]^2\hat{C}_3(a) = 1. \tag{2.91}$$

Hence, when $p = 0$, the group G must satisfy Eqs. (2.91) as well as Eqs. (2.79), i.e.,

$$\hat{G}: \begin{cases} \bar{w} = w, \quad \bar{U} = [\hat{C}_1(a)]^{-2}U, \quad \bar{v} = [\hat{C}_1(a)]^{-1}v, \\ \bar{y} = \hat{C}_1(a)y, \quad \bar{x} = x + r\ln\hat{C}_1(a). \end{cases} \tag{2.92}$$

Next, three *additional* absolute invariants will be determined for groups of the form $\tilde{G}$ and $\hat{G}$, respectively. With $C_1(a_0)$ as the group identity, the symbol of any group $\tilde{G}$ is

$$\tilde{Q} = (p-2)U\frac{\partial}{\partial U} - v\frac{\partial}{\partial v} + y\frac{\partial}{\partial y} + p\left(x + \frac{r}{p}\right)\frac{\partial}{\partial x}. \tag{2.93}$$

The function $\tilde{g}_i(w, U, v, x, y)$ is an absolute invariant of any group satisfying Eq. (2.89) if $\tilde{Q}\tilde{g}_i = 0$. Furthermore, if $\tilde{\lambda}_i$, $i = 1, 2, 3, 4$ are independent solutions of $\tilde{Q}\tilde{\lambda}_i \equiv 0$, then a general solution of $\tilde{Q}\tilde{g}_i \equiv 0$ is given by $\tilde{g}_i = \tilde{\Gamma}_i(\tilde{\lambda}_1, \tilde{\lambda}_2, \tilde{\lambda}_3, \tilde{\lambda}_4)$, where $\tilde{\Gamma}_i$ is arbitrary. By elementary separation of variables, it is easily demonstrated that

$$\tilde{\lambda}_1 = w, \qquad \tilde{\lambda}_2 = U[x + r/p]^{(2/p)-1},$$
$$\tilde{\lambda}_3 = v[x + r/p]^{1/p}, \qquad \text{and} \qquad \tilde{\lambda}_4 = y[x + r/p]^{-1/p}$$

are independent solution of $\tilde{Q}\tilde{\lambda}_i \equiv 0$. Thus for $i = 1, 2, 3$,

$$\tilde{g}_i(w, U, v, x, y) = \tilde{\Gamma}_i[w, U(x + r/p)^{(2/p)-1}, v(x + r/p)^{1/p}, y(x + r/p)^{-1/p}]. \tag{2.94}$$

Of course, many choices in Eq. (2.94) are possible. As a specific choice, we select functionally independent quantities $\tilde{g}$, $\tilde{g}_1$, $\tilde{g}_2$, and $\tilde{g}_3$ to be

$$\tilde{\eta} = \tilde{g}(x, y) = y[x + r/p]^{-1/p}, \tag{2.95}$$
$$\tilde{g}_1(w, U, v, x, y) = \tilde{\lambda}_1 = w,$$
$$\tilde{g}_2(w, U, v, x, y) = \tilde{\lambda}_2 = v[x + r/p]^{1/p},$$
$$\tilde{g}_3 = \tilde{\lambda}_3 = U[x + r/p]^{(2/p)-1}.$$

Now these, together with the relations from the Morgan theorem

$$\tilde{g}_1 = w = \tilde{F}_1(\tilde{\eta}), \tag{2.96}$$

$$\tilde{g}_2 = v[x + r/p]^{1/p} = \tilde{F}_2(\tilde{\eta}), \tag{2.97}$$

$$\tilde{g}_3 = U[x + r/p]^{(2/p)-1} = \tilde{F}_3(\tilde{\eta}), \tag{2.98}$$

permit the differential equations (2.71) to be transformed into ordinary differential equations.

Since $U = U(x)$, it follows that $U[x + r/p]^{(2/p)-1}$ is a function of x alone. Since $\tilde{F}_3 = \tilde{F}_3[y(x + r/p)^{-1/p}]$ depends on both x and y, it follows that this must be a constant, say U_0. Consequently,

$$\tilde{U}(x) = U(x) = U_0[x + r/p]^{1-(2/p)}, \tag{2.99}$$

that is, $U(x)$ may not be prescribed arbitrarily but must satisfy Eq. (2.99) in order to have a solution to the problem invariant under any group of the form $\tilde{G}$ [Eqs. (2.89)].

With Eqs. (2.95)–(2.99), we find the following ordinary differential equations in $\tilde{\eta}$ for Eqs. (2.70):

$$\begin{gathered}(\nu/U_0)\tilde{F}_1'' + (\tilde{\eta}/p)\tilde{F}_1\tilde{F}_1' - \tilde{F}_2\tilde{F}_1'/U_0 - (1 - 2/p)[(\tilde{F}_1)^2 - 1] = 0, \\ (1 - 2/p)\tilde{F}_1 - (\tilde{\eta}/p)\tilde{F}_1' + \tilde{F}_2'/U_0 = 0,\end{gathered} \tag{2.100}$$

while the auxiliary conditions, Eqs. (2.71), become

$$\tilde{F}_1(0) = 0, \qquad \tilde{F}_1(\infty) = 1, \qquad \tilde{F}_2(0) = 0,$$
$$\tilde{F}_1, \tilde{F}_2 \quad \text{analytic on} \quad \tilde{\eta} > 0. \tag{2.101}$$

One similarity representation of the problem, Eqs. (2.70) and (2.71), is provided by Eqs. (2.99)–(2.101). Equation (2.99) together with a solution of Eqs. (2.100) subject to Eq. (2.101) would be a similarity solution of the problem.

Based on the groups of $\hat{G}$, with $p = 0$, alternative similarity representations independent of the former may be found. From the symbol

$$\hat{Q} = -2U\frac{\partial}{\partial U} - v\frac{\partial}{\partial v} + y\frac{\partial}{\partial y} + r\frac{\partial}{\partial x},$$

of any such group $\hat{G}$, we have the solutions of $\hat{Q}\hat{\lambda}_i = 0$,

$$\hat{\lambda}_1 = w, \qquad \hat{\lambda}_2 = U\exp[2x/r + 2k],$$
$$\hat{\lambda}_3 = v\exp[x/r + k], \qquad \hat{\lambda}_4 = y\exp[-x/r - k].$$

Consequently, for $i = 1, 2, 3$,

$$\hat{g}_i(w, U, v, x, y) = \hat{\Gamma}_i[w, U\exp(2x/r + 2k), v\exp(x/r + k), y\exp(-x/r - k)],$$

where $\hat{\Gamma}_i$ is arbitrary, is a general solution to $\hat{Q}\hat{g}_i = 0$. Since $\hat{g}_i$ are absolute invariants, we may choose a specific set to be

$$\hat{\eta} = \hat{g}(x, y) = y\exp[-x/r - k],$$
$$\hat{g}_1 = \hat{\lambda}_1 = w = \hat{F}_1(\hat{\eta}),$$
$$\hat{g}_2 = \hat{\lambda}_2 = v\exp[x/r + k] = \hat{F}_2(\hat{\eta}),$$
$$\hat{g}_3 = \hat{\lambda}_3 = U\exp[2x/r + 2k] = \hat{F}_3(\hat{\eta}).$$

Again, since $\hat{F}_3(\hat{\eta})$ depends upon x and y and $U = U(x)$, it follows that this must be constant, say U_0. Thus

$$\hat{U}(x) = U(x) = U_0\exp[-2x/r - 2k].$$

Equations (2.70) transform into

$$[\nu/U_0]\hat{F}_1'' + [\hat{\eta}/r]\hat{F}_1\hat{F}_1' - \hat{F}_2\hat{F}_1'/U_0 + (2/r)(\hat{F}_1^{\,2} - 1) = 0,$$
$$2\hat{F}_1 + \hat{\eta}\hat{F}_1' - (r/U_0)\hat{F}_2' = 0,$$

and the auxiliary conditions, Eqs. (2.71), become $\hat{F}_1(0) = 0$, $\hat{F}_2(0) = 0$, $\hat{F}_1(\infty) = 1$, $\hat{F}_1$, $\hat{F}_2$ analytic on $\hat{\eta} > 0$.

Before closing this section, we must discuss the results obtained and the complications that can occur when a more general system than Eqs. (2.72) is employed. Woodard [42] has explored a variety of alternatives. For example, suppose the boundary layer equations without a pressure gradient

$$uu_x + vu_y = u_{yy}, \qquad u_x + v_y = 0, \tag{2.102}$$

are considered with the transformation class

$$\begin{aligned} \bar{x} &= f^x(x, y; a), & \bar{y} &= f^y(x, y; a), \\ \bar{u} &= C_4(a)u + C_5(x, y, u, v; a), & \bar{v} &= C_6(a)v + C_7(x, y, u, v; a). \end{aligned} \tag{2.103}$$

The required invariance conditions and group properties generate the group

$$\begin{aligned} \bar{x} &= C_6^{-2}C_4 x, & \bar{y} &= C_6^{-2}\{C_6 y + (C_6 - C_4)x\}, \\ \bar{u} &= C_4 u, & \bar{v} &= C_6 v + (C_6 - C_4)u. \end{aligned} \tag{2.104}$$

If, from the general form of the invariants, we select the special class

$$\begin{aligned} \eta &= yx^{1/(m-2)} + x^{(m-1)/(m-2)}, \\ ux^{m/(2-m)} &= f(\eta), \\ vu^{-1/m} + u^{(m-1)/m} &= h(\eta), \end{aligned} \tag{2.105}$$

with $m = C_4'(a)/C_6'(a)$, then it follows that

$$v = h(\eta)u^{1/m} - u.$$

When the most general forms are employed, special care must be taken to assure that the group properties are satisfied. Satisfaction of the closure property and invariance under the group restricts the functions of the parameters, $C_4(a)$ and $C_6(a)$ herein, in a usually nontrivial way.

Additional research in the application of these techniques has been carried out by Moran and Gaggioli [43] for real boundary layers, in a generalization of dimensional analysis by the same authors [44] and for diffusion and other transport problems by Woodard [42].

2.7 SIMILARITY FORMALISM WITH MULTIPARAMETER GROUPS

Here, we wish to call attention to an error in the similarity discussion of Volume I. This error was brought to our attention by Gaggioli and Moran [38] (see also Moran and Gaggioli [40]). On page 141, Section 4.5 of Volume I, we essentially assert that the one parameter group $\bar{x} = a^{\alpha}x$, $\bar{y} = a^{\alpha}y$, $\bar{z} = a^{\beta}z$, $\bar{u} = a^{\gamma}u$ can be reduced to an ordinary differential equation in the variables (invariants)

$$\eta_1 = z/(bx + cy)^{\beta/\alpha}, \qquad f(\eta_1) = u/(bx + cy)^{\gamma/\alpha}. \tag{2.106}$$

However, this *does not constitute a complete set of absolute invariants, and the assertion is invalid.* As is readily seen, another independent invariant is

$$\eta_2 = x/y, \tag{2.107}$$

and therefore in Eq. (2.106), $f = f(\eta_1, \eta_2)$. Moreover, in this particular case, Eqs. (2.106) and (2.107) can be easily shown to be absolute invariants of the above group. *But one should use the formalism of this chapter to obtain the absolute invariants.* For this problem we must solve $Q_S\eta_i = 0$, where Q_S is the symbol of the subgroup formed from the first three transformations. Thus we should solve

$$x\frac{\partial \eta_i}{\partial x} + y\frac{\partial \eta_i}{\partial y} + \frac{\beta}{\alpha} z\frac{\partial \eta_i}{\partial z} = 0, \tag{2.108}$$

with the result that $\eta = G(\eta_1, \eta_2)$, G arbitrary, provides the *two* absolute invariants of S. An alternative set

$$\hat{\eta}_1 = zy^{-\beta/\alpha}, \qquad \hat{\eta}_2 = x/y,$$

is also possible.

> *One must be especially careful to note that a similarity representation is not guaranteed to evolve from an incomplete set of invariants.*

In many examples, a formalism for multiparameter groups is useful. The machinery for some elementary cases with two parameters is contained in Volume I, page 142. Herein, we shall provide the basic theorems for multiparameter systems with theorems from Eisenhart [2] as discussed by Moran and Gaggioli [41].

Let

$$G: \{\bar{z}^i = f^i(z^1, z^2, \ldots, z^m; a_1, \ldots, a_r); r \leqslant m; i = 1, \ldots, m\} \tag{2.109}$$

be an r-parameter continuous group. Then the *symbols* of G are the operators defined by

$$Q_\alpha = \sum_{j=1}^{m} \xi_\alpha^{\ j}(z^1,\ldots, z^m)(\partial/\partial z^j), \qquad \alpha = 1,\ldots, r, \tag{2.110}$$

where the elements of the matrix $\xi = [\xi_\alpha^{\ j}]$ are defined by [compare the one parameter case of Section 2.5, Eq. (2.60)]

$$\xi_\alpha^{\ j} = \left[\sum_{p=1}^{m} \psi_p^{\ j}(\partial f^p/\partial a_\alpha)\right](z; a^0), \tag{2.111}$$

where a^0 signifies the group identity, and the elements of $\psi = [\psi_p^{\ j}]$ are specified by

$$\sum_{p=1}^{m} \psi_p^{\ j}(z; a)(\partial f^p/\partial z^k)(z; a) = \delta_k^{\ j}, \qquad j, k = 1, 2,\ldots, m, \tag{2.112}$$

$\delta_k^{\ j} = 1$, if $j = k$, and zero otherwise.

In terms of these definitions, the pertinent results are stated by the following theorem.

Theorem 2.7-1. (*a*) *A function $F(z^1,\ldots, z^m)$ is an absolute invariant of G, Eq.* (2.109), *if and only if it is a solution of*

$$Q_\alpha F = 0. \tag{2.113}$$

(*b*) *The group G possesses $m - \rho$*† *and only $m - \rho$ functionally independent absolute invariants, where $\rho < m$ is the* **rank** *of the matrix $\xi = [\xi_\alpha^{\ j}]$, $j = 1,\ldots, m$; $\alpha = 1,\ldots, r$. (If $r = 1$, it is immediate that $\rho = 1$.)*

(*c*) *If $F_i(z^1,\ldots, z^m)$, $i = 1,\ldots, m - \rho$ is a set of functionally independent solutions of Eq.* (2.113) *and if $F(z^1,\ldots, z^m)$ is any other solution of Eq.* (2.113), *then*

$$F = G\{F_1, F_2, \ldots, F_{m-\rho}\},$$

where G is an arbitrary differentiable function.

Consequently, a group

$$\Gamma_r \begin{cases} S_r: & \bar{x}^i = f^i(x^1,\ldots, x^m; a_1,\ldots, a_r), & i = 1,\ldots, m, \\ & \bar{y}_j = f_j(y_1,\ldots, y_n; a_1,\ldots, a_r), & j = 1,\ldots, n, \end{cases}$$

† In most cases, $\rho = r$, r the number of parameters, although this must be verified.

will possess $m + n - \rho$ functionally independent absolute invariants. We shall discuss only those systems Γ_r such that the subgroup S_r has $m - \rho$ functionally independent absolute invariants denoted $\eta_j(x^1,..., x^m) = \eta_j(\bar{x}^1,..., \bar{x}^m)$, $j = 1,..., m - \rho$ and, additionally, there are n absolute invariants

$$g_k(y_1 ,..., y_n; x^1,..., x^m) = g_k(\bar{y}_1 ,..., \bar{y}_n; \bar{x}^1,..., \bar{x}^m)$$

so selected that the rank of the Jacobian

$$\partial[g_1 ,..., g_n]/\partial[y_1 ,..., y_n]$$

is n.

For the boundary layer equations

$$\begin{aligned} uu_x + vu_y + wu_z - \nu u_{yy} - UU_x - WU_z &= 0, \\ uw_x + vw_y + ww_z - \nu w_{yy} - UW_x - WW_z &= 0, \\ u_x + v_y + w_z &= 0, \end{aligned} \tag{2.114}$$

with boundary conditions

$$\begin{gathered} u(x, 0, z) = v(x, 0, z) = w(x, 0, z) = 0, \\ \lim_{y\to\infty} u = U(x, z), \qquad \lim_{y\to\infty} w = W(x, z), \end{gathered} \tag{2.114a}$$

one might attempt the development of a similarity representation and solution with a class of two-parameter transformation groups of the form

$$G \begin{cases} S\colon & \bar{x} = C_1 x + k_1 , \quad \bar{y} = C_2 y + k_2 , \quad \bar{z} = C_3 z + k_3 , \\ & \bar{u} = C_4 u + k_4 , \quad \bar{v} = C_5 v + k_5 , \quad \bar{w} = C_6 w + k_6 , \\ & \bar{U} = C_7 U + k_7 , \quad \bar{W} = C_8 W + k_8 , \end{cases} \tag{2.115}$$

where C_i , k_i $i = 1, 2,..., 8$ are functions of the two parameters, a_1 and a_2 .

When conditions of invariance of the differential equations, Eqs. (2.114), and the auxiliary conditions, Eqs. (2.114a), are invoked, we find

$$G' \begin{cases} S'\colon & \bar{x} = C_4 C_2^2 x + k_1 , \quad \bar{y} = C_2 y, \quad \bar{z} = C_6 C_2^2 z + k_3 , \\ & \bar{u} = C_4 u, \quad \bar{v} = v/C_2 , \quad \bar{w} = C_6 w, \\ & \bar{U} = C_4 U, \quad \bar{W} = C_6 W, \end{cases} \tag{2.115a}$$

and the invariants follow from Theorem 2.7-1. In particular, we obtain, for the subgroup S', the two equations

$$[\alpha_1 x + \alpha_2] \frac{\partial \eta}{\partial x} + \alpha_3 y \frac{\partial \eta}{\partial y} + [\alpha_4 z + \alpha_5] \frac{\partial \eta}{\partial z} = 0,$$
$$[\beta_1 x + \beta_2] \frac{\partial \eta}{\partial x} + \beta_3 y \frac{\partial \eta}{\partial y} + [\beta_4 z + \beta_5] \frac{\partial \eta}{\partial z} = 0, \tag{2.116}$$

where

$$\alpha_1 = [\partial C_1/\partial a_1](a_1{}^0, a_2{}^0) = [[\partial C_4/\partial a_1] + 2[\partial C_2/\partial a_1]](a_1{}^0, a_2{}^0),$$
$$\alpha_2 = [\partial k_1/\partial a_1](a_1{}^0, a_2{}^0),$$
$$\alpha_3 = [\partial C_2/\partial a_1](a_1{}^0, a_2{}^0),$$
$$\beta_1 = [\partial C_4/\partial a_2](a_1{}^0, a_2{}^0),$$
$$\vdots$$

η is an invariant of S' if and only if it satisfies *both* equations.

In matrix form, Eqs. (2.116) become

$$\begin{bmatrix} \alpha_1 x + \alpha_2, & \alpha_3 y, & \alpha_4 z + \alpha_5 \\ \beta_1 x + \beta_2, & \beta_3 y, & \beta_4 z + \beta_5 \end{bmatrix} \begin{bmatrix} \eta_x \\ \eta_y \\ \eta_z \end{bmatrix} = 0,$$

and this has one and only one solution if the coefficient matrix has rank 2. This has rank 2 whenever at least one of its two-by-two submatrices has a nonzero determinant. This condition is met whenever *at least one* of the following hold:

$$\lambda_{31} x + \lambda_{32} \neq 0, \qquad \lambda_{34} z + \lambda_{35} \neq 0,$$
$$\lambda_{14} xz + \lambda_{15} x + \lambda_{24} z + \lambda_{25} \neq 0, \qquad \lambda_{ij} = \alpha_i \beta_j - \alpha_j \beta_i . \tag{2.117}$$

For convenience in our subsequent analysis, we rewrite Eqs. (2.116) in the notation given by Eqs. (2.117). Thus

$$[\lambda_{31} x + \lambda_{32}] \frac{\partial \eta}{\partial x} + [\lambda_{34} z + \lambda_{35}] \frac{\partial \eta}{\partial z} = 0,$$
$$[\lambda_{31} x + \lambda_{32}] y \frac{\partial \eta}{\partial y} - [\lambda_{14} xz + \lambda_{15} x + \lambda_{24} z + \lambda_{25}] \frac{\partial \eta}{\partial z} = 0, \tag{2.118}$$

and differences between the groups S' are now reflected by differences among the λ's.

The first of Eqs. (2.118) has the general solution

$$\eta = f[y, \xi(x, z)], \qquad \xi = \lambda_{35}x - \lambda_{32}z. \tag{2.119}$$

However, to obtain a solution to the system Eqs. (2.118) the second equation must also be satisfied. Upon setting Eq. (2.119) into the second of Eqs. (2.118), we have

$$[\lambda_{31}x + \lambda_{32}]y\frac{\partial f}{\partial y} + \lambda_{32}[\lambda_{14}xz + \lambda_{15}x + \lambda_{24}z + \lambda_{25}]\frac{\partial f}{\partial \xi} = 0. \tag{2.120}$$

A variety of special cases, all of which have been tabulated by Moran and Gaggioli [41], now appear. We shall examine only one, that is, with $\lambda_{31} = 0$ and $\lambda_{32} \neq 0$. The coefficient of $\partial f/\partial \xi$ is independent of y. Thus for $f = f(y, \xi)$ it is necessary for that coefficient to depend only upon ξ. Consequently, it is necessary that $\lambda_{14} = 0$ and $\lambda_{32}\lambda_{15} = -\lambda_{24}\lambda_{35}$, whereupon this *special* case of Eq. (2.120) becomes

$$y\frac{\partial f}{\partial y} + \left[\left(\frac{\lambda_{15}}{\lambda_{35}}\right)\xi + \lambda_{25}\right]\frac{\partial f}{\partial \xi} = 0. \tag{2.121}$$

The general solution of Eq. (2.121) is

$$f = F[yh(\xi)], \tag{2.122}$$

where $h(\xi)$ satisfies

$$\left[\left(\frac{\lambda_{15}}{\lambda_{35}}\right)\xi + \lambda_{25}\right]\frac{d\ln h}{d\xi} = 1.$$

With $\lambda_{15} \neq 0$, $h(\xi)$ is found to be

$$\begin{aligned} h(\xi) &= [(\lambda_{15}/\lambda_{35})\xi + \lambda_{25}]^{\lambda_{35}/\lambda_{15}} \\ &= [\lambda_{15}x + \lambda_{24}z + \lambda_{25}]^{\lambda_{35}/\lambda_{15}}. \end{aligned} \tag{2.123}$$

Thus with Eqs. (2.119), (2.122), and (2.123), it follows that for those groups S' with $\lambda_{14} = \lambda_{31} = \lambda_{34} = 0$, $\lambda_{32} \neq 0$, $\lambda_{35} \neq 0$, $\lambda_{15} \neq 0$, $\lambda_{32}\lambda_{15} = -\lambda_{24}\lambda_{35}$, absolute invariants are of the form

$$\eta = F[y(x + Az + B)^k]. \tag{2.124}$$

In addition, a *complete set* for a group G' not only contains an $\eta(x, y, z)$ but also five functionally invariant g's. The procedure to obtain the g's is parallel to that employed in obtaining the η's. For the group G',

Eq. (2.115a), five independent solutions $g(x, y, z, u, v, w, U, W)$ are to be obtained for the symbol equations

$$\alpha_6 u \frac{\partial g}{\partial u} - \alpha_3 v \frac{\partial g}{\partial v} + \alpha_7 w \frac{\partial g}{\partial w} + \alpha_6 U \frac{\partial g}{\partial U} + \alpha_7 W \frac{\partial g}{\partial W}$$
$$+ [\alpha_1 x + \alpha_2] \frac{\partial g}{\partial x} + \alpha_3 y \frac{\partial g}{\partial y} + [\alpha_4 z + \alpha_5] \frac{\partial g}{\partial z} = 0,$$

$$\beta_6 u \frac{\partial g}{\partial u} - \beta_3 v \frac{\partial g}{\partial v} + \beta_7 w \frac{\partial g}{\partial w} + \beta_6 U \frac{\partial g}{\partial U} + \beta_7 W \frac{\partial g}{\partial W}$$
$$+ [\beta_1 x + \beta_2] \frac{\partial g}{\partial x} + \beta_3 y \frac{\partial g}{\partial y} + [\beta_4 z + \beta_5] \frac{\partial g}{\partial z} = 0,$$

where

$$\alpha_6 = [\partial C_4/\partial a_1](a_1{}^0, a_2{}^0) = \alpha_1 - 2\alpha_3,$$
$$\beta_6 = [\partial C_4/\partial a_2](a_1{}^0, a_2{}^0) = \beta_1 - 2\beta_3,$$
$$\vdots$$

Moran and Gaggioli [45, 46] have suggested that under certain circumstances the aforementioned reduction in number of variables may be employed to reduce differential equations to algebraic equations. The basic result is embodied in the following theorem:

Theorem 2.7-2. *Let groups G with the form*

$$G \begin{cases} S\colon & \bar{x}^i = F^i(x^1,\ldots, x^m; y_1,\ldots, y_n; a_1,\ldots, a_m) \\ & \bar{y}_j = F_j(x^1,\ldots, x^m; y_1,\ldots, y_n; a_1,\ldots, a_m) \end{cases}$$
$$i = 1,\ldots, m; \quad j = 1, 2,\ldots, n,$$

possess n and only n functionally independent absolute invariants *$g_j(y_1,\ldots, y_n; x^1,\ldots, x^m)$, $j = 1,\ldots, n$, which are differentiable in each argument and*

$$\partial[g_1,\ldots, g_n]/\partial[y_1,\ldots, y_n] \neq 0.$$

If and only if, for some set of differentiable functions I_j, $y_j = I(x^1,\ldots, x^m)$ becomes $\bar{y}_j = I_j(\bar{x}^1,\ldots, \bar{x}^m)$ when transformed under G, then

$$g_j(y_1,\ldots, y_n; x^1,\ldots, x^m) = k_j(\text{const}), \qquad j = 1,\ldots, n. \tag{2.125}$$

In order to apply Theorem 2.7-2, it is first necessary to determine the functions g_j. This methodology has been presented in this and preceding sections. Once this is done the implicit solution for the y's is obtained from Eq. (2.125).

2.8 INFINITESIMAL TRANSFORMATIONS

The application of infinitesimal transformation groups to the solution of partial differential equations was first discussed by Lie [1] and considerably later by Ovsjannikov [47] and Müller and Matschat [48]. A general discussion and application to the linear and nonlinear diffusion equations has been carried out by Bluman [49] and Bluman and Cole [50]. Other nonlinear problems are contained in the work of Woodard [42], who also develops a useful extension which applies to simultaneous equations.

The basic theory and application of infinitesimal transformation groups to ordinary differential equations is found in the work of Cohen [13] and a condensed form is found in the work of Page [26] or Ames [3]. We shall briefly sketch the fundamental concepts herein but rest our primary discussion upon applications to partial differential equations.

Let $u = \theta(x, t)$ be a solution of the partial differential equation

$$Lu = 0, \tag{2.126}$$

defined over a region R in the x, t-plane shown in Fig. 2-1. Further, we suppose boundary conditions $\beta_j(u, x, t) = 0$ are given on curves $\Gamma_j(x, t) = 0$ and that the problem has a unique solutions defining a single surface $u = \theta(x, t)$ in the u, x, t-space. To this problem, we apply a group of transformations, with parameter ϵ,

$$\begin{aligned} x' &= x'(x, t, u; \epsilon), \\ t' &= t'(x, t, u; \epsilon), \\ u' &= u'(x, t, u; \epsilon), \end{aligned} \tag{2.127}$$

which maps the u, x, t-space into itself. Under Eq. (2.127), we have R transforming into R' and $u = \theta(x, t)$ into $u' = \theta'(x', t')$, in general. But, specifically we wish to invoke *invariance conditions* which ensure that:

(1) Equation (2.126) is invariant under the transformation specified by Eq. (2.127); and, (2.128)

(2) The boundary conditions and boundary curves are left invariant, that is $R' = R$ and $\beta_j(u', x', t) = 0$ on $\Gamma_j(x', t')$ for each j. (2.129)

Consequently, if a unique solution to Eq. (2.126) exists over R with the associated auxiliary conditions, the solution surface must be invariant,

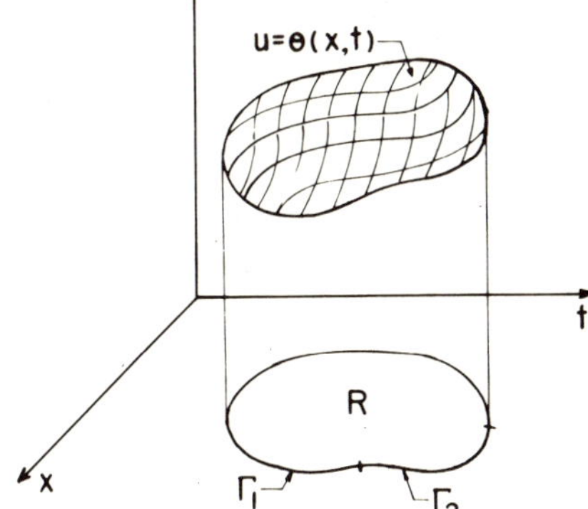

FIG. 2-1. Integration domain and solution surface.

that is, $u'[x, t, \theta(x, t); \epsilon] = \theta(x', t')$. In principle, the functional form of θ can then be deduced.

More specifically, we now consider the following infinitesimal transformations:

$$\begin{aligned} x' &= x + \epsilon X(x, t, u) + O(\epsilon^2), \\ t' &= t + \epsilon T(x, t, u) + O(\epsilon^2), \\ u' &= u + \epsilon U(x, t, u) + O(\epsilon^2). \end{aligned} \tag{2.130}$$

The infinitesimal form of the invariance condition of the solution surface becomes, with Eq. (2.130),

$$\theta[x + \epsilon X, t + \epsilon T] = \theta(x, t) + \epsilon U(x, t, u) + O(\epsilon^2). \tag{2.131}$$

Upon expanding the left-hand side of Eq. (2.131) and equating $O(\epsilon)$ terms, there results

$$X(x, t, \theta)(\partial\theta/\partial x) + T(x, t, \theta)(\partial\theta/\partial t) = U(x, t, \theta), \tag{2.132}$$

which is the *equation of an invariant surface for* θ. The characteristic (Lagrange) equations resulting from Eq. (2.132) are

$$dx/X = dt/T = d\theta/U. \tag{2.133}$$

These are solvable in principle. In particular, if X/T is independent of θ, we obtain the similarity representation

$$\eta(x, t) = \text{const}, \qquad u(x, t) = F[x, t, \eta, f(\eta)], \tag{2.134}$$

where the first relation is the integral of the first equality in Eq. (2.133).

$\eta(x, t) = \text{const}$ defines "similarity curves" in x, t-space. The dependence of F upon η involves a function $f(\eta)$ which is the solution to some ordinary differential equation obtained by setting the second relation of Eq. (2.134) into Eq. (2.126).

To ascertain those infinitesimal transformations which can be admitted, we must study the invariance of $Lu = 0$. It is more convenient to calculate the partial derivatives of L in coordinates (x', t') along a surface $u = \theta(x, t)$. Along that surface

$$x' = x'(x, t), \qquad t' = t'(x, t), \tag{2.135}$$

so that

$$\begin{aligned}
\partial x/\partial x' &= 1 - \epsilon[X_x + X_u\theta_x] + O(\epsilon^2),\\
\partial x/\partial t' &= -\epsilon[X_t + X_u\theta_t] + O(\epsilon^2),\\
\partial t/\partial t' &= 1 - \epsilon[T_t + T_u\theta_t] + O(\epsilon^2),\\
\partial t/\partial x' &= -\epsilon[T_x + T_u\theta_x] + O(\epsilon^2).
\end{aligned} \tag{2.136}$$

With Eqs. (2.136) we can now calculate the transformation between various partial derivatives beginning with [see Eq. (2.130)]

$$\theta'(x', t') = \theta(x, t) + \epsilon U(x, t, \theta) + O(\epsilon^2).$$

Thus

$$\partial\theta'/\partial x' = \theta_x + \epsilon[U_x + (U_u - X_x)\theta_x - T_x\theta_t - X_u\theta_x{}^2 - T_u\theta_x\theta_t] + O(\epsilon^2), \tag{2.137}$$

and the second derivative becomes

$$\begin{aligned}
\partial^2\theta'/\partial x'^2 = \theta_{xx} + \epsilon[&U_{xx} + (2U_{xu} - X_{xx})\theta_x - T_{xx}\theta_t + (U_{uu} - 2X_{xu})\theta_x{}^2\\
&- T_{xu}\theta_x\theta_t - X_{uu}\theta_x{}^3 - T_{uu}\theta_x{}^2\theta_t + (U_u - 2X_x)\theta_{xx}\\
&- 2T_x\theta_{xt} - 3X_u\theta_{xx}\theta_x - T_u\theta_{xx}\theta_t - 2T_u\theta_{xt}\theta_x] + O(\epsilon^2).
\end{aligned} \tag{2.138}$$

For the time derivatives, similar expressions are formed by interchanging the roles of x and t and X and T. Thus

$$\partial\theta'/\partial t' = \theta_t + \epsilon[U_t + (U_u - T_t)\theta_t - X_t\theta_x - T_u\theta_t{}^2 - X_u\theta_t\theta_x] + O(\epsilon^2), \tag{2.139}$$

and

$$\begin{aligned}
\partial^2\theta'/\partial t'^2 = \theta_{tt} + \epsilon[&U_{tt} + (2U_{tu} - T_{tt})\theta_t - X_{tt}\theta_x + (U_{uu} - 2T_{tu})\theta_t{}^2\\
&- X_{tu}\theta_x\theta_t - T_{uu}\theta_t{}^3 - X_{uu}\theta_t{}^2\theta_x + (U_u - 2T_t)\theta_{tt}\\
&- 2X_t\theta_{xt} - 3T_u\theta_{tt}\theta_t - X_u\theta_{tt}\theta_x - 2X_u\theta_{xt}\theta_t] + O(\epsilon^2).
\end{aligned} \tag{2.140}$$

For a given equation $Lu = 0$, we search for those infinitesimals (X, T, U) for which the fact that $\theta(x, t)$ is a solution of $L\theta = 0$ implies that $\theta'(x', t')$ is also a solution† of $L'\theta' = 0$. This requirement together with the invariance conditions, Eqs. (2.128) and (2.129), will ensure that the solution is invariant. There are at least two methods to follow—the "classical" and "nonclassical."

2.9 CLASSICAL DETERMINATION OF INFINITESIMAL TRANSFORMATIONS

The classical method *only makes use of the given equation* $Lu = 0$ and thus involves setting $L'u'$ proportional to Lu. This provides a set of conditions on X, T, U without the use of the invariant surface condition, Eq. (2.133). The invariant surface condition is employed later to find the functional form of the solution. As a vehicle to explain this method we use the nonlinear diffusion equation

$$[D(u)u_x]_x - u_t = 0, \tag{2.141}$$

first examined in this context by Ovsjannikov [47, 51] and later by Bluman [49].

With Eqs. (2.137)–(2.139), we have

$$\begin{aligned}
&\frac{\partial}{\partial x'}\left(D(u')\frac{\partial u'}{\partial x'}\right) - \frac{\partial u'}{\partial t'} \\
&\quad = \frac{\partial}{\partial x}\left(D(u)\frac{\partial u}{\partial x}\right) - \frac{\partial u}{\partial t} \\
&\qquad + \epsilon\{u_x[X_t + 2D'(u)U_x - D(u)X_{xx} + 2D(u)U_{xu}] \\
&\qquad + [D(u)U_{xx} - U_t] - 2u_{tx}D(u)T_x - 2u_xu_{tx}D(u)T_u \\
&\qquad + u_tu_x[X_u - 2D'(u)T_x - 2D(u)T_{xu}] \\
&\qquad - u_x^2u_t[2D'(u)T_u + D(u)T_{uu}] - u_x^3[2D'(u)X_u + D(u)X_{uu}] \\
&\qquad - u_{xx}u_tD(u)T_u - 3u_xu_{xx}D(u)X_u + u_t[T_t - U_u - D(u)T_{xx}] \\
&\qquad + u_t^2T_u + u_x^2[UD''(u) - 2D'(u)X_x + 2D'(u)U_u - 2D(u)X_{xu} + D(u)U_{uu}] \\
&\qquad + u_{xx}[D(u)U_u + UD'(u) - 2D(u)X_x]\} + O(\epsilon^2).
\end{aligned} \tag{2.142}$$

† L' designates L with the primed coordinates replacing the unprimed system.

After substituting $u_t = D(u)u_{xx} + D'(u)u_x^2$ into that portion of Eq. (2.142) in the braces, the *classical method* consists in equating to zero terms with the same derivative of u, i.e., the coefficients of u_{tx}, $u_x u_{tx}$,..., and the terms free of derivatives of u, etc. Setting the coefficients of u_{tx} and $u_x u_{tx}$ equal to zero, we find

$$T_x = 0, \qquad T_u = 0,$$

respectively, so that

$$T = T(t). \tag{2.143}$$

Equating to zero the coefficient of $u_x u_{xx}$, we see that

$$X_u = 0. \tag{2.144}$$

Continuing to equate to zero, successively, the coefficients of u_x, u_{xx}, u_x^2, etc., and employing Eqs. (2.143) and (2.144), we are led to the relations

$$X_t + 2D'(u)U_x - D(u)X_{xx} + 2D(u)U_{xu} = 0, \tag{2.145}$$

$$D(u)T'(t) + UD'(u) - 2DX_x = 0, \tag{2.146}$$

$$UD''(u) - 2D'(u)X_x + D'(u)U_u + D(u)U_{uu} + D'(u)T'(t) = 0, \tag{2.147}$$

$$D(u)U_{xx} - U_t = 0. \tag{2.148}$$

These together with Eqs. (2.143) and (2.144) assure the invariance condition.

Equation (2.146) implies that

$$U = [D(u)/D'(u)][2X_x - T'(t)], \tag{2.149}$$

and setting this into Eq. (2.148) gives

$$2D(u)X_{xxx} - (2X_x - T'(t))_t = 0.$$

Since neither X nor T are functions of u, it follows that $X_{xxx} = 0$ and $2X_x - T'(t) = h(x)$. Consequently,

$$\begin{aligned} X &= \frac{x}{2}T'(t) + \alpha x^2 + \beta x + \gamma, \\ T &= T(t), \\ U &= \frac{D(u)}{D'(u)}[4\alpha x + 2\beta], \end{aligned} \tag{2.150}$$

where α, β, γ are arbitrary constants.

Next we substitute Eq. (2.150) into Eq. (2.147), whereupon it reduces to

$$[D/D']'' = 0. \tag{2.151}$$

Thus, if one of α, $\beta \neq 0$,

$$D(u) = a(u + b)^c, \tag{2.152}$$

where a, b, c are arbitrary constants. That this nonlinear diffusion coefficient permits a similarity solution is well known.

Now upon setting Eqs. (2.150) into Eq. (2.145), we find

$$\frac{x}{2} T''(t) + 2\alpha D(u) \left[7 - \frac{4D(u)D''}{(D')^2}\right] = 0. \tag{2.153}$$

For arbitrary $D(u)$, Eq. (2.153) can only hold if $T''(t) = 0$ and $\alpha = 0$, whereupon

$$T(t) = 2A + 2Bt, \tag{2.154}$$

where A and B are arbitrary constants.

However, if $\alpha \neq 0$, we find an additional group which corresponds to the fixed function $D(u) = a(u + b)^{-4/3}$ which satisfies

$$7[D'(u)]^2 - 4D(u)D''(u) = 0.$$

Let us now summarize the three cases and employ the equation of the invariant surface, Eq. (2.133), to find the functional forms:

Case 1. *$D(u)$ arbitrary* ($\alpha = 0$, $\beta = 0$).

$$X = Bx + \gamma, \qquad T = 2A + 2Bt, \qquad U = 0.$$

With $B \neq 0$, Eq. (2.133), becomes

$$dx/(x + \gamma') = dt/[2(A' + t)] = du/0.$$

The similarity variable, obtained by integrating the first equation, is

$$\eta = (x + \gamma')/(A' + t)^{1/2},$$

with

$$u = F(\eta), \qquad D(u) = D(F),$$

and the resulting ordinary differential equation is

$$\frac{\eta}{2}\frac{dF}{d\eta} + \frac{d}{d\eta}\left[D(F)\frac{dF}{d\eta}\right] = 0.$$

Case 2. $D(u) = a(u + b)^c$ ($\alpha = 0$, $\beta \neq 0$).

$$X = (\beta + B)x + \gamma, \qquad T = 2A + 2Bt, \qquad U = (2\beta/c)(u + b).$$

With $B \neq 0$, Eq. (2.133), becomes

$$\frac{dx}{(1 + \beta')x + \gamma'} = \frac{dt}{2(A' + t)} = \frac{du}{2(\beta'/c)(u + b)},$$

with the similarity variable,

$$\eta = \frac{[x + \gamma'/(1 + \beta')]}{(A' + t)^{(\beta'+1)/2}}, \qquad \text{and} \qquad u + b = (A + t)^{\beta'/c} F(\eta).$$

The resulting ordinary differential equation is

$$\frac{\beta' + 1}{2} \eta \frac{dF}{d\eta} - \frac{\beta'}{c} F(\eta) + \lambda \frac{d}{d\eta} \left\{ [F(\eta)]^c \frac{dF}{d\eta} \right\} = 0.$$

Case 3. $D(u) = a(u + b)^{-4/3}$ ($\alpha \neq 0$, $\beta \neq 0$).

$$X = (\beta + B)x + \alpha x^2 + \gamma, \qquad T = 2A + 2Bt, \qquad U = -\tfrac{3}{2}(u + b)(2\alpha x + \beta).$$

With $B \neq 0$ and $(\beta + 1)^{-4/3} = 4\alpha\gamma$, the similarity variable is

$$\eta = \frac{\exp[-2/(2\alpha' x + \beta' + 1)]}{(A' + t)^{1/2}},$$

$$u + b = \frac{F(\eta)}{\{x + [(\beta' + 1)/2\alpha]\}^3} \exp[-3/(2\alpha' x + \beta' + 1)].$$

The resulting ordinary differential equation is

$$F'' - \tfrac{4}{3}(F')^2 F^{-1} - \tfrac{3}{4} F \eta^{-2} + (\alpha^2/2a) \eta F^{4/3} F' = 0.$$

Bluman and Cole [50] discuss the linear diffusion equation in great detail including sketches of the similarity curves for a number of cases.

2.10 NONCLASSICAL DETERMINATION OF INFINITESIMAL TRANSFORMATIONS

The nonclassical procedure, introduced by Bluman and Cole [50], makes use of both the given equation $Lu = 0$ and the invariant surface condition, Eq. (2.132):

$$X(x, t, u)(\partial u/\partial x) + T(x, t, u)(\partial u/\partial t) = U(x, t, u).$$

Now Eq. (2.132) really possesses only two independent infinitesimals, since it can be divided through by X, T, or U. Assuming $T \neq 0$, we divide by T and write

$$\bar{X} = X/T, \qquad \bar{U} = U/T. \tag{2.155}$$

Upon dropping the bars the *condition for the invariant surface* now reads

$$u_t = U - Xu_x . \tag{2.156}$$

Using Eq. (2.156) and its implications on other derivatives, which when combined with $Lu = 0$ may be simplified, we discuss the nonclassical method as applied to Burgers' equation (see Woodard [42])

$$Lu = u_t + uu_x - u_{xx} = 0. \tag{2.157}$$

From Eq. (2.156) it follows that

$$u_{tx} = U_x + U_u u_x - Xu_{xx} - X_x u_x - X_u(u_x)^2.$$

Since u satisfies Eq. (2.157), $u_{xx} = (u_t + uu_x)$, and using Eq. (2.156) we find, after collecting terms,

$$u_{tx} = (U_x - XU) + (U_u + X^2 - X_x - Xu)u_x - X_u(u_x)^2. \tag{2.158}$$

Now using Eqs. (2.156) and (2.158) we examine the transformed equations in the following format:

$$\begin{aligned}
u'u'_{x'} - uu_x &= \epsilon\{(uU_x) + (U + uU_u - uX_x)u_x \\
&\quad + (-X_u u)(u_x)^2\} + O(\epsilon^2), \\
u'_{t'} - u_t &= \epsilon\{(U_t + UU_u) + (-XU_u - X_t - X_u U)u_x \\
&\quad + (X_u X)u_x{}^2\} + O(\epsilon^2) \\
-[u'_{x'x'} - u_{xx}] &= \epsilon\{[-U_{xx} - UU_u + 2UX_x] \\
&\quad + [X_{xx} - 2U_{xu} - (u - X)(U_u - 2X_x) + 3X_u U]u_x \\
&\quad + [2X_{xu} - U_{uu} + 3X_u(u - X)]u_x{}^2 + X_{uu}(u_x)^3\} + O(\epsilon^2).
\end{aligned}$$

When the three foregoing equations are added and the invariance conditions invoked, the left-hand side vanishes. For the right-hand side to vanish to $O(\epsilon^2)$, we require the coefficients of u_x , $u_x{}^2$, $u_x{}^3$, and terms not involving u_x to vanish. Since the higher powers of u_x often involve simpler forms, we usually simplify the analysis by considering them first. Thus for the u_x^3 *coefficient*, we must have $X_{uu} = 0$, so that

$$X = C_2(x, t)u + C_1(x, t). \tag{2.159}$$

For the u_x^2 coefficient,

$$U_{uu} = 2X_{xu} + 2uX_u - 2XX_u,$$

which, upon application of Eq. (2.159), becomes

$$U_{uu} = 2[(C_2)_x + C_2(1 - C_2)u - C_2C_1],$$

so that

$$U = B(x, t)u + D(x, t) + u^2[(C_2)_x - C_2C_1] + \tfrac{1}{3}u^3C_2(1 - C_2). \tag{2.160}$$

The coefficients of u_x and $(u_x)^0$ when equated to zero are, respectively,

$$U - X_t - (2U_{xu} - X_{xx}) + uX_x - 2XX_x = 0, \tag{2.161}$$

$$U_t + uU_x - U_{xx} + 2UX_x = 0. \tag{2.162}$$

Determination of the general similarity solution to the Burgers equation has been changed to the study of the nonlinear equations (2.161) and (2.162) together with Eqs. (2.159) and (2.160). However, it is not generally feasible to construct the general solution of these equations. Rather, classes of special solutions must be examined, each of which generates a similarity solution of the original equation. *Any* solution to the system reduces Burgers' equation to an ordinary differential equation.

Once B, C_1, C_2, and D are determined, the characteristic differential equations corresponding to Eq. (2.156) become

$$dx/X = dt/1 = du/U. \tag{2.163}$$

The similarity variable $\eta(x, t) =$ constant is the integral of the first equality of Eq. (2.163). Once η is known explicitly the functional form is found, for example, by replacing x by $x(t, \eta)$ and integrating the second of Eqs. (2.163).

We shall describe the analysis for one case. If it is assumed that $C_2 = 0$, then Eqs. (2.159) and (2.160) become

$$X = C_1(x, t) = A(x, t), \qquad U = B(x, t)u + D(x, t). \tag{2.164}$$

Substituting Eqs. (2.164) into Eqs. (2.161) and (2.162) gives

$$Bu + D - A_t - 2B_x + A_{xx} + uA_x - 2AA_x = 0, \qquad (2.165)$$

$$B_t u + D_t + u(B_x u + D_x) - (B_{xx} u + D_{xx}) + 2(Bu + D)A_x = 0. \qquad (2.166)$$

Since A, B, and D are independent of u, we eliminate dependence upon u in the foregoing equations by equating coefficients of u and u^2 to zero. Thus in Eq. (2.165) we set

$$B = -A_x \qquad (2.167)$$

and in Eq. (2.166)

$$B_t + D_x - B_{xx} + 2BA_x = 0, \qquad (2.168)$$

$$B_x = 0. \qquad (2.169)$$

Equation (2.169) implies that $B = B(t)$. From Eq. (2.167) we have

$$A = -B(t)x + E(t). \qquad (2.170)$$

Consequently, Eq. (2.168) becomes

$$B_t + D_x - 2B^2 = 0,$$

which implies that $D_x = F(t) = 2[B(t)]^2 - B'(t)$. Then

$$D = F(t)x + G(t). \qquad (2.171)$$

As a consequence of Eqs. (2.167)–(2.171), Eqs. (2.165) and (2.166) now reduce to

$$D - A_t - 2AA_x = 0, \qquad \text{and} \qquad D_t + 2DA_x = 0.$$

The remaining analysis consists of employing these together with Eqs. (2.170), (2.171), and $B = B(t)$ to find B, E, F, and G. We record some of the results below:

Case 1. $E = 0$.

$$X = A = x/(2t + m), \qquad m \text{ const},$$

$$U = -u/(2t + m).$$

Similarity:

$$\eta = x/(2t + m)^{1/2},$$

$$f(\eta) = u(2t + m)^{1/2},$$

$$f'' + f'(\eta - f) + f = 0.$$

Case 2. $E = -RB(R$ const). With $G = b[(b^2/2)(t+d)^2 + c]^{-1}$, b, c, d const,

$$X = A = -(G'/2G)(x + R),$$
$$U = (G'/2G)u + G[(b/2)x + 1].$$

Similarity:

$$\eta = (t+d)/(x+R), \qquad \text{with} \quad c = 0,$$
$$f(\eta) = (t+d)^{-1}(u - 1/\eta),$$
$$\eta^2 f'' + 2\eta f' + ff' = 0.$$

Integral:

$$f = a_2 \tanh[(a_2/2)(a_3 - \eta^{-1})].$$

Solution:

$$u = a_2(t+d)^{-1} \tanh\{(a_2/2)[a_3 - (x+R)/(t+d)]\} + (x+R)/(t+d).$$

Case 3. *No assumption on E.*

$$X = (t+d)^{-1}(x + \tfrac{1}{2}) + N(t+d)^{-2}, \qquad d, N \text{ const},$$
$$U = -(t+d)^{-1}u + (t+d)^{-2}(x + \tfrac{1}{2}).$$

Similarity:

$$\eta = [(x + \tfrac{1}{2}) + \tfrac{1}{2}N(t+d)^{-1}]/(t+d),$$
$$f(\eta) = [u - \eta - \tfrac{1}{2}N(t+d)^{-2}](t+d).$$

Solution: In Bessel functions (see Woodard [42]).

2.11 THE NONCLASSICAL METHOD AND SIMULTANEOUS EQUATIONS

A convenient form for treating simultaneous equations has been introduced and applied by Woodard [42]. We shall discuss it using the boundary layer equations. Since the basic expansions, Eqs. (2.136)–(2.140), have been developed in (t, x) variables, it is convenient to write the equations in that notation. Thus our system becomes

$$uu_t + vu_x = u_{xx}, \qquad u_t + v_x = 0. \tag{2.172}$$

If an auxiliary (stream) function is introduced, the first equation becomes third order. This requires the computation of higher-order derivatives in addition to our basic expansions. An alternative is to expand Eqs. (2.130) to

$$u' = u + \epsilon U(x, t, u) + O(\epsilon^2), \tag{2.173a}$$

$$v' = v + \epsilon V(x, t, u, v) + O(\epsilon^2), \tag{2.173b}$$

with the two associated invariant surface conditions

$$Xu_x + u_t = U, \qquad Xv_x + v_t = V.$$

We wish to especially note that $U = U(x, t, u)$ while V depends upon x, t, u, and v!

Woodard [42] shows clearly how this method applies. It is easily generalized.

2.12 SOME SIMILARITY LITERATURE

In addition to the papers discussed in the previous sections, we record here some literature of importance to specific problems.

A series of papers by Müller and Matschat [48, 52, 53] employ transformation groups to study possible similarity solutions for the steady flow of a gas. Their problems concern the equations

$$\rho(uu_x + vu_y) + p_x = 0, \qquad \rho(uv_x + vv_y) + p_y = 0, \qquad u_x + v_y = 0,$$

and other steady gas flow situations. Möhring [54] and Ames [3] apply some elementary groups to the gas dynamics problems. Shock waves in one-dimensional plasmas are examined by similarity analysis by Friedhoffer [27] and for a spherical shock wave in a gas by Latter [25].

Three-dimensional laminar compressible boundary layers in general orthogonal coordinates are discussed from the similarity vantage by Fong [55]. For two-dimensional boundary layer flow over curved surfaces, Murphy [56] develops the basic equations and discusses similarity solutions.

For non-Newtonian fluids, similarity discussions based upon group methods have been carried out by Lee and Ames [17] and Hansen and Na [18]. The references in the aforementioned papers describe some alternative methods. Rotem [57] considers the boundary layer solutions for pseudoplastic fluids whose equations are

$$uu_x + vu_y = y^{-p}(\partial/\partial y)\{y^p[(\partial u/\partial y)^2]^{(n-1)/2}(\partial u/\partial y)\},$$
$$u_x + y^{-p}(\partial/\partial y)(y^p v) = 0.$$

In addition to the work of Schultz [24] similar analyses for wave propagation problems have concerned Bykhovskii [58] who examined a propagating wave in water flow on a sloping channel. Rosen [28] uses finite transformation groups in his investigation of the nonlinear wave equation

$$\phi_{tt} + 2\phi\phi_t - \phi_{xx} = 0.$$

Lee and Chou [59] consider the wave equation

$$u_{tt} - F(u_x/u_t)u_{xx} = 0,$$

while Nariboli [60] examines the nonlinear heat equation

$$u_{xx} + ku_x/x = f(u)u_t\,,$$

the plane transonic flow equation,

$$u_{xx} = u_y u_{yy}\,,$$

and the boundary layer equations, all by group methods.

Irmay [61] in an excellent extensive review of the nonlinear diffusion equation with a gravity term discusses the use of similarity in that subject. In addition, he discusses many alternative methods of analysis. An associated problem is examined by Silberg [62]. Lastly, we mention the work of Abbott [63], who considers the concept of *generalized similarity*. This is applied in fluid mechanics to attempt answers to such questions as "Is there any basis of comparison between compressible and incompressible flow problems, axisymmetric and planar flow?" This definition is in contrast to the sense in which we have used the term, that is, in terms of independent variables of a problem.†

2.13 TRANSFORMATION OF BOUNDARY-VALUE PROBLEMS INTO INITIAL-VALUE PROBLEMS—SINGLE EQUATIONS

When a similarity representation is obtained by the methods of this chapter, or by any other procedure, the resulting problem is usually a boundary-value problem with the new independent variable ranging from 0 to ∞. In Volume I we discussed Klamkin's [72] generalization

† Self-similar solutions for two-dimensional unsteady isentropic flow of a polytropic gas have been investigated by Mackie [64], Pogodin *et al.* [65], Suchkov [66], Ermolin and Sidorov [67], and Levine [68–70]. See also Ianenko [71].

of the brilliant idea of Blasius (see Goldstein [73]) for the transformation of the boundary-value problem

$$y''' + yy'' = 0, \qquad y(0) = y'(0) = 0, \quad y'(\infty) = 2,$$

into the *pair* of initial-value problems

$$F''' + FF'' = 0, \qquad F(0) = F'(0) = 0, \quad F''(0) = 1, \tag{2.174}$$

and

$$y''' + yy'' = 0, \qquad y(0) = y'(0) = 0, \quad y''(0) = \{2/F'(\infty)\}^{3/2}. \tag{2.175}$$

In principle, there is no need to solve the second problem, since $F(\eta)$ has been determined from the problem given by Eq. (2.174) and $y = \lambda^{1/3}F(\lambda^{1/3}\eta)$, $2 = \lambda^{2/3}F'(\infty)$. However, if y is to be obtained at the *same* uniformly spaced values of η as $F(\eta)$, then it is generally easier and more accurate to solve Eq. (2.175) than to interpolate the values of $\lambda^{1/3}F(\lambda^{1/3}\eta)$ from $F(\eta)$. Consideration of the size of η to approximate to $\eta = \infty$ is due to Rubel [74].

The original methods of Klamkin [72] were applicable to ordinary differential equations or systems of them which were invariant under certain groups of homogeneous linear transformations. The boundary conditions were specified as homogeneous at the origin and some finite value at infinity. Subsequently, Na [75, 76] noted that the method was applicable to *finite intervals* and also to equations which were invariant under other groups of transformations. All the boundary conditions at the initial point were taken to be homogeneous by Klamkin [72] and Na [75, 76]. Klamkin [77] has shown that this is unnecessary. We describe his analyses for a general second-order equation over an infinite domain and over a finite domain.

Let a second-order equation over $0 \leqslant \eta < \infty$ be

$$\sum_{\substack{m,n\\r,s}} A_{mnrs}(y'')^m(y')^n y^r \eta^s = 0, \tag{2.176}$$

subject to the (more general) boundary conditions

$$y'(0) = ay(0) + b, \qquad y^{(e)}(\infty) = k. \tag{2.177}$$

Here m, n, r, and s are arbitrary indices, A_{mnrs} are arbitrary constants, and e is an arbitrary integer (usually 0, 1, or 2). In what follows, we are tacitly assuming the existence and uniqueness of the initial-value problems which will have implications for the basic boundary-value problem. If Eq. (2.176) is multiple valued for y'', a particular branch is specified and the analysis is carried through for that branch.

Let λ, μ be two parameters and assume that y can be expressed in the form

$$y = \lambda F(\mu\eta), \tag{2.178}$$

where $F(\eta)$ also satisfies Eq. (2.176) but is subject to the *initial conditions*

$$F(0) = F'(0) = 1. \tag{2.179}$$

For both y and $F(\eta)$ to satisfy Eq. (2.176), the equation must be invariant under the simple two-parameter group

$$\bar{\eta} = \mu\eta, \qquad \bar{y} = \lambda^{-1}y. \tag{2.180}$$

This implies certain restrictions on the indices m, n, r, and s which are obtained by setting Eq. (2.178) into Eq. (2.176), that is,

$$\sum_{\substack{m,n\\r,s}} A_{mnrs}[F''(\mu\eta)]^m[F'(\mu\eta)]^n F^r(\mu\eta)^s \lambda^c \mu^d = 0, \tag{2.181}$$

where

$$c = m + n + r, \qquad d = 2m + n - s. \tag{2.182}$$

For invariance under Eq. (2.180), c and d must be constant for all sets of indices m, n, r, and s. Then Eq. (2.176) reduces to

$$\sum_{m,n} A_{mn}(y'')^m(y')^n y^{c-m-n}\eta^{2m+n-d} = 0. \tag{2.183}$$

It now follows that

$$\begin{gathered} y(0) = \lambda, \qquad y'(0) = a\lambda + b = \lambda\mu, \\ k = \lambda\mu^e F^{(e)}(\infty). \end{gathered} \tag{2.184}$$

After solving the initial-value problem for F, thereby obtaining $F^{(e)}(\infty)$ (perhaps using an analysis similar to that of Rubel [74]), λ and μ are determined from the simultaneous equations of Eqs. (2.184) which then provide the information to specify $y(0)$ and $y'(0)$. Thus we have converted the boundary-value problem into two initial-value problems. Both can be calculated, or we can determine y from $y = \lambda F(\mu\eta)$ by interpolation of $F(\eta)$.

The existence and uniqueness of $F(\eta)$, $0 \leqslant \eta < \infty$, has been tacitly assumed. Consequently, from Eq. (2.178), the existence and uniqueness of y depends upon that for μ and λ. Eliminating λ in Eqs. (2.184) it follows that the equation for μ is

$$k'(\mu - a) = \mu^e,$$

where $k' = k/bF^{(e)}(\infty)$. Depending upon the relative values of a, e, and k', there can be zero, one, or two solutions for μ. Once μ has been determined, λ is obtained from $\lambda = b/(\mu - a)$. Clearly, y may not have a unique solution.

If the interval is *finite*, a similar analysis is possible. Thus if the second boundary condition in Eq. (2.177) is

$$y^{(e)}(L) = k, \qquad [2.177a]$$

then we have two equations in λ and μ,

$$y'(0) = a\lambda + b = \lambda\mu, \qquad k = \lambda\mu^e F^{(e)}(\mu L),$$

as before. From these, we find $\lambda = b/(\mu - a)$ and

$$[k/bF^{(e)}(\mu L)](\mu - a) = \mu^e.$$

If we replace the boundary conditions, Eqs. (2.177), by

$$y(0) = a, \qquad y^{(e)}(\infty) = k, \qquad (2.185)$$

then we can employ one parameter μ. With $y = F(\mu\eta)$, the requirement that F also satisfy Eq. (2.176) leads to the invariance condition

$$2m + n - s = \text{const.}$$

Since $y(0) = a$, $F(0) = a$. If we now take $F'(0) = 1$, then $y'(0) = \mu$ which is found from

$$k = \mu^e F^{(e)}(\infty), \qquad e \neq 0.$$

For $e = 0$, an anomaly occurs probably due to impossible boundary conditions.

For a finite interval, with

$$y(0) = a, \qquad y^{(e)}(L) = k,$$

we can proceed as in the immediately preceding paragraph. The boundary condition at the terminal point, η_T (L or ∞) can be replaced by

$$\sum_i B_i y^{(i)}(\eta_T) = k.$$

The special cases

$$\frac{d}{d\eta}\left(T^m \frac{dT}{d\eta}\right) + \beta T = 0, \qquad T(0) = 0, \quad T(L) = 0,$$

and

$$(d^2T/d\eta^2) + \beta e^T = 0, \qquad (dT/d\eta)(0) = 0, \quad T(1) = 0,$$

are treated by Na [75] with the one-parameter groups

$$(\text{linear})\ \bar{\eta} = a^{\alpha}\eta, \qquad \bar{T} = a^{\beta}T,$$

and

$$(\text{nonlinear})\ \bar{\eta} = e^{\alpha_1 a}\eta, \qquad \bar{T} = T + \alpha_2 a,$$

respectively.

Klamkin [77] also considers third-order equations of the form

$$\sum_{\substack{m,n,r\\ s,t}} A_{mnrst}(y''')^m(y'')^n(y')^r(y)^s\eta^t = 0,$$

with $y = \lambda F(\mu\eta)$. The analysis is essentially the same as that described above.

2.14 TRANSFORMATION OF BOUNDARY-VALUE PROBLEMS INTO INITIAL-VALUE PROBLEMS—SIMULTANEOUS EQUATIONS

In Klamkin [72], it was also shown how to treat simultaneous equations employing a typical example of two second-order equations in two variables. For broader classes of boundary conditions, the same author [77] (see also Ames [3]) has demonstrated the procedure employing two simultaneous equations that are third order in y and second in z. Such examples actually occur. Greenspan and Carrier [78] study the similarity representation for the flow of a viscous, electrically conducting fluid past a semi-infinite flat plate in the presence of a magnetic field and find the equations

$$f''' + ff'' - \beta gg'' = 0,$$

$$g'' + \epsilon(fg' - f'g) = 0, \qquad \beta \leq 1,\dagger$$

$$f(0) = f'(0) = 0, \quad f'(\infty) = 2; \qquad g(0) = 0, \quad g'(\infty) = 2.$$

These equations are invariant under the one-parameter group,

$$\bar{x} = \lambda x, \qquad \bar{f} = \lambda^{-1}f, \qquad \bar{g} = \lambda^{-1}g,$$

† There can be no solutions if $\beta > 1$, as shown by Reuter and Stewartson [79].

but this is not sufficient to convert our boundary-value problem into an initial-value problem, since there are two conditions at infinity. Actually a two-parameter group of transformations is necessary, but no two-parameter groups of homogeneous linear transformations exist.

More generally, we consider the two equations (here $y_i = d^i y/d\eta^i$)

$$\sum_i A_i y_3^{m_i} y_2^{n_i} y_1^{p_i} y^{q_i} z_2^{r_i} z_1^{s_i} z^{t_i} \eta^{u_i} = 0,$$
$$\sum_i \bar{A}_i y_3^{\bar{m}_i} y_2^{\bar{n}_i} y_1^{\bar{p}_i} y^{\bar{q}_i} z_2^{\bar{r}_i} z_1^{\bar{s}_i} z^{\bar{t}_i} \eta^{\bar{u}_i} = 0, \tag{2.186}$$

with the boundary conditions†

$$y(0) = a, \qquad y'(0) = 0, \qquad \sum_i a_i y^{(i)}(L) = k,$$
$$z(0) = 0, \qquad \sum_i b_i z^{(i)}(L) = l.$$

The occurrence of one nonhomogeneous condition at the initial point and two at the terminal point suggests the need for a *three*-parameter group of transformations. With λ, μ, and ν as arbitrary parameters, we take

$$y = \lambda F(\mu\eta), \qquad z = \nu G(\mu\eta), \tag{2.187}$$

and ask that F and G also satisfy Eqs. (2.186). This invariance condition imposes the following six requirements for each value of the index i:

$$\begin{aligned} m_i + n_i + p_i + q_i &= \text{const}, \\ \bar{m}_i + \bar{n}_i + \bar{p}_i + \bar{q}_i &= \text{const}, \\ r_i + s_i + t_i &= \text{const}, \\ \bar{r}_i + \bar{s}_i + \bar{t}_i &= \text{const}, \\ 3m_i + 2n_i + p_i + 2r_i + s_i - u_i &= \text{const}, \\ 3\bar{m}_i + 2\bar{n}_i + \bar{p}_i + 2\bar{r}_i + \bar{s}_i - \bar{u}_i &= \text{const}, \end{aligned} \tag{2.188}$$

If Eqs. (2.188) hold, let us set

$$F(0) = 1, \qquad F'(0) = 0, \qquad F''(0) = 1,$$
$$G(0) = 0, \qquad G'(0) = 1.$$

† We can replace the finite interval by an infinite interval.

As a consequence of Eqs. (2.187), it follows that

$$y(0) = a = \lambda, \qquad y''(0) = \lambda\mu^2, \qquad z'(0) = \nu\mu,$$

$$\lambda \sum_i a_i \mu^i F^{(i)}(\mu L) = k, \tag{2.189}$$

$$\nu \sum_i b_i \mu^i G^{(i)}(\mu L) = l. \tag{2.190}$$

In principle, we can determine μ from Eq. (2.189) and then ν from Eq. (2.190). Then y and z are given by Eq. (2.187).

Suppose $y(0) = 0$. Then in the above, we would let $\lambda = 1$, change $F(0) = 1$ to $F(0) = 0$, and keep everything else the same. In this case, we can use the two-parameter group

$$y = \lambda F(\mu\eta), \qquad z = \lambda^\alpha \mu^\beta G(\mu\eta),$$

and use the two extra constants α and β to relieve the severe restrictions Eq. (2.188). Additional boundary conditions are treated by Klamkin [77]. Na [76] extends the nonlinear spiral group ($\bar{x} = e^{\alpha a} x, \bar{y} = y + \beta a$) to N general classes of boundary-value transformations.

At this point we remark that these methods can be used on similar systems of equations of any order and any number of dependent variables subject to a considerable variety of boundary conditions. For the method to apply, the system of equations has to be invariant under a group of transformations with an appropriate number of parameters. For homogeneous linear transformations this number will generally correspond to the sum of the number of conditions at the terminal point plus the number of nonhomogeneous or mixed conditions at the initial point. The condition $y'(0) = ay(0)$ is a mixed one, since it contains more than one derivative. Although homogeneous it requires an extra parameter.

Lastly we remark that there are problems where the previous method will not apply directly unless the boundary conditions are first transformed into a suitable form. Such examples are given by Klamkin [77].

References

1. Lie, S., *Arch. Math. (Kristiana)* **6**, 328 (1881). *See also Math. Ann.* **25**, 71 (1885).
2. Eisenhart, L. P., "Continuous Groups of Transformations." Dover, New York, 1961.
3. Ames, W. F., "Nonlinear Ordinary Differential Equations in Transport Processes." Academic Press, New York, 1968.
4. Dickson, L. E., *Ann. Math.* [2] **25**, 287 (1924).
5. Birkhoff, G., "Hydrodynamics," 2nd ed. (1st ed., 1950). Princeton Univ. Press, Princeton, New Jersey, 1960.

6. Mikusinski, J., "Operational Calculus." Pergamon, Oxford, 1959.
7. Feller, W., *Ann. Math.* [2] **55**, 468 (1952).
8. Yosida, K., "Lectures on Semigroup Theory and Its Application to Cauchy's Problem in Partial Differential Equations." Tata Inst. Fund. Research, Bombay, 1957.
9. Aris, R., *Ind. Eng. Chem. Fundam.* **3**, 28 (1964).
10. Morgan, A. J. A., *Quart. J. Math. Oxford Ser.* **2**, 250 (1952).
11. Michal, A. D., *Proc. Nat. Acad. Sci. USA* **37**, 623 (1952).
12. Manohar, R., Some Similarity Solutions of Partial Differential Equations of Boundary Layers, Tech. Summary Rep. #375, MRC, Univ. of Wisconsin, Madison, Wisconsin, 1963.
13. Cohen, A., "An Introduction to the Lie Theory of One-Parameter Groups." Stechert, New York, 1931.
14. Birkhoff, G., and MacLane, S., "A Survey of Modern Algebra," 2nd. ed. Macmillan, New York, 1953.
15. Ames, W. F., *Ind. Eng. Chem. Fundam.* **4**, 72 (1965).
16. Na, T. Y., and Hansen, A. G., *Internat. J. Nonlinear Mech.* **2**, 373 (1967).
17. Lee, S. Y., and Ames, W. F., *AIChE J.* **12**, 700 (1966).
18. Hansen, A. G., and Na, T. Y., *ASME Paper* No. 67-WA/FE-2 (1967).
19. Hansen, A. G., "Similarity Analyses of Boundary Value Problems in Engineering." Prentice-Hall, Englewood Cliffs, New Jersey, 1964.
20. Na, T. Y., Abbott, D. E., and Hansen, A. G., "Similarity Analysis of Partial Differential Equations." Univ. of Michigan Tech. Rep. NASA Contract 8-20065 (1967).
21. Acrivos, A., Shah, M. J., and Peterson, E. E., *AIChE J.* **6**, 312 (1960).
22. Gutfinger, C., and Shinnar, R., *AIChE J.* **10**, 631 (1964).
23. Kapur, J. N., *Phys. Soc. Japan J.* **17**, 1303 (1962). *See also* Kapur, J. N., and Srivastava, R. C., *Z. Angew. Math. Phys.* **14**, 383 (1963).
24. Schultz, A. B., *Int. J. Solids Structures* **4**, 799 (1968).
25. Latter, R., *J. Appl. Phys.* **26**, 954 (1959).
26. Page, J. M., "Ordinary Differential Equations with an Introduction to Lie's Theory of the Group of One Parameter." Macmillan, New York, 1897.
27. Friedhoffer, J. A., A Study of the Magnetogasdynamic Equations as Applied to Shock and Blast Waves, Ph.D. Dissertation, Univ. of Delaware, Newark, Delaware (1968).
28. Rosen, G., *J. Math. Phys.* **45**, 235 (1966).
29. Vicario, A. A., Jr., Longitudinal Wave Propagation along a Moving Threadline, Ph.D. Dissertation, Univ. of Delaware, Newark, Delaware (1968).
30. Strumpf, A., On a Class of Transformations Leading to Similar Solutions of the Steady Two-Dimensional Navier–Stokes Equations, Ph.D. Dissertation, Stevens Institute of Technology, Hoboken, New Jersey (1964).
31. v. Krzywoblocki, M. Z., and Roth, H., *Comment. Math. Univ. St. Paul.* **13** (1964).
32. v. Krzywoblocki, M. Z., and Roth, H., *Comment. Math. Univ. St. Paul.* **14** (1966).
33. v. Krzywoblocki, M. Z., and Roth, H., *Comment. Math. Univ. St. Paul.* **15** (1966).
34. Hellums, J. D., and Churchill, S. W., *Chem. Eng. Progr. Symp. Ser.* (No. 32) **57**, 75 (1961).
35. Hellums, J. D., and Churchill, S. W., *AIChE J.* **10**, 110 (1964).
36. Moran, M. J., A Unification of Dimensional Analysis and Similarity Analysis via Group Theory, Ph.D. Dissertation, Univ. of Wisconsin, Madison, Wisconsin (1967).
37. Gaggioli, R. A., and Moran, M. J., "Groups Theoretic Techniques for the Similarity Solution of Systems of Partial Differential Equations with Auxiliary Conditions." Math. Res. Center (Univ. of Wisconsin) Tech. Summary Rep. No. 693 (1966).

38. Gaggioli, R. A., and Moran, M. J., "Similarity Analyses of Compressible Boundary Layer Flows via Group Theory," Math. Res. Center (Univ. of Wisconsin) Tech. Summary Rep. No. 838 (1967).
39. Moran, M. J., and Gaggioli, R. A., *SIAM J. Appl. Math.* **16**, 202 (1968).
40. Moran, M. J., and Gaggioli, R. A., *AIAA J.* **6**, 2014 (1968).
41. Moran, M. J., and Gaggioli, R. A., *J. Engrg. Math.* **3**, 151 (1969).
42. Woodard, H. S., Similarity Solutions for Partial Differential Equations Generated by Finite and Infinitesimal Groups, Ph.D. Dissertation, University of Iowa, Iowa City, Iowa, 1971.
43. Moran, M. J., and Gaggioli, R. A., "Similarity for a Real Gas Boundary Layer Flow." Math. Res. Center (Univ. of Wisconsin) Tech. Summary Rep. No. 919 (1968).
44. Moran, M. J., and Gaggioli, R. A., "A Generalization of Dimensional Analysis." Math. Res. Center (Univ. of Wisconsin) Tech. Summary Rep. No. 927 (1968).
45. Moran, M. J., and Gaggioli, R. A., "On the Reduction of Differential Equations to Algebraic Equations." Math. Res. Center (Univ. of Wisconsin) Tech. Summary Rep. No. 925 (1968).
46. Moran, M. J., and Gaggioli, R. A., *SIAM J. Math. Anal.* **1**, 37 (1970).
47. Ovsjannikov, L. V., "Gruppovye svoystva differentsialny uravneni," Novosibirsk (1962). ["Group Properties of Differential Equations," (G. Bluman, transl.) (1967).] Available from California Inst. of Technology Library, Pasadena, California.
48. Müller, E. A., and Matschat, K., "Miszellaneen der Angewandten Mechanik," p. 190. Akademie-Verlag, Berlin, 1962.
49. Bluman, G. W., Construction of Solutions to Partial Differential Equations by the use of Transformation Groups. Ph.D. Thesis, California Inst. of Technology (1967).
50. Bluman, G. W., and Cole, J. D., *J. Math. Mech.* **18**, 1025 (1969).
51. Ovsjannikov, L. V., *Dokl. Akad. Nauk CCCP* **125**, 492 (1959).
52. Müller, E. A., and Matschat, K., *Z. Angew. Math. Mech.* **41**, 41 (1961).
53. Müller, E. A., and Matschat, K., *Proc. Eleventh Int. Cong. Appl. Math.* (Munich), 1061 (1964).
54. Möhring, W., *Z. Angew. Math. Mech.* **46**, 208 (1966).
55. Fong, M. C., *AIAA J.* **2**, 2205 (1964).
56. Murphy, J. S., *AIAA J.* **3**, 2043 (1965).
57. Rotem, Z., *Chem. Eng. Sci.* **21**, 618 (1966).
58. Bykhovskii, E. B., *Prikl. Mat. Mekh.* **30**, 303 (1966).
59. Lee, S-y. and Chou, D. C., "Some Generalizations of Simple Waves—Similarity Solutions." Univ. of Iowa, Iowa City, Iowa, 1971.
60. Nariboli, G. A., *Appl. Sci. Res.* **22**, 449 (1970).
61. Irmay, S., "Solutions of the Nonlinear Diffusion Equation with a Gravity Term in Hydrology," Proc. of the Wageningen Symp. (Internat. Assoc. of Scientific Hydrology) 478 (1966).
62. Silberg, P. A., "A Class of Periodic Solutions to the Two Dimensional Cubic Differential Equation." Northrop Nortronics, Norwood, Massachusetts, 1968.
63. Abbott, D. E., "The Generalized Similarity Method." IEEE Simulation and Modeling Conf., Pittsburgh, Pennsylvania, 1967.
64. Mackie, A. G., *Proc. Cambridge Philos. Soc.* **64**, 1099 (1968).
65. Pogodin, I. A., Suchkov, V. A., and Ianenko, N. N., *J. Appl. Math. Mech.* **22**, 256 (1958).
66. Suchkov, V. A., *J. Appl. Math. Mech.* **27**, 1132 (1963).
67. Ermolin, E. V., and Sidorov, A. F., *J. Appl. Math. Mech.* **30**, 412 (1966).
68. Levine, L. E., *Proc. Cambridge Philos. Soc.* **64**, 1151 (1968).

69. Levine, L. E., Self Similar Solutions of the Equations Governing the Two-Dimensional, Unsteady Motion of a Polytropic Gas." Univ. Maryland Tech. Note BN 549 (1968).
70. Levine, L. E., *Quart. Appl. Math.* **27**, 399 (1969).
71. Ianenko, N. N., *Dokl. Akad. Nauk. SSSR* **109**, 44 (1956).
72. Klamkin, M. S., *SIAM Rev.* **4**, 43 (1962).
73. Goldstein, S., "Modern Developments in Fluid Dynamics," Vol. 1, p. 135. Oxford Univ. Press, London and New York, 1957.
74. Rubel, L. A., *Quart. Appl. Math.* **13**, 203 (1955).
75. Na, T. Y., *SIAM Rev.* **9**, 204 (1967).
76. Na, T. Y., *SIAM Rev.* **10**, 85 (1968).
77. Klamkin, M. S., "Transformation of Boundary Value Problems into Initial Value Problems." Publication Preprint of Scientific Research Staff, Ford Motor Company, Dearborn, Michigan (Oct. 1969).
78. Greenspan, H. P., and Carrier, G. F., *J. Fluid Mech.* **6**, 77 (1959).
79. Reuter, G. E. H., and Stewartson, K., *Phys. Fluids* **4**, 276 (1961).

CHAPTER **3**

Approximate Methods

3.0 INTRODUCTION

Approximate methods are interpreted here as analytical procedures for developing solutions which are close, in some sense, to the exact solution of the nonlinear problem. Thus numerical solutions are excluded, since they result in tables or graphs rather than functional forms. Experience and intuition can often be employed to select a reasonable and sometimes quite accurate first guess, from which it is possible to proceed to successively improved approximations. Moreover, the analytical form of the approximate solution is often more useful than solutions generated by numerical integration, since it displays the parameters of the problem in general form. Approximate solutions usually require less computation time to generate.

Chapters 5 and 6 of Volume I were devoted to approximate methods. The material of this chapter is intended to be supplementary. Additional material will be included on weighted residual methods, regular and singular perturbation, maximum operations, and other iterative methods. Some methods from specific fields rest heavily upon physical knowledge, while others are completely analytical. Examples and references employing these procedures will be included wherever possible.

3.1 WEIGHTED RESIDUAL METHODS (WRM)

A general discussion of weighted residual (Galerkin,† collocation, subdomain, moments,‡ least squares) methods is found in Chapter 5 of Volume I. A later review and historical development of the field has been published by Finlayson and Scriven [1, 2].

To set the stage for this supplementary discussion, we consider the differential equation

$$u_t = N(u), \qquad \mathbf{x} \in V, \quad t > 0, \tag{3.1}$$

for $u = u(\mathbf{x}, t)$, where $N(\cdot)$ denotes a general differential operator in the space derivatives of u, $\mathbf{x}$ is a vector of space variables, V is a (one, two, three)-dimensional domain with boundary S. Furthermore, we suppose Eq. (3.1) is subject to the initial and boundary conditions

$$\begin{aligned} u(\mathbf{x}, 0) &= u_0(\mathbf{x}), \qquad && \mathbf{x} \in V, \\ u(\mathbf{x}, t) &= f_S(\mathbf{x}, t), \qquad && \mathbf{x} \in S. \end{aligned} \tag{3.2}$$

Basically there are three variations of any of the assorted weighted residual methods (WRM). To review these we employ the vehicle of the *assumed trial solution*

$$u_T(\mathbf{x}, t) = u_S(\mathbf{x}, t) + \sum_{i=1}^{N} C_i(t)\, u_i(\mathbf{x}, t), \tag{3.3}$$

where the approximating functions, u_i, are prescribed and satisfy§

$$u_S = f_S, \qquad u_i = 0, \quad \mathbf{x} \text{ on } S. \tag{3.4}$$

Consequently, u_T satisfies the boundary conditions, but not the initial condition or equation, for all functions $C_i(t)$. It is not necessary that the trial solution be linear in the C_i. Indeed, we shall see that for some nonlinear problems it may be useful to assume trial solutions of a more general form than Eq. (3.3).

† Called the *Bubnov–Galerkin* method by Mikhlin [3] and others in the recent Russian literature.

‡ Sometimes called the *integral method* or the *Karman–Pohlhausen method* in the first approximation when the weighting function is unity. For a review see Goodman [4].

§ The rationale of prescription is discussed in Volume I.

The differential equation residual, R_{E}, and initial residual, R_{I},

$$R_{\mathrm{E}}(u_{\mathrm{T}}) \equiv N(u_{\mathrm{T}}) - (u_{\mathrm{T}})_t \tag{3.5}$$

$$R_{\mathrm{I}}(u_{\mathrm{T}}) \equiv u_0(\mathbf{x}) - u_S(\mathbf{x}, 0) - \sum_{i=1}^{N} C_i(0)\, u_i(\mathbf{x}, 0), \tag{3.6}$$

are measures of how well the trial function u_{T} satisfies the equation and initial conditions, respectively. With increasing N, one hopes that the residuals will become smaller. The exact solution is obtained when both residuals are identically zero. To develop the approximation of this ideal we select N weighting functions† w_j, $j = 1, 2, \ldots, N$, and introduce the spatial average (inner product or weighted integral)

$$(w, v) \equiv \int_V wv\, dV. \tag{3.7}$$

Upon setting the weighted integrals of the equation residual R_{E} equal to zero,

$$(w_j, R_{\mathrm{E}}(u_{\mathrm{T}})) \equiv 0, \qquad j = 1, 2, \ldots, N, \tag{3.8}$$

we obtain N simultaneous nonlinear ordinary differential equations for the $C_j(t)$, $j = 1, 2, \ldots, N$. In a similar way, when the weighted integrals of the initial residual R_{I} is set equal to zero, we generate the initial conditions $C_j(0)$ for the preceding differential equations.

Once the C_j's are determined by this *interior method*, the approximate solution is obtained by substituting these into Eq. (3.3). Successive approximations are obtained by increasing N and repeating the process.

In the interior method the trial solution is selected to satisfy the boundary conditions but not the differential equation or initial conditions. The converse situation can also be treated either for initial-value or boundary-value problems. In this *boundary method*, trial solutions are selected which satisfy the differential equation but not the boundary conditions. This procedure replaces the spatial average, Eq. (3.7), by an average over the boundary.

An intermediate situation exists. In so-called *mixed methods* the trial solution does not satisfy either the differential equations or boundary conditions. Schuleshko [5] treated mixed methods by requiring the

---

† Galerkin: $w_j = u_j$ of Eq. (3.3),

Collocation: $w_j = \delta(\mathbf{x} - \mathbf{x}_j)$, δ = Dirac delta,

Subdomain: $w_j = \begin{cases} 1, & x \in V_j, \\ 0, & x \notin V_j, \end{cases}$

Least squares: $w_j = \partial R(u_T)/\partial C_j$.

differential equation residual to be orthogonal to one set of weighting functions using Eq. (3.7), while the boundary residual is simultaneously made orthogonal to another set of functions using an appropriate surface integral as the inner product. With N weighting functions, this leads to $2N$ conditions, yet in general only N conditions can be satisfied by the N independent C_j. For this procedure to work, some of the conditions must be discarded. This was done by Snyder *et al.* [6].

Bolotin [7], Mikhlin [3], and Finlayson [8] note that when the Galerkin method is employed, the dilemma of the previous paragraph can be resolved by adding the differential equation residuals to the boundary residuals. The combination is accomplished in such a way that the differential equation residual, when integrated by parts, cancels identical terms of the boundary residual. The situation is analogous to the treatment of *natural*† boundary conditions in the variational calculus (see, e.g., Mikhlin [3] or Hildebrand [9]). In fact, only boundary conditions analogous to natural boundary conditions can be treated this way. This *combination of residuals* for more general problems is important in establishing the equivalence between the Galerkin method and several variational methods (see Finlayson [8] and Finlayson and Scriven [1]).

A variety of additional modifications are possible. One of these requires

† The Euler equation of the problem

$$\delta \int_{x_1}^{x_2} \int_{y_1}^{y_2} F(x, y, u, u_x, u_y, u_{xx}, u_{xy}, u_{yy})\, dx\, dy = 0, \tag{3.9}$$

where x_1, x_2, y_1, and y_2 are constants is

$$\frac{\partial^2}{\partial x^2}\left(\frac{\partial F}{\partial u_{xx}}\right) + \frac{\partial^2}{\partial x\,\partial y}\left(\frac{\partial F}{\partial u_{xy}}\right) + \frac{\partial^2}{\partial y^2}\left(\frac{\partial F}{\partial u_{yy}}\right) - \frac{\partial}{\partial x}\left(\frac{\partial F}{\partial u_x}\right) - \frac{\partial}{\partial y}\left(\frac{\partial F}{\partial u_y}\right) + \frac{\partial F}{\partial u} = 0, \tag{3.10}$$

with no restrictions on the variation of u, u_x, or u_y provided the *natural* boundary conditions, given below, hold:

$$\left[\frac{\partial}{\partial x}\left(\frac{\partial F}{\partial u_{xx}}\right) + \frac{\partial}{\partial y}\left(\frac{\partial F}{\partial u_{xy}}\right) - \frac{\partial F}{\partial u_x}\right]\Bigg|_{x_1}^{x_2} = 0,$$

$$\left[\frac{\partial}{\partial y}\left(\frac{\partial F}{\partial u_{yy}}\right) + \frac{\partial}{\partial x}\left(\frac{\partial F}{\partial u_{xy}}\right) - \frac{\partial F}{\partial u_y}\right]\Bigg|_{y_1}^{y_2} = 0, \tag{3.11}$$

$$\frac{\partial F}{\partial u_{xx}}\Bigg|_{x_1}^{x_2} = \frac{\partial F}{\partial u_{yy}}\Bigg|_{y_1}^{y_2} = 0.$$

that the approximating functions satisfy *derived* or *secondary boundary conditions* which are obtained by insisting that the differential equation be satisfied on the boundary. This was first done by Duncan [10] but it is also used in the Karman–Pohlhausen method when solving boundary layer problems (see Goodman [4]). Other conditions such as continuity of the velocity and certain of its derivatives are employed as well. Additional compatibility conditions are required to assure good results when the integral method is applied to MHD boundary layer flow. In those problems Hugelman and Haworth [11] and Hugelman [12] find that all trial solutions should satisfy the derivative of the differential equation normal to the surface at both the solid surface and the edge of the boundary layer.

The method reducing a partial differential equation to ordinary differential equations just described, can be applied to boundary-value or eigenvalue problems. The spatial averages are taken over all but one of the independent variables, and the approximate solution is obtained by solving the resulting set of ordinary differential equations.

The WRM can be combined with other methods. Indeed Collatz [13] discusses a combination of WRM with an iterative scheme. Kaplan [14], Kaplan and Bewick [15], and Kaplan *et al.* [16] coupled WRM with finite differences to reduce computing time in large nuclear reactor problems. More recently Galerkin-type numerical methods have appeared.

Finlayson and Scriven [2] give 187 references to applications of weighted residual methods. In the next section, we describe several intriguing modifications not available at the time of that excellent review.

3.2 NOVEL APPLICATIONS OF WRM IN FLUID MECHANICS

In Section 3.1 we remarked that trial solutions linear in the C_i [Eq. (3.3)] are not necessary. Following a linearization employed first by Oseen [17, 18] and improved by Lewis and Carrier [19] and Carrier [20], the use of collocation is examined by Schetz [21, 22] and Schetz and Jannone [23, 24]. Their goal was the development of relatively simple methods useful in obtaining functional forms for the approximate solution to nonlinear boundary-layer problems.

Consider the boundary-layer equations

$$u_x + v_y = 0, \tag{3.12}$$

$$uu_x + vu_y = \nu u_{yy}, \tag{3.13}$$

with

$$u(0, y) = u_e, \quad y > 0; \qquad u(x, 0) = 0, \quad x \geqslant 0;$$
$$\lim_{y \to \infty} u(x, y) = u_e, \quad x \geqslant 0; \qquad v(x, 0) = 0, \quad x \geqslant 0. \tag{3.14}$$

Since $v(x, 0) = 0$, this will be called the *impermeable case* as there is no surface mass transfer. In this situation one can approximate the convective derivative on the left-hand side of Eq. (3.13) as

$$uu_x + vu_y \approx \tilde{u}(x)\, u_x. \tag{3.15}$$

Excellent results are reported by Schetz with

$$\tilde{u} = C_1 u_e(\text{const}), \tag{3.16}$$

where C_1 is determined by *single-point collocation*. The actual procedure is to find the solution of the approximate momentum equation

$$C_1 u_e \bar{u}_x = \nu \bar{u}_{yy}, \tag{3.17}$$

subject to the boundary conditions on u, which are given in Eq. (3.14). The standard auxiliary (stream) function $\psi(x, y)$ is now introduced by

$$u(x, y) = \psi_y, \qquad v(x, y) = -\psi_x, \tag{3.18}$$

which identically satisfies the continuity equation and puts the momentum equation in the form

$$\psi_y \psi_{xy} - \psi_x \psi_{yy} = \nu \psi_{yyy}. \tag{3.19}$$

The approximation to the stream function is now found directly from the solution of Eq. (3.17) using

$$\bar{\psi}(x, y) = \int_0^y \bar{u}(x, y)\, dy.$$

This expression is then inserted into Eq. (3.19) and any remainder is forced to zero at one point,† thus evaluating C_1. The actual selection of the collocation point remains strictly arbitrary, but Schetz found that the simple rule of dividing the region of interest into roughly equal parts yielded consistently good results.

† We discuss the case for several points subsequently.

To illustrate, note that the solution of Eq. (3.17) is

$$\bar{u}(x, y) = u_e \operatorname{erf}[yC_1^{1/2} \operatorname{Re}_x^{1/2}/2x] = u_e \operatorname{erf} \eta. \tag{3.20}$$

Since erf(2.000) = 1.000, it is reasonable to assert that for this computation the region of interest ("boundary layer thickness") is $\eta = 2.000$. Dividing this into equal parts gives the collocation point as $\eta = 1.0$. The value of C_1 so determined is 0.4861. This yields a very good approximation to the velocity profile but predicts a skin friction coefficient which is 18% too large.

If we wish to extend the one-parameter approximation for improved results, the velocity profile can be expressed as

$$\bar{u} = u_e \operatorname{erf}[\eta + C_2^{-1} \eta^2 + C_3^{-1} \eta^3 + \cdots],$$

still satisfying all boundary conditions. Collocating at $\eta = 0.60$ and $\eta = 1.20$ with two parameters C_1 and C_2, leads to a substantial improvement over the single-parameter result and an excellent approximation to the exact result with a skin friction coefficient which is only 9% too low. On the basis of these results, it appears that an excellent approximate solution is obtainable by collocating with a very limited number of parameters when the functional form of the approximating expression is related to the solution of a linearized approximation to the boundary-layer equations. Of course, other WRM could be used with the trial function [Eq. (3.20) here] chosen in this manner. Schetz compares the two-point collocation with two WRM methods employing weighting functions, $\exp[-\eta^2]$ and erfc η. He finds that the required additional computation may not be worth the small improvement.

An analogous procedure can be applied in the case of a porous wall. In that situation, the boundary conditions [Eq. (3.14)] include $v(x, 0) = v_0(x)$. Taking $\tilde{u}(x) = C_1 u_e$ and $\tilde{v}(x) = C_2 v_0(x)$, our approximate (linearized) momentum equation becomes

$$C_1 u_e \bar{u}_x + C_2 v_0(x)\, \bar{u}_y = \nu \bar{u}_{yy}\,. \tag{3.21}$$

The problem now is to find solutions to Eq. (3.21) for the physical boundary conditions and then determine the two parameters C_1 and C_2 so as to force this solution of an approximate boundary-layer equation to satisfy the exact boundary-layer equation (stream function form) at two points in the flow field. The procedures are essentially the same as those discussed above, but the lengthy details preclude inclusion herein (see Schetz [25]).

The *classical linearization*, leading to Eq. (3.17), is in essence a *linearization operating upon the convective terms*. If $\tilde{u} = \tilde{u}(x)$, then

$$\tilde{u}(x)\,\bar{u}_x = \nu \bar{u}_{yy}\,,$$

becomes

$$\bar{u}_\xi = \bar{u}_{yy}\,, \tag{3.22}$$

if we set

$$\xi = \int_0^x \frac{\nu\, dx'}{\tilde{u}(x')}\,.$$

An alternative procedure, called *von Mises linearization*, has been discussed by Hamielec *et al.* [26]. When the von Mises transformation $(x, y) \to (x, \psi)$ is applied to the momentum equation, Eq. (3.13), we obtain

$$\partial u/\partial x = \nu(\partial/\partial\psi)[u(\partial u/\partial\psi)]. \tag{3.23}$$

If we linearize by setting $u \approx \hat{u}(x)$ so that

$$(\partial/\partial\psi)[u(\partial u/\partial\psi)] \approx \hat{u}(x)(\partial^2 u/\partial\psi^2),$$

then Eq. (3.23) is approximated by the linear equation

$$(\partial\bar{u}/\partial x) = \nu\hat{u}(x)(\partial^2\bar{u}/\partial\psi^2). \tag{3.24}$$

Upon setting

$$s = \int_0^x \nu\hat{u}(x')\, dx',$$

Eq. (3.23) becomes

$$\partial\bar{u}/\partial s = \partial^2\bar{u}/\partial\psi^2 \tag{3.25}$$

[compare Eq. (3.22)], which is an equation possessing all the attractive aspects of the corresponding linearized equation in the physical x, y-plane, but which certainly is not the same type of linearization. To ascertain the corresponding equation in the physical plane, we recall the transformations

$$\left.\frac{\partial}{\partial x}\right|_y = \left.\frac{\partial}{\partial x}\right|_\psi - v\left.\frac{\partial}{\partial\psi}\right|_x, \qquad \left.\frac{\partial}{\partial y}\right|_x = u\left.\frac{\partial}{\partial\psi}\right|_x.$$

With these it follows that the corresponding equation is

$$u\frac{\partial\bar{u}}{\partial x} + v\frac{\partial\bar{u}}{\partial y} = \nu\frac{\partial}{\partial y}\left(\frac{\hat{u}}{\bar{u}}\frac{\partial\bar{u}}{\partial y}\right). \tag{3.26}$$

In the physical plane, the corresponding equation contains the *exact* nonlinear convective derivative but a *modified nonlinear viscous term*! In addition, the von Mises linearization introduces a singularity at the wall where $\bar{u} \to 0$. This is ironic. The wall shear is defined as

$$\tau_W = \mu \, \partial \bar{u}/\partial y \,|_{y=0} ,$$

which becomes

$$\tau_W = \mu(\bar{u} \, \partial \bar{u}/\partial \psi)$$

after applying the von Mises transformation. Thus the occurrence of the singularity creates difficulty with regard to predicted shear at the wall.

To obtain any meaningful results, we must have the solution of Eq. (3.25) behave as

$$\bar{u}(s, \psi) \approx \psi^{1/2} F(s), \qquad \text{as} \quad \psi \to 0,$$

that is, (3.27)

$$\tau_W = \mu[\tfrac{1}{2} \psi^{1/2} F(s) \, \psi^{-1/2} F(s)] = \tfrac{1}{2} \mu F^2(s), \qquad \text{as} \quad \psi \to 0.$$

But Eq. (3.25) has the solution

$$\bar{u}(s, \psi) = u_e \operatorname{erf}[\psi/(4s)^{1/2}],$$

whose asymptotic form as $\psi \to 0$ is

$$\bar{u}(s, \psi) \sim (2u_e/\pi^{1/2})[\psi/(4s)^{1/2}].$$

Thus,

$$\tau_W \sim (4\mu u_e^2/\pi) \lim_{\psi \to 0} [\psi/4s] = 0,$$

which is physically untrue! We must therefore modify the von Mises linearization to insure that the solution is physically meaningful.

The foregoing difficulty is avoided by adopting an inner–outer expansion, similar to the von Karman–Millikan procedure described in Volume I (page 222). An inner expansion is matched to an outer (von Mises) linearized solution to obtain a complete velocity profile.† A free parameter, associated with the location of the matching point, is determined to yield the exact Blasius value of the skin friction coefficient or it can be determined by WRM.

An *inner expansion*, taken to be valid near the wall ($0 \leqslant \psi \leqslant \psi_0$), is assumed in the form

$$\tilde{u}_I(s, \psi) \equiv \bar{u}(s, \psi)/u_e = \psi^{1/2} F(s), \tag{3.28}$$

† As an alternative, we could use the classical linearization described previously.

thereby ensuring proper behavior for shear at the wall. The outer solution, $\tilde{u}_O$, valid for $\psi \geqslant \psi_0$, is obtained from the von Mises linearized equation

$$\partial\tilde{u}_O/\partial s = \partial^2\tilde{u}_O/\partial\psi^2, \tag{3.29}$$

subject to

$$\begin{aligned} &\tilde{u}_O(0, \psi) = 1, \qquad \lim_{\psi\to\infty} \tilde{u}_O(s, \psi) = 1, \\ &\tilde{u}_O(s, \psi_0) = \tilde{u}_I(s, \psi_0) = \psi_0^{1/2}F(s). \end{aligned} \tag{3.30}$$

The latter condition reflects the continuity of the velocity at ψ_0, which like $F(s)$, is as yet unspecified. The problem specified by Eqs. (3.29) and (3.30) has the solution

$$\tilde{u}_O = \operatorname{erf}\left(\frac{\psi - \psi_0}{(4s)^{1/2}}\right) + \frac{(\psi - \psi_0)\,\psi_0^{1/2}}{2\pi^{1/2}} \int_0^s F(\lambda) \frac{\exp[-(\psi - \psi_0)^2/4(s - \lambda)]}{(s - \lambda)^{3/2}}\, d\lambda.$$

The function $F(s)$ can be determined by requiring *continuity of the shear as well as the velocity* at the matching point $\psi = \psi_0$. A simple computation shows that

$$F(s) = 2(\psi_0/\pi s)^{1/2}, \tag{3.31}$$

whereupon,

$$\begin{aligned} \tilde{u}_I &= [4(\Gamma_0/\pi)]^{1/2}\, \Gamma^{1/2}, \qquad 0 \leqslant \psi \leqslant \psi_0, \\ \tilde{u}_O &= \operatorname{erf}[\Gamma - \Gamma_0] + (4\Gamma_0/\pi^{1/2}) \exp[-(\Gamma - \Gamma_0)^2], \qquad \psi \geqslant \psi_0, \end{aligned}$$

and we have used $\Gamma = \psi/(4s)^{1/2}$, $\Gamma_0 = \psi_0/(4s)^{1/2}$.

Now taking Γ_0 as a constant, the matching point ψ_0 is located at a fixed position within a velocity profile plotted as a function of a similarity variable $\Gamma = \psi/2s^{1/2}$.

3.3 WRM IN TRANSPORT PHENOMENA—SOME RECENT LITERATURE

a. Galerkin's Method

The choice of approximating functions is the most crucial step in applying WRM. Determination of a good, if not the best, trial function is an outstanding problem. Certainly any symmetry properties should be exploited, and in problems of conventional types it is usually convenient to use the interior method. Approximating functions are therefore chosen to satisfy the boundary conditions. Kantorovich and Krylov [27] (see also Volume I) demonstrate how to construct complete sets of functions which vanish on a boundary of complicated shape. Snyder and Stewart [28] combine this scheme and symmetry arguments to develop approximating functions for Galerkin's method applied to Newtonian creeping flow through a regular packed bed of spheres. Other *applications of Galerkin's method* include Kawaguti's [29] study of the critical Reynolds number for flow past a sphere, a study by Hays [30] of Couette and Poiseuille flow, the work of Biot [31] and Biot and Agrawal [32] on heat transfer and ablation, and the research of Richardson [33] on unsteady heat conduction with a nonlinear boundary condition. Donnelly *et al.* [34] include a variety of applications of WRM and variational procedures which are closely associated. In particular, we mention the research of Hays [35] who applied the Galerkin method to time-dependent and steady heat conduction, wherein the thermal conductivity is temperature dependent. A similar equation was studied by Collings [36] in which Galerkin's method is favorably compared with other WRM.

Finlayson [37] employs five WRM to study nonlinear transfer of heat and mass. Grotch [38] develops an approximate solution for the nonlinear boundary value problem

$$Ay'' - y' + By^n = 0,$$
$$y - Ay' = 1, \quad \text{at} \quad z = 0, \quad y' = 0, \quad \text{at} \quad z = 1,$$

by means of Galerkin's method.

The original work of Kawaguti [29] was aimed at the approximate description of the entire flow field through use of trial stream function polynomials. He employed two trial stream functions and determined the unknown coefficients by Galerkin's method for low Reynolds number flows, $\text{Re} < 70$. These solutions were extended by Hamielec and Johnson [39] and Hamielec *et al.* [40] to include the entire laminar flow regime past spheres. We sketch those results which are compared to finite difference calculations by Hamielec *et al.* [40]. Consider the Navier–Stokes equations for viscous, incompressible axisymmetric flow in spherical coordinates. If ψ is the stream function, we have

$$\tfrac{1}{2}\,\text{Re}[\psi_r(E^2\psi/r^2 \sin^2\theta)_\theta - \psi_\theta(E^2\psi/r^2 \sin^2\theta)_r] \sin\theta = E^4\psi, \tag{3.32}$$

where

$$E^2 = \frac{\partial^2}{\partial r^2} + \frac{\sin\theta}{r^2}\frac{\partial}{\partial\theta}\left(\frac{1}{\sin\theta}\frac{\partial}{\partial\theta}\right).$$

Upon introducing the vorticity, ω, by means of

$$\omega r \sin\theta = E^2\psi, \tag{3.33}$$

Eq. (3.32) becomes

$$\tfrac{1}{2}\,\mathrm{Re}[\psi_r(\omega/r\sin\theta)_\theta - \psi_\theta(\omega/r\sin\theta)_r]\sin\theta = E^2[\omega r\sin\theta], \tag{3.34}$$

and the velocity components are related to the stream function by

$$V_\theta = \psi_r/r\sin\theta, \qquad V_r = -\psi_\theta/r^2\sin\theta.$$

Perhaps motivated by the Stokes solution of $E^4\psi = 0$, the authors studying this problem chose the trial stream function as

$$\psi = \Big(\tfrac{1}{2}r^2 + \sum_{i=1}^{4} A_i/r^i\Big)\sin^2\theta + \Big(\sum_{i=1}^{4} B_i/r^i\Big)\sin^2\theta\cos\theta, \tag{3.35}$$

and determined the constants A_i, B_i, $i = 1, 2, 3, 4$, by Galerkin's method.

A modification of the classical Galerkin procedure termed the *N-parameter integral formulation of Galerkin–Kantorovich–Dorodnitsyn* was introduced by Dorodnitsyn [41] to solve the hypersonic blunt-body problem by what has become known as the *strip method*. Later Dorodnitsyn [42, 43] published several solutions of the Golubev [44] boundary-layer formulation after employing a simple transformation of the dependent variable. The technique was utilized to solve the compressible boundary-layer equations with and without wall suction by Pavlovskii [45, 46] and Liu [47, 48]. The N-parameter method has been applied by Abbott and Bethel [49] to a number of laminar boundary-layer problems (compressible and incompressible, axisymmetric and planar, adiabatic and nonadiabatic walls). Herein we shall emphasize the mathematical technique. Consequently the model used will be that of steady incompressible planar flow.

With a given, but arbitrary, external velocity distribution $U_1(x)$ the boundary-layer equations are

$$UU_X + VU_Y = U_1U_{1X} + \nu U_{YY}, \qquad U_X + V_Y = 0, \tag{3.36}$$

where X and Y are surface and normal coordinates, U and V are the X and Y components of velocity, respectively, and ν is the kinemetic fluid viscosity. We suppose the boundary conditions to be $Y = 0$: $U = 0$, $V = 0$; $Y \to \infty$: $U(X, Y) \to U_1(X)$. Before proceeding to the N-parameter formulation, a Goertler [50] type transformation is employed.

With $\mathrm{Re}_\infty = U_\infty l/\nu$, where U_∞ and l are arbitrary constant reference quantities, let

$$\begin{gathered} u = U/U_1\,, \qquad v = (\mathrm{Re}_\infty)^{1/2}\, V/U_1\,, \qquad u_1 = U_1/U_\infty\,, \\ \partial\xi = (U_1/U_\infty)\,\partial X/l, \qquad \eta = (\mathrm{Re}_\infty)^{1/2}\,(U_1/U_\infty)\,Y/l, \\ w = v + \eta u u_1'/u_1\,, \qquad u_1' = u_{1\xi}\,, \end{gathered} \tag{3.37}$$

whereupon Eqs. (3.36) become[†]

$$uu_\xi + wu_\eta = (1 - u^2)\,u_1'/u_1 + u_{\eta\eta}\,, \tag{3.38a}$$

$$u_\xi + w_\eta = 0, \tag{3.38b}$$

with the boundary conditions $u(\xi, 0) = w(\xi, 0) = 0$, and $u(\xi, \eta) \to 1$ as $\eta \to \infty$.

The N-parameter method is now applied to Eqs. (3.38). Let $H_i(u)$, $i = 1, 2, \ldots, N$ be linearly independent weighting functions with properties to be discussed in the sequel. The first step in the development of the system of N linearly independent integral equations is to combine the continuity equation [Eq. (3.38b)] and the momentum equation by multiplying the former by u and adding the result to the latter. Thus we find

$$2uu_\xi + uw_\eta + wu_\eta = (1 - u^2)\,u_1'/u_1 + u_{\eta\eta}\,,$$

or

$$(\partial u^2/\partial\xi) + [\partial(uw)/\partial\eta] - (1 - u^2)\,u_1'/u_1 - (\partial^2 u/\partial\eta^2) = 0, \tag{3.39}$$

which is in *divergence* form. (It has been standard practice in the Russian literature to develop approximate solutions from the divergence form of the system. An interesting comment on the necessity of this procedure for some finite difference methods is given by Babenko [51].) Ultimately the dependent variable w can be removed by imposing a suitable side condition on the weighting function. For simplicity, Eq. (3.39) can be symbolically represented as

$$F[u(\xi, \eta), w(\xi, \eta)] = 0. \tag{3.40}$$

† The usefulness of solving boundary-layer equations in ξ, η coordinates lies in the ease with which the initial conditions can be specified in terms of similarity solutions. A computational advantage occurs because of the simple manner in which the ordinary differential equations resulting from the N-parameter method in the transformed coordinates are found to reduce to algebraic equations for similarity solutions.

The second step in the analysis is to multiply Eq. (3.40) by a member of the set of linearly independent functions of u, $\{H_i(u), i = 1, 2,..., N\}$,

$$H_i(u) F[u, w] = 0, \tag{3.41}$$

and then to integrate the result over the domain of η,

$$\int_0^\infty H_i(u) F[u, w]\, d\eta = 0, \qquad i = 1, 2,..., N. \tag{3.42}$$

This set of N integral conditions is of Galerkin type.

Next, Eq. (3.41) is expanded, by substitution for $F[u, w]$ and calculating the derivatives, to give

$$uH_iu_\xi + uH_iu_\xi + wH_iu_\eta + uH_iw_\eta - H_i(1 - u^2)\, u_1'/u_1 - H_iu_{\eta\eta} = 0. \tag{3.43}$$

A function $h_i(u)$ is now introduced by means of the definition

$$\frac{dh_i}{du} = H_i(u), \tag{3.44}$$

so that

$$\frac{\partial h_i}{\partial \xi} = H_i \frac{\partial u}{\partial \xi} \qquad \text{and} \qquad \frac{\partial h_i}{\partial \eta} = H_i \frac{\partial u}{\partial \eta}.$$

Substituting $h_i(u)$ into Eq. (3.43) and noting that $uH_i(u_\xi + w_\eta) = 0$, we find by a second application of the continuity equation, Eq. (3.38b), that

$$u \frac{\partial h_i}{\partial \xi} + h_i \frac{\partial u}{\partial \xi} + w \frac{\partial h_i}{\partial \eta} + h_i \frac{\partial w}{\partial \eta} - h_i'(1 - u^2) \frac{u_1'}{u_1} - h_i' \frac{\partial^2 u}{\partial \eta^2} = 0.$$

Combining the partial derivatives in the above equation yields the explicit form of Eq. (3.41):

$$\frac{\partial(h_iu)}{\partial \xi} + \frac{\partial(h_iw)}{\partial \eta} - h_i' \left\{(1 - u^2) \frac{u_1'}{u_1} + \frac{\partial^2 u}{\partial \eta^2}\right\} = 0. \tag{3.45}$$

An explicit form of Eq. (3.42) is obtained by integrating Eq. (3.45) over the domain of η,

$$\int_0^\infty \frac{\partial(h_iu)}{\partial \xi}\, d\eta + h_iw \Big|_0^\infty - (u_1'/u_1) \int_0^\infty h_i'(1 - u^2)\, d\eta - \int_0^\infty h_i' \frac{\partial^2 u}{\partial \eta^2}\, d\eta = 0. \tag{3.46}$$

Since the boundary condition for an impermeable wall requires

$w(\xi, 0) = 0$, it is clear that the dependent variable w can be eliminated from Eq. (3.46) by requiring that

$$\lim_{\substack{\eta\to\infty \\ u\to 1}} h_i(u) = 0. \tag{3.47}$$

Finally an application of Leibnitz's rule (on differentiation) to Eq. (3.46) gives the final form

$$\frac{d}{d\xi}\int_0^\infty h_i(u)\, u\, d\eta - \frac{u_1'}{u_1}\int_0^\infty h_i'(u)(1-u^2)\, d\eta + h_i'(0)\left.\frac{\partial u}{\partial\eta}\right|_{\eta=0}$$
$$+ \int_0^\infty h_i''(u)\left(\frac{\partial u}{\partial\eta}\right)^2 d\eta = 0, \qquad i = 1, 2, \ldots, N, \tag{3.48}$$

of the integral conditions.† Strictly speaking $H_i(u)$ are the weighting functions but since only their integrals, $h_i(u)$, appear, we shall refer to $h_i(u)$ as the weighting functions.

If there is a single weighting function $h(u) = 1 - u$, Eq. (3.48) becomes

$$\frac{d}{d\xi}\int_0^\infty u(1-u)\, d\eta + \frac{u_1'}{u_1}\int_0^\infty (1-u^2)\, d\eta = \left.\frac{\partial u}{\partial\eta}\right|_{\eta=0},$$

which is the *Karman momentum integral equation* in the present coordinates.

If a set of weighting functions is taken in the form

$$h_i(u) = (1-u)^i, \qquad i = 1, 2, \ldots, N,$$

Eq. (3.48) takes the *form proposed by Golubev* [44]. Choosing the set in the form

$$h_i(u) = (1-u^i)/i, \qquad i = 1, 2, \ldots, N,$$

yields the *moment of momentum equations* derived by Truckenbrodt [52].

---

† Dorodnitsyn [42] obtained Eq. (3.48) by multiplying Eq. (3.38b) (continuity) by $h_i(u)$ and Eq. (3.38a) (momentum) by $h_i'(u)$ and adding. One immediately obtains Eq. (3.48) with the side condition Eq. (3.47) necessary for the elimination of w. The derivation above shows the relation to the Galerkin WRM. The weighting function $H_i(u)$ appears in the form of the integral, that is as $h_i(u)$.

The Dorodnitsyn [41] *strip method*, actually not employed by Dorodnitsyn on boundary-layer problems, is obtained by selecting discontinuous weighting functions,

$$h_i(u) = \begin{cases} 0, & u < u_{i-1}, \\ 1, & u_{i-1} \leqslant u \leqslant u_i, \\ 0, & u > u_i, \quad \text{with} \quad u_0 = 0, \quad u_N = 1. \end{cases}$$

Pallone [53] used the variation

$$h_i(u) = \begin{cases} 0, & u < u_i, \\ 1, & u_i \leqslant u \leqslant 1, \end{cases}$$

on the preceding form. The computations of Medvedev [54], and Abbott and Bethel [49] indicate that the use of discontinuous functions leads to less rapidly convergent solutions than those obtained with continuous weighting functions.

The set of integral conditions, Eq. (3.48), with $u(\xi, \eta)$ as the dependent variable, was obtained from the original boundary-layer equations without introducing any approximations. It is therefore an exact restatement of the original problem. In the present form, it has some undesirable features. When one seeks an approximating function of the form, say,

$$u_{\mathrm{T}}(\xi, \eta) = \sum_{n=1}^{N} B_n(\xi)\, \psi_n(\xi, \eta), \tag{3.49}$$

each member of the prescribed set $\{\psi_n(\xi, \eta)\}$, which is complete with respect to $u(\xi, \eta)$, must satisfy the η boundary conditions of $u(\xi, \eta)$. The selection of such functions is often not straightforward because of the asymptotic nature of $u(\xi, \eta)$ as $\eta \to \infty$. Not only is the integration over a semi-infinite domain, but the set probably must include error, exponential, or other transcendental functions which may be difficult to handle. While Hanson and Richardson [55] found a transcendental approximation to be of no great difficulty in boundary-layer theory, most authors have preferred the course we are to take.

Dorodnitsyn [42], in an attempt to surmount these problems, introduced the *inverse slope*,

$$\theta(\xi, u) = (\partial u/\partial \eta)^{-1}, \tag{3.50}$$

as a new dependent variable by transforming the independent variables from (ξ, η) to (ξ, u). Clearly θ is the inverse shear stress in the standardized coordinate system. This transformation replaces a dependent variable which is bounded and defined over a semi-infinite region with one which is defined over a finite interval, but has a pole as $u \to 1$.

Applying the transformation† of the independent variables to Eqs. (3.48) gives

$$\frac{d}{d\xi}\int_0^1 h_i u\theta\, du - \frac{u_1'}{u_1}\int_0^1 h_i'(1-u^2)\theta\, du + \frac{h_i'(0)}{\theta(\xi,0)} + \int_0^1 [h_i''/\theta]\, du = 0,$$

$$i = 1,\ldots, N, \qquad (3.51)$$

with the boundary condition

$$\theta(\xi, u) \to \infty, \qquad \text{as} \quad u \to 1. \qquad (3.52)$$

In this form, the limits of integration are finite, and the problem of integrand approximation arises linearly in each integral with the function $\theta(\xi, u)$.

At this point, we make our approximation for the function $\theta(\xi, u)$ in the form

$$\theta(\xi, u) \approx \theta_N(\xi, u) = \sum_{n=1}^{N} C_n(\xi)\,\phi_n(\xi, u), \qquad (3.53)$$

where:

(i) The set $\{\phi_n(\xi, u)\}n = 1, 2,\ldots$ is mathematically complete with respect to the function being approximated;
(ii) each $\phi_n(\xi, u)$ satisfies the boundary condition, Eq. (3.52); and
(iii) the dependence upon u of each $\phi_n(\xi, u)$ is *explicitly chosen.*

Substitution of Eq. (3.53) into Eq. (3.51) and subsequent integration over u transforms the integral conditions to a system of ordinary differential equation in the $C_n(\xi)$.

---

† The general Dorodnitsyn transformation replaces the independent variables (τ, ξ, η) by $(\bar{\tau}, \bar{\xi}, u)$ and the dependent variable u by Θ, where

$$\bar{\tau} = \tau, \qquad \bar{\xi} = \xi, \qquad u = u(\tau, \xi, \eta), \qquad \Theta = \Theta(\bar{\tau}, \bar{\xi}, u) = (\partial u/\partial\eta)^{-1},$$

and

$$\eta = \int_0^u \Theta(\bar{\tau}, \bar{\xi}, u)\, du.$$

The transformation equations are

$$\frac{\partial}{\partial\tau} = \frac{\partial}{\partial\bar{\tau}} + \frac{\partial u}{\partial\tau}\frac{\partial}{\partial u}, \qquad \frac{\partial}{\partial\xi} = \frac{\partial}{\partial\bar{\xi}} + \frac{\partial u}{\partial\xi}\frac{\partial}{\partial u},$$

and

$$\frac{\partial}{\partial\eta} = \frac{\partial u}{\partial\eta}\frac{\partial}{\partial u}.$$

If the system of integral conditions, Eqs. (3.51), is written in the operator notation

$$G_i[\theta(\xi, u)] = 0, \qquad i = 1, 2, ..., N, \tag{3.54}$$

then with θ_T replacing θ, it follows that

$$G_i[\theta_T(\xi, u)] = R_i \neq 0, \qquad i = 1, 2, ..., N, \tag{3.55}$$

where R_i is the residual. If θ_T is the exact solution, R_i is zero, thus its size is, in some sense, an indication of the "nearness" of θ_T to θ. When the set $\{\phi_n\}$ is complete with respect to the solution $\theta(\xi, u)$, that is,

$$\lim_{N\to\infty} \sum_{n=1}^{N} C_n(\xi)\, \phi_n(\xi, u) = \lim_{N\to\infty} \theta_N(\xi, u) = \theta(\xi, u),$$

and

$$G_i[\lim_{N\to\infty} \theta_N(\xi, u)] = G_i[\theta(\xi, u)] = 0, \qquad i = 1, 2, ...$$

One *presumes* the converse of this also to be true. Thus if the coefficients are determined from the system

$$G_i[\theta_N(\xi, u)] = 0, \qquad i = 1, 2, ..., N,$$

then $\lim_{N\to\infty}\theta_N(\xi, u) = \theta(\xi, u)$, and the approximate solution converges to the exact.

From the preceding discussion, it is apparent that the requirement of mathematical completeness of the set $\{\phi_n(\xi, u)\}$ is a necessary condition for the convergence of the approximate solution to the exact. Presuming completeness, each successive order of approximation will yield a better solution, and the convergence will be uniform. If, on the other hand, the selected set is not complete but yields $\theta_N(\xi, u)$ which is in qualitative agreement with $\theta(\xi, u)$, then the solution will converge to some "best possible" solution of unknown accuracy. In this case, uniform convergence cannot be expected *a priori* and in fact will usually be lacking. *One cannot infer convergence to the exact solution from the decrease of the differences between the approximate solutions for successive orders of approximations.* When the approximating function θ_N is not known to be complete, the only means of testing the approximate result is to compare it with exact or accurate numerical solutions. Unfortunately the mathematical procedures for selecting complete sets of functions for many engineering problems, such as the present one, are nonexistent.

Abbott and Bethel [49] observe that in the process of selecting the prescribed functions, $\phi_n(\xi, u)$, considerable guidance is available from the physical meaning of the dependent variable

$$\theta(\xi, u) = \left(\frac{\partial u}{\partial \eta}\right)^{-1} = \frac{u_1^2(\xi)(\mathrm{Re}_\infty)^{1/2}}{l(\partial u/\partial y)} \sim \frac{u_1^2(\xi)}{\tau(\xi, u)}. \qquad (3.56)$$

At any particular streamwise section, say ξ_0,

$$\theta(\xi_0, u) \sim [\tau(\xi_0, u)]^{-1}.$$

The qualitative physical behavior of η and the shear stress τ for $\xi = \xi_0$ is shown in Fig. 3-1. From these we can infer the qualitative form of

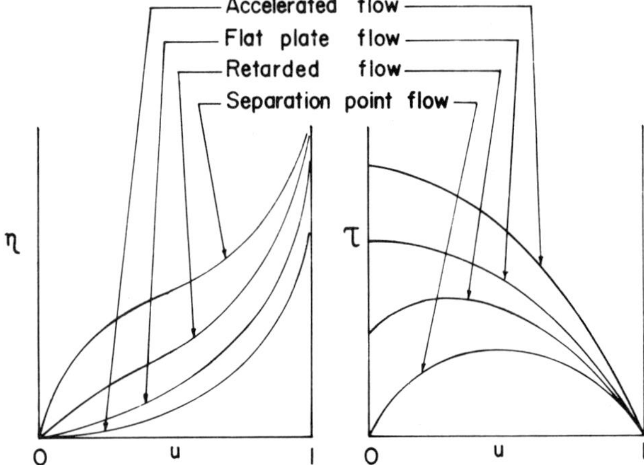

FIG. 3-1. Qualitative behavior of velocity and shear stress.

$\theta(\xi_0, u)$, which is shown in Fig. 3-2. Clearly there is a singularity in $\theta(\xi_0, u)$ as $u \to 1$ for all cases and as $u \to 0$ for the separation case. These must be accounted for in selecting the functions $\phi_n(\xi, u)$.

Dorodnitsyn [42] used the family

$$\phi_n(\xi, u) = \phi_n(u) = u^{n-1}/(1 - u),$$

which generates the approximating function

$$\theta_N(\xi, u) = \sum_{n=1}^{N} C_n(\xi)\, u^{n-1}/(1 - u).$$

This provides for satisfaction of the boundary conditions in the accelerated flow and mildly retarded regimes but does not admit a pole at the

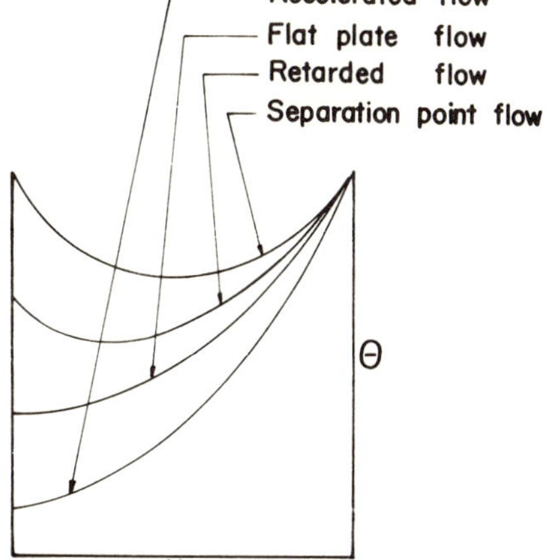

FIG. 3-2. Qualitative behavior of inverse slope profiles.

separation point. Abbott and Bethel [49] sought an increase in accuracy with the trial family

$$\theta_{N,M}(\xi, u) = \sum_{n=1}^{N} C_n(\xi)\, u^{n-1}/(1 - u^M),$$

but found it to be of academic interest only.

In the case of retarded flows, several authors (Dorodnitsyn [42] and Nielsen *et al.* [56]) provide for the occurrence of a possible separation point with retarded flows by selecting[†]

$$\theta_{N,\alpha} = \sum_{n=1}^{N-1} C_n(\xi)\, u^{n-1}/(1 - u)[\alpha(\xi) + u]^{1/2}, \tag{3.57}$$

where $\alpha(\xi) \to 0$ as the separation point ($u = 0$) is approached. We shall give the details for this case only.

Using Eq. (3.57) with $h_i(u) = 1 - u^i$, $i = 1, 2,\ldots, N$, in Eq. (3.48) the following system is obtained:

$$\sum_{n=1}^{N-1} \Big[\sum_{p=1}^{i} Z\alpha(2, n + p)\Big] \frac{dC_n}{d\xi} - \frac{1}{2}\frac{d\alpha}{d\xi} \sum_{n=1}^{N-1} \Big[\sum_{p=1}^{i} Z_\alpha(1, n + p)\Big] C_n$$

$$+ i\frac{u_1'}{u_1} \sum_{n=1}^{N-1} [Z_\alpha(2, i + n - 1) + Z_\alpha(2, i + n)]\, C_j$$

$$- \frac{\alpha^{1/2}}{C_1}\delta_{il} - i(i - 1)\, I_\alpha(i) = 0, \qquad i = 1, 2,\ldots, N, \tag{3.58}$$

[†] Note that $(N - 1)$ functions $C_n(\xi)$, $n = 1, 2,\ldots (N - 1)$, and $\alpha(\xi)$ are to be determined from the N integral relations.

where

$$Z_\alpha(k, l) = \int_0^l u^{l-1}(\alpha + u)^{k-5/2}\, du = \sum_{m=1}^{l} K_\alpha(l, m)\, W_\alpha(k + l - m),$$

with

$$W_\alpha(q) = \int_0^1 (\alpha + u)^{q-1/2}\, du = [(\alpha + 1)^{q-3/2} - \alpha^{q-3/2}]/(q - \tfrac{3}{2})$$

$$K_\alpha(l, m) = \begin{cases} 1 & \text{for} \quad m = 1, \\ -K_\alpha(l, m-1)\left[\dfrac{l - m + 1}{m - 1}\right]\alpha & \text{for} \quad m \neq 1. \end{cases}$$

Also $\delta_{il} = 1$ for $i = l$ and 0 otherwise, and

$$I_\alpha(i) = \int_0^1 \left[u^{i-2}(1 - u)(\alpha + u)^{1/2} / \sum_{n=1}^{N-1} C_n u^{n-1}\right] du.$$

Actual calculations are reported by Abbott and Bethel [49].

Application of the preceding method to time dependent laminar boundary layers is provided by Koob and Abbott [57]. Also Koob and Abbott [58] discuss viscous flow over an impulsively accelerated semi-infinite plate by WRM. Deiwert and Abbott [59, 60] and Abbott *et al.* [61] consider application of WRM to turbulent boundary layers.

Calculation by the *method of lines*, discussed briefly in Volume I, has been employed by a number of authors. Smith and Clutter [62], and Koob and Abbott [57] use this computational method in boundary-layer calculations, while Hicks and Wei [63] treat parabolic partial differential equations with two point-boundary conditions by this method.

b. Integral Method

Since this procedure is so often used, we give only a few applications. Goodman [4] discusses the application of this method to nonlinear heat transfer problems, Acrivos *et al.* [64] use the integral method to develop solutions of the two-dimensional boundary-layer equations for a power-law non-Newtonian fluid. Power-law fluids in the inlet flow of a channel were considered by Kapur and Gupta [65], while Rajeshwari and Rathna [66] study the flow of non-Newtonian viscoelastic and viscoinelastic fluids near a stagnation point with this integral method of Pohlhausen.

Simultaneous development of velocity and temperature distributions have been treated in this manner by Lemlich and Steinkamp [67] and Siegel and Sparrow [68].

c. Subdomain Method

The *strip method* of Dorodnitsyn [41] is of the subdomain type and was applied by him on hypersonic problems. Belotserkovskii and Chuskin [69] describe the background of the procedure, while Carter and Gill [70] employ it to develop solutions for free and forced convection in vertical and horizontal conduits. Launder [71] asserts that this method is of Pohlhausen type in his boundary-layer calculations. Coupled heat and mass transfer problems are discussed by Finlayson [37]. In the same paper he compares five WRM (subdomain, collocation, moments, Galerkin's, and least squares).

d. Collocation

In addition to the examples of Volume I and the work of Schetz [22–25], we add here a reference to the researches of Conway [72] on boundary-value problems, Jeffreys [73] for cellular convection, and Kadner [74] for his general discussion. Villadsen and Stewart [75] develop new collocation methods which utilize orthogonal polynomial expansions fitted by collocation techniques. By appropriate choices of the weight function in the orthogonality relation, their methods yield accuracy comparable to the Galerkin procedure.

e. Least Squares

Finlayson [37] is an up-to-date reference on the treatment of coupled heat and mass transfer. The least squares method is discussed at some length in a monograph by Becker [76]. Lumping this method with the related variational methods he lists criteria which he maintains a good method must satisfy. His list includes the following.

(i) Errors should be minimized in some sense.
(ii) The functional should be positive definite.
(iii) The procedure should be capable of treating initial-value problems, as well as others.

Becker asserts that the least squares method is the best general criterion of WRM and he illustrates the advantages of the method by solving a nonlinear time dependent partial differential equation which governs the fuel depletion in a nuclear reactor.

f. Method of Moments

An example of the application of moments was given in Volume I. Here, we add the work of Yamada [77], on approximate integration of the boundary layer equations, Prager [78] in transport processes with chemical reactions, and Poots [79] to heat conduction in a problem possessing a two-dimensional solidification front.

3.4 WRM IN DYNAMICS AND SOLID MECHANICS

The application of WRM, especially that of Galerkin, has been especially popular in developing approximate solutions to problems of solid mechanics. After the introduction of the Galerkin method in 1915 it was employed on a variety of problems, before the convergence of the method was studied (see Section 3.5 for a discussion of the state of the theory for WRM). Many of these early problems concerned the solution of ordinary differential equations or linear partial differential equations. Among the early nonlinear problems we find the work of Lourie and Tchekmarev [80] on nonlinear vibrations and that of Panov [81] on nonlinear problems in elasticity.

A plethora of applications has been generated in the years 1950–1970. Early in this period, Schwesinger [82] published one-term approximations of forced nonharmonic vibrations, and Eringen [83] studied transverse impact on beams and plates. The flutter of two-dimensional panels was studied by Fung [84], while Tanner [85] estimated bearing loads using Galerkin's method. As mentioned in Volume I, Nowinski and his students have employed Galerkin's procedure in a variety of studies arising from static and dynamic elastic systems. In particular, we note Nowinski [86] on the nonlinear transverse vibrations of orthotropic cylindrical shells and Nowinski and Woodall [87] on the finite vibrations of a free rotating anisotropic membrane. As a typical dynamic application, we present the essence of Nowinski's [88] study of the nonlinear transverse vibrations of a spinning disk.

Let a flat circular elastic plate of radius a and uniform thickness h rotate about its axis of symmetry with a constant angular velocity ω. The origin of a plane polar coordinate system r, θ, embedded in and rotating with the disk, is located at the center of the middle plane of the disk. The disk is subjected to a normal load $g(r, \theta, t)$, which includes the weight, and the transverse deflections are assumed to be large that is, of the order of the magnitude of the thickness h. While a geometrical nonlinearity enters into the problem due to the large deflection assumption, it is postulated that Hooke's law is valid in its isotropic form. With σ_r, σ_θ, and $\tau_{r\theta}$ representing the radial, hoop, and shear stresses, respectively, the well known *stress equilibrium* equations for the symmetric case are

$$\begin{aligned} \frac{\partial \sigma_r}{\partial r} + \frac{1}{r}\frac{\partial \tau_{r\theta}}{\partial \theta} + \frac{\sigma_r - \sigma_\theta}{r} + \rho^* \omega^2 r &= 0, \\ \frac{1}{r}\frac{\partial \sigma_\theta}{\partial \theta} + \frac{\partial \tau_{r\theta}}{\partial r} + 2\frac{\tau_{r\theta}}{r} &= 0, \end{aligned} \tag{3.59}$$

where ρ^* is the mass density of the disk.†

To the equations of equilibrium we adjoin the nonlinear strain-displacement relations

$$\begin{aligned} \epsilon_r &= \frac{\partial u}{\partial r} + \frac{1}{2}\left(\frac{\partial w}{\partial r}\right)^2, \\ \epsilon_\theta &= \frac{u}{r} + \frac{1}{r}\frac{\partial v}{\partial \theta} + \frac{1}{2r^2}\left(\frac{\partial w}{\partial \theta}\right)^2, \\ \gamma_{r\theta} &= \frac{1}{r}\frac{\partial u}{\partial r} + \frac{\partial v}{\partial r} - \frac{v}{r} + \frac{1}{r}\frac{\partial w}{\partial r}\frac{\partial w}{\partial \theta}. \end{aligned} \tag{3.60}$$

These are directly obtainable from any book on elasticity by adding the nonlinear terms. Clearly u, v, and w denote radial, hoop, and axial components of the displacement of the particles of the disk, respectively, and no symmetry of the state of strain has been postulated. Equations (3.59) are identically satisfied if we derive the stress components from a stress function ϕ defined by means of

$$\begin{aligned} \sigma_r &= \frac{1}{r}\frac{\partial \phi}{\partial r} + \frac{1}{r^2}\frac{\partial^2 \phi}{\partial \theta^2} - \frac{1}{2}\rho^* \omega^2 r^2, \\ \sigma_\theta &= \frac{\partial^2 \phi}{\partial r^2} - \frac{1}{2}\rho^* \omega^2 r^2, \\ \tau_{r\theta} &= -\frac{\partial}{\partial r}\left(\frac{1}{r}\frac{\partial \phi}{\partial \theta}\right). \end{aligned} \tag{3.61}$$

The governing field equations are furnished, in the present case, by the equation of motion in the direction of the axis of the spin, and by the

† In this development, the inertia terms arising from the in-plane motion of the particles of the plate are neglected, since the vibrations take place principally in the transverse direction. This has been verified by Chu and Herrmann [89], Chu [90], and more recently by Kuo [91] using a *perturbation procedure*. To first order, this approximation is valid.

only identically nonvanishing equation of compatibility of the in-plane deformations. The former equation may be obtained, for the asymmetric case under investigation, by transforming from rectangular cartesian to the polar coordinates the corresponding well-known von Karman equation (see Volume I) and adding the inertia term. This gives, finally,

$$\frac{D}{h}\nabla^4 w + \rho^* \frac{\partial^2 w}{\partial t^2} = \frac{q(r, \theta, t)}{h} + \frac{\partial^2 w}{\partial r^2}\left(\frac{1}{r}\frac{\partial \phi}{\partial r} + \frac{1}{r^2}\frac{\partial^2 \phi}{\partial \theta^2}\right)$$
$$+ \left(\frac{1}{r}\frac{\partial w}{\partial r} + \frac{1}{r^2}\frac{\partial^2 w}{\partial \theta^2}\right)\frac{\partial^2 \phi}{\partial r^2} - 2\frac{\partial}{\partial r}\left(\frac{1}{r}\frac{\partial w}{\partial \theta}\right)\frac{\partial}{\partial r}\left(\frac{1}{r}\frac{\partial \phi}{\partial \theta}\right)$$
$$- \frac{1}{2}\rho^*\omega^2 r^2 \nabla^2 w - \rho^*\omega^2 r \frac{\partial w}{\partial r}, \tag{3.62}$$

where $D = Eh^3/12(1 - \nu^2)$ is the bending rigidity of the plate; $w(r, \theta, t)$ is the deflection; t, time; and ∇^2 the two-dimensional Laplace operator in polar coordinates. The equation of compatibility of large deformations in polar coordinates and in the asymmetric case may be derived from the corresponding classical equation

$$\frac{1}{r^2}\frac{\partial^2 \epsilon_r^{\text{lin}}}{\partial \theta^2} - \frac{1}{r}\frac{\partial^2 (r\gamma_{r\theta}^{\text{lin}})}{\partial r\, \partial \theta} + \frac{1}{r^2}\frac{\partial}{\partial r}\left(r^2 \frac{\partial \epsilon_\theta^{\text{lin}}}{\partial r}\right) - \frac{1}{r}\frac{\partial \epsilon_r^{\text{lin}}}{\partial r} = 0, \tag{3.63}$$

by a substitution $\epsilon_r^{\text{lin}} = \epsilon_r - \bar{\epsilon}_r$, and so on, where ϵ_r, and so on, denote the left-hand members of Eqs. (3.60), and the barred symbols, the nonlinear terms in the right-hand members of these equations. Using the linear isotropic stress–strain relations and Eqs. (3.61), we obtain

$$\nabla^4 \phi = -\frac{1}{r}\left(\frac{\partial w}{\partial r} + \frac{1}{r^2}\frac{\partial^2 w}{\partial \theta^2}\right)\frac{\partial^2 w}{\partial r^2} + \frac{1}{r^2}\left(\frac{\partial^2 w}{\partial r\, \partial \theta}\right)^2 - \frac{2}{r^3}\frac{\partial^2 w}{\partial r\, \partial \theta}\frac{\partial w}{\partial \theta} + \frac{1}{r^4}\left(\frac{\partial w}{\partial \theta}\right)^2. \tag{3.64}$$

Faced with the integration of the two simultaneous nonlinear equations, Eqs. (3.62) and (3.64), together with appropriate boundary conditions, an approximate method of integration comes to mind. Motivated by exact solutions to linear problems, we could choose the deflection function as a double series,

$$w(r, \theta, t) = \sum_i F_i(r)\, \tau_i(t) \cos(n\theta + \alpha), \tag{3.65}$$

but, to avoid the laborious details, we actually select

$$w(r, \theta, t) = F(r)\, \tau(t) \cos(n\theta + \alpha), \tag{3.66}$$

where α is a constant, and $\tau(t)$ is a function to be determined. The function $F(r)$ may be represented in the form of a power series

$$F(r) = r^n \sum_{j=0}^{\infty} C_j r^{2j}, \tag{3.67}$$

similar to that describing the radial profile of the deflection surface in the linear case (see, e.g., Prescott [92, Eq. (18.24)]). In the linear case, for normal modes of vibration, the series, Eq. (3.67), reduces to a finite number of terms, and their sum vanishes for values of r associated with the nodal circles. In what follows we confine our attention to two terms of the series and assume

$$w(r, \theta, t) = A_0 \tau(t)\, r^n(1 + gr^2) \cos(n\theta + \alpha), \tag{3.68}$$

where A_0 is the amplitude of the vibration and g is to be determined from the problem's auxiliary conditions.

At first glance, this drastic truncation would appear to restrict the generality of the solution severely. However, Eq. (3.68) does permit solutions to a variety of interesting problems. Some of these, together with the appropriate boundary conditions, are listed below and depicted in Fig. 3-3:

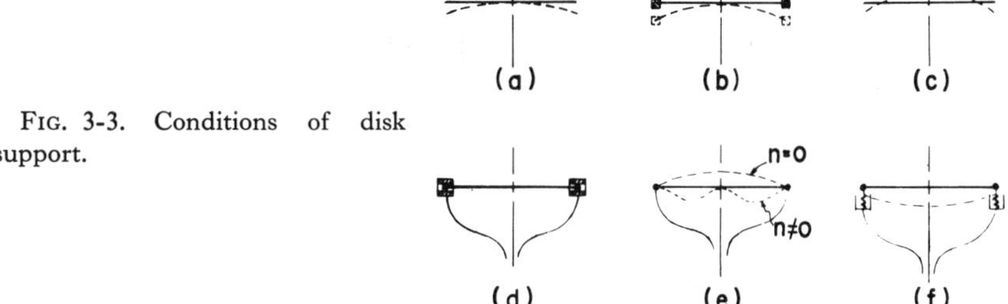

FIG. 3-3. Conditions of disk support.

(*a*) *Free flexible disk with no nodal circle and any number n of nodal diameters.* In this case, we may set $g = 0$, thus obtaining

$$w = A_0 \tau r^n \cos(n\theta + \alpha).$$

The edge of the disk may be either free from radial and shear stress, in conformity with the boundary conditions $\sigma_r = \tau_{r\theta} = 0$ for $r = a$ (Fig. 3-3a) or subjected to a varying radial tension exerted, for instance, by a heavy ring attached to the edge of and rotating with the disk (Fig. 3-3b).

(*b*) *Free flexible disk with no nodal diameter* ($n = 0$) *and one nodal circle.* Since no external force acts on the disk in the axial direction, it follows that the total momentum of the disk in this direction is constant, and may be assumed zero. This yields

$$w = A_0\tau[1 - 2(r^2/a^2)].$$

Note that the radius of the existing nodal circle equals $0.707a$ just as in the linear case. The edge of the disk, for instance, may be taken as free from external tractions (Fig. 3-3c).

(*c*) *Free flexible disk with one nodal diameter* ($n = 1$) *and one nodal circle.* The angular momentum of the whole disk about the nodal diameter being zero during the vibration, the parameter g becomes equal to $(-1.5/a^2)$, in conformity with the value of g furnished in the linear case:

$$w = A_0\tau r[1 - \tfrac{3}{2}(r^2/a^2)] \cos(\theta + \alpha).$$

The boundary conditions may be specified in the present case as in that discussed previously.

(*d*) *Flexible disk supported at the edge with an arbitrary number of nodal diameters.* In this particular case $g = -1/a^2$, and

$$w = A_0\tau r^n[1 - (r^2/a^2)] \cos(n\theta + \alpha).$$

Apparently, lack of an additional free parameter excludes the possibility of assuming the existence of a nodal circle. The edge of the disk may be taken as stress free or, say, prevented against radial displacements: $u = 0$ for $r = a$ (Figs. 3-3d and e). In the latter case an initial tension of an arbitrary magnitude may also be superposed.

(*e*) *Stiff disk.* In this case we have in Eq. (3.68) too few free parameters at our disposal in order to satisfy the boundary conditions of a plate subjected to bending. However, it is possible to assume that the disk is free and be satisfied with a pointwise vanishing of the radial bending moment at the edge of the disk, as well as with an overall vanishing of the transverse force (say, over two adjacent bays bounded by nodal half-rays). One may also postulate that the edge of the disk is elastically supported in such a fashion that, for instance, the bending moment at the edge of the disk vanishes, and the deflection at the edge is proportional to the transverse force (with a suitably adjusted coefficient of proportionality, e.g., corresponding to the Winkler-type foundation) (Fig. 3-3f). Clearly, the edge may be either freely movable in the radial direction or prevented from such a motion.

We now proceed to the determination of the stress function $\phi(r, \theta, t)$ and of the membrane stresses in terms of the function ϕ. For this purpose, let us substitute Eq. (3.68) into the right-hand member of the governing Eq. (3.64) and find a particular integral of this equation suitable for the boundary-value problem under investigation. The desired representation is

$$\phi = K \left\{ \frac{a_1}{c_1} r^{2n} + \frac{a_2}{c_2} g^2 r^{2(n+2)} + \frac{a_3}{c_3} g r^{2(n+1)} + \frac{b_2}{c_2 + d_2} g^2 r^{2(n+2)} \cos 2(n\theta + \alpha) \right\} + \frac{1 - \nu}{32} \omega^2 \rho^* r^4 + Ar^2 + (Cr^{2n} + Dr^{2(n+1)} \cos 2(n\theta + \alpha), \qquad K \equiv \frac{E}{2} A_0{}^2 \tau^2, \tag{3.69}$$

where A, C, and D are constants to be determined later from the in-plane boundary conditions, and

$$\begin{aligned}
a_1 &= 2n^2(n-1)^2, \qquad a_2 = 2(n+1)^2(n^2-2), \qquad a_3 = 4n^2(n^2-1), \\
b_2 &= -4(n+1)^2 \\
c_1 &= 2n\{(2n-1)[2(n-1)(2n-1)-1]+1\}, \\
c_2 &= 2(n+2)\{(2n+3)[2(n+1)(2n+3)-1]+1\}, \\
c_3 &= 2(n+1)\{(2n+1)[2n(2n+1)-1]+1\}, \\
d_1 &= -16n^2(n-1)^2, \qquad d_2 = -16n^2(n+1)(n+5), \\
d_3 &= -16n^2(n+1)^2.
\end{aligned} \tag{3.70}$$

With these in mind, and upon the substitution of Eq. (3.69) into Eq. (3.61), one obtains the following final equations for the components of the membrane stresses:

$$\sigma_r = K \left\{ \frac{2na_1}{c_1} r^{2(n+1)} + \frac{2(n+2)\, a_2}{c_2} g^2 r^{2(n+1)} + \frac{2(n+1)\, a_3}{c_3} g r^{2n} \right\} - \frac{\nu + 3}{8} \omega^2 \rho^* r^2 + 2A + \left\{ K \frac{2b_2}{c_2 + d_2} (n - 2n^2 + 2)\, g^2 r^{2(n+1)} + 2Cn(1-2n)\, r^{2(n-1)} + 2D(n - 2n^2 + 1)\, r^{2n} \right\} \cos 2(n\theta + \alpha), \tag{3.71}$$

$$\sigma_\theta = K\left\{\frac{2n(2n-1)\,a_1}{c_1}\,r^{2(n-1)} + \frac{2(n+2)(2n+3)\,a_2}{c_2}\,g^2r^{2(n+1)}\right.$$

$$\left. + \frac{2(n+1)(2n+1)\,a_3}{c_3}\,gr^{2n}\right\} - \frac{1+3\nu}{8}\,\omega^2\rho^* r^2 + 2A$$

$$+ \left\{K\,\frac{2(n+2)(2n+3)\,b_2}{c_2+d_2}\,g^2r^{2(n+1)} + 2Cn(2n-1)\,r^{2(n-1)}\right.$$

$$\left. + 2D(n+1)(2n+1)\,r^{2n}\right\}\cos 2(n\theta+\alpha), \tag{3.72}$$

$$\tau_{r\theta} = 2n\left\{K\,\frac{(2n+3)\,b_2}{c_2+d_2}\,g^2r^{2(n+1)} + C(2n-1)\,r^{2(n-1)}\right.$$

$$\left. + D(2n+1)\,r^{2n}\right\}\sin 2(n\theta+\alpha). \tag{3.73}$$

We now apply the procedure of Galerkin to the remaining field Eq. (3.62). Upon substitution of Eqs. (3.68) and (3.69) into Eq. (3.62), one multiplies both members of the latter equation by the spatial part $r^n(1+gr^2)\cos(n\theta+\alpha)$ of Eq. (3.68) and integrates the result over the domain of the plate. A rather lengthy calculation yields the following nonlinear differential equation of the second order for the time function $\tau(t)$:

$$(d^2\tau/dt^2) + \alpha^*\tau + \beta^*\tau^3 = 0, \tag{3.74}$$

provided the loading function $q(r, \theta, t)$ is independent of the space coordinates (uniform loading). Equation (3.74) includes the following notation (with μ as coefficient depending on the prescribed boundary conditions):

$$\alpha^* = -\frac{\omega^2}{2s}\left(s_4 + s_1\,\frac{a^2}{8}\,g\mu\right),$$

$$\beta^* = -\frac{EA_0{}^2a^{2(n-2)}}{\rho^* s}\left(\frac{s_1 gA^*}{a^{2(n-2)}} + s_2C^* + s_3a^2D^* + \frac{1}{2}\,s_5\right),$$

$$A = A^*EA_0{}^2\tau^2 + \frac{\omega^2\rho^* a^2}{16}\,\mu,$$

$$C = C^*EA_0{}^2\tau^2,$$

$$D = D^*EA_0{}^2\tau^2,$$

$$s = \frac{1}{2(n+1)} + \frac{a^2}{n+2}\,g + \frac{a^4}{2(n+3)}\,g^2,$$

$$s_1 = 4(n+1)\left(\frac{1}{n+1} + \frac{a^2}{n+2}g\right),$$

$$s_2 = 2n^2\left((1-2n)\,a^2g + \frac{(n+1)(1-2n)}{2n+1}\,a^4g^2 - (n-1)\right),$$

$$s_3 = -\,n(n-1)(2n+1) + 2(1+n-2n^3)\,a^2g + (1+2n-n^2-2n^3)\,a^4g^2.$$

$$\begin{aligned} s_4 = &-\frac{n}{n+1}\left\{1 + \frac{n-1}{4}(1-\nu)\right\} \\ &+ \left\{\left(n+1-\frac{n^2}{2}\right)(1-\nu) - 4(n+1)\right\}\frac{a^2}{n+2}g \\ &+ \left\{\frac{(n+1)(4-n)}{4}(1-\nu) - (3n+4)\right\}\frac{a^4}{n+3}g^2, \end{aligned}$$

$$\begin{aligned} s_5 = {} & n(1-n)\{(n+1)(a_3/c_3) - 2(n-1)(a_1/c_1)\}\,a^2g \\ & + [2/(2n+1)]\{n(1-n^2)(n+2)(a_2/c_2) \\ & \qquad + n(1-n)(2n^2+3n+1)[b_2/(c_2+d_2)] \\ & \qquad + 2(n+1)(1+2n+n^2-n^3)(a_3/c_3) \\ & \qquad + n^2(1+n)(3-n)(a_1/c_1)\}\,a^4g^2 \\ & + (n+1)^{-1}\{2(n+2)(3n-n^3+2)(a_2/c_2) \\ & \qquad + 2(2+4n+n^2-3n^3-2n^4)[b_2/(c_2+d_2)] \\ & \qquad + (n+1)^2(2n-n^2+2)(a_3/c_3)\}\,a^6g^3 \\ & + [(n+1)/2(2n+3)]\{4(n+2)(4+n-n^2)(a_2/c_2) \\ & \qquad + [(n+2)(n-2n^2+2) + (2n+3)(4-3n^2)] \\ & \qquad \cdot [b_2/(c_2+d_2)]\} \\ & \cdot a^8g^4 - [2n^2(n-1)^2/(2n-1)](a_1/c_1). \end{aligned}$$

Let us suppose that $\alpha^*, \beta^* > 0$. Then the solution of the differential Eq. (3.74) is given by the Jacobian elliptic function

$$\tau(t) = \operatorname{cn}(\omega^* t, k),$$

provided the following normalized initial conditions for the time function are satisfied:

$$\tau(0) = 1, \qquad [d\tau/dt]_{t=0} = 0,$$

and the notation

$$\omega^* = (\alpha^* + \beta^*)^{1/2}, \qquad k^2 = \beta^*/[2(\alpha^* + \beta^*)],$$

is used (with k as modulus). Clearly, $\operatorname{cn}(\omega^* t, k)$ is a periodic function with the period $T^* = 4K/\omega^*$, where $K(k) = F(k, \pi/2)$ is the complete elliptic integral of the first kind. Since the modulus k and the parameter ω^* involve β^*, which is dependent on the amplitude A_0, it follows that the periodic time of the nonlinear vibrations depends on the amplitude, a fact which is well known.

The complete details for the free nonlinear vibrations having two nodal diameters and no nodal circles is developed in the cited paper. This mode, in which the nodal diameters are orthogonal, is most interesting since it corresponds, in the linear case, to the gravest mode of vibration.

Evensen [93] has made an interesting theoretical and experimental investigation of the nonlinear flexural vibration of rings. His work on the ring was motivated by certain discrepancies that he had noted (Evensen [93]) between his analogous results for a cylindrical shell and those previously obtained by Chu [90] and Nowinski [86]. These authors both theoretically concluded that the nonlinearity was of a hardening type, while Evensen's theoretical and experimental results indicate a softening type of nonlinearity (decrease of frequency with increase in vibration amplitude). Dowell [94] has removed Evensen's assumption of zero midplane strain, so we sketch his slightly more general analyses.

The equations of motion are taken to be

$$D\left[\frac{\partial^2}{\partial y^2} + \frac{1}{R^2}\right]\left[\frac{\partial^2 w}{\partial y^2} + \frac{w}{R^2}\right] + \frac{\sigma_y}{R} - \frac{\partial}{\partial y}\left[\sigma_y \frac{\partial w}{\partial y}\right] + \rho h \frac{\partial^2 w}{\partial t^2} = q, \tag{3.75}$$

and

$$\partial \sigma_y / \partial y = 0, \tag{3.76}$$

where

$$\sigma_y = Eh\epsilon_{yy}\,|_{z=0} = Eh\left[\frac{\partial v}{\partial y} + \frac{w}{R} + \frac{1}{2}\left(\frac{\partial w}{\partial y}\right)^2\right], \tag{3.77}$$

and

$$q = (F_n/\pi R)\cos(ny/R)\cos\omega t.$$

From Eq. (3.76) it follows that $\sigma_y = \sigma_y(t)$. The trial solution is selected as

$$w(y, t) = A_n(t)\cos(ny/R) + B_n(t)\sin(ny/R) + A_0(t), \tag{3.78}$$

with the consequence that

$$\sigma_y = Eh\{\tfrac{1}{4}(n/R)^2[A_n{}^2 + B_n{}^2] + A_0/R\}.$$

Galerkin's method is then employed to find the ordinary differential equations for A_0, A_n, and B_n.

Advani [95–97], using Nowinski's work as his reference, considers the set of coupled equations

$$\frac{\partial^2 w}{\partial r^2}\left\{\frac{1}{r}\frac{\partial \phi}{\partial r}+\frac{1}{r^2}\frac{\partial^2 \phi}{\partial \theta^2}\right\}+\left\{\frac{1}{r}\frac{\partial w}{\partial r}+\frac{1}{r^2}\frac{\partial^2 w}{\partial \theta^2}\right\}\frac{\partial^2 \phi}{\partial r^2}$$

$$-2\frac{\partial}{\partial r}\left(\frac{1}{r}\frac{\partial w}{\partial \theta}\right)\frac{\partial}{\partial r}\left(\frac{1}{r}\frac{\partial \phi}{\partial \theta}\right)-\frac{1}{2}\rho\omega^2 r^2 h\nabla^2 w$$

$$-\rho\omega^2 rh\frac{\partial w}{\partial r}=\rho h\frac{\partial^2 w}{\partial t^2},$$

and (3.79)

$$\nabla^4\phi = Eh\left\{-\left[\frac{1}{r}\frac{\partial w}{\partial r}+\frac{1}{r^2}\frac{\partial^2 w}{\partial \theta^2}\right]\frac{\partial^2 w}{\partial r^2}+\frac{1}{r^2}\left(\frac{\partial^2 w}{\partial r\,\partial \theta}\right)^2\right.$$

$$\left.-\frac{2}{r^3}\left(\frac{\partial^2 w}{\partial r\,\partial \theta}\right)\frac{\partial w}{\partial \theta}+\frac{1}{r^4}\left(\frac{\partial w}{\partial \theta}\right)^2\right\}+2(1-\nu)\,\rho\omega^2 h,$$

where ϕ is the stress function defined in Eqs. (3.61). By direct substitution, exact harmonic solutions to Eqs. (3.79) are found in the form

$$\begin{aligned} w &= Ar^2\sin 2(\theta\pm ct),\\ \phi &= Br^4\cos 4(\theta\pm ct)+[2hEA^2+(1-\nu)\,\rho\omega^2 h](r^4/32)-Cr^2, \end{aligned}\tag{3.80}$$

where the wave velocity c is related to the wave amplitude by the relation

$$c^2-\frac{5-\nu}{8}\omega^2=\frac{E}{4\rho}A^2-\frac{12B}{\rho h}.\tag{3.81}$$

Actually two nonsuperposable solutions are contained in Eq. (3.80). The wave governed by $\theta + ct$ is a backward wave, since it travels in the direction of decreasing θ. The wave governed by $\theta - ct$ is the forward wave. When waves "run backward" as fast as the membrane spins forward, stationary waves will occur. These may be obtained by setting $c = \omega$ in Eq. (3.81), whereupon

$$\omega^2=\frac{8}{3+\nu}\left(\frac{E}{4\rho}A^2-\frac{12B}{\rho h}\right).$$

Note that the linear theory of a spinning membrane with a free edge ($B = 0$) excludes the possibility of standing waves.

Lastly we mention the research of Woodall [98] on the large-amplitude oscillations of a thin elastic beam. For the case of a simply supported beam he obtains and compares solutions obtained by regular perturbation, Galerkin's method, and by finite differences. The regular perturbation will be described later in this chapter and the finite difference method in Chapter 4.

3.5 COMMENTS ON WRM THEORY

In Section 5.18 of Volume I we present some of the available convergence theorems and error bounds for WRM. Herein we add some additional references to specific, generally linear, problems with the remark that very little is known about the convergence of WRM for nonlinear problems *without a corresponding variational principle.*

Some 25 years elapsed between the introduction of Galerkin's method (1915) and study of the convergence of the procedure. Repman [99][†] was probably the first to prove convergence of solutions obtained by Galerkin's method though only for a Fredholm integral equation. Petrov [100] examined convergence of the Galerkin method for eigenvalues of the Orr–Sommerfeld equation of hydrodynamic stability theory (fourth-order ordinary differential equations). Keldysh [101] treated general ordinary differential equations and second-order elliptic partial differential equations. The proofs of Keldysh were simplified by Mikhlin [3], especially for the equations

$$-\sum_{i,j=1}^{m} \frac{\partial}{\partial x_i}\left(A_{ij}\frac{\partial u}{\partial x_j}\right) + \sum_{i=1}^{m} B_i \frac{\partial u}{\partial x_i} + cu = f. \tag{3.82}$$

The essential result is that the first derivatives of the Galerkin approximate solution converge in the mean to the first derivatives of the exact solution. Convergence proofs for certain eigenvalue problems in Taylor–Dean hydrodynamic stability have been developed by Sani [102] and Di Prima and Sani [103] who employ a generalized Galerkin method in their studies.

For problems of the unsteady state, Faedo [104] employs Galerkin's procedure on a hyperbolic equation thereby inspiring Green [105] to prove uniform convergence of the method when applied to the parabolic equation

$$u_{xx} - u_t - g(x, t)u = f(x, t). \tag{3.83}$$

Hopf [106] uses the Galerkin method to prove the existence of *weak* solutions to the time dependent Navier–Stokes equations and Il'in *et al.*

[†] All of these studies refer to linear problems unless otherwise stated.

[107] carry out a similar analysis for the unsteady-state transport equation

$$\sum_{i,j=1}^{m} \frac{\partial}{\partial x_i}\left(A_{ij}\frac{\partial u}{\partial x_j}\right) + \sum_{i=1}^{m} B_i \frac{\partial u}{\partial x_i} + Cu - \frac{\partial u}{\partial t} = F(x, t), \tag{3.84}$$

with a known velocity field. Meister [108] has applied the Galerkin method to the Taylor stability problem with time-dependent disturbances. Finlayson and Scriven [109] have generated improvable, pointwise upper and lower bounds (error bounds!) for the solution to Eq. (3.84).

Convergence theorems for the collocation method are given by Kadner [110], while the method of moments is treated by Kravchuk [111]. Considerably more information exists for boundary-value problems treated by the method of least squares. Mikhlin [3] proves conditions which insure that the method of least squares gives a sequence of approximate solutions convergent in the mean to the exact solution and the mean-square error can be determined. He also notes that the least squares method converges more slowly than the Ritz (variational) method (when the latter is applicable).

For nonlinear problems, the state of the theory both in convergence criteria and error bounds continues to be extremely limited. Indeed, for nonlinear problems without a corresponding variational principle we find only a handful of results. In particular Krasnosel'ski [112] presents theorems for the Galerkin method applied to nonlinear integral equations. Glansdorff [113] announces a theorem on the convergence of the so-called local potential method (which is identical in application to the Galerkin procedure) which he applies to the steady heat conduction equation with temperature dependent thermal conductivity. For a discussion of the local potential method as a disguised Galerkin method, see Finlayson and Scriven [114]. Whenever the Rayleigh–Ritz (variational) and Galerkin methods coincide, the convergence proofs for the Rayleigh–Ritz method imply convergence of the Galerkin method also. Thus the Galerkin-convergence proofs given by Kantorovich and Krylov [27] apply only to specific problems with a maximum or minimum principle. The results on linear problems, discussed previously, apply whether or not a corresponding variational principle exists.†

When the same trial functions are used, and the Ritz‡ or Rayleigh–Ritz

† Claims have been made that completeness of the set of approximating functions is sufficient for convergence. The proofs given by Mikhlin [3] show clearly that this is not enough!

‡ The term "Ritz method" is used when discussing stationary principles, while "Rayleigh–Ritz" is used when either a minimum or maximum principle is intended.

method is applicable, the resulting calculation is identical to Galerkin's calculation. This equivalence still persists when the trial functions do not satisfy the natural boundary conditions (see, e.g., Mikhlin [3], Bolotin [7], or Finlayson and Scriven [1]) which they need not do in the Rayleigh–Ritz method. The boundary residual is either added to or subtracted from the equation residual, and the calculations are equivalent to those of the Ritz or Rayleigh–Ritz method.

The important difference between the Rayleigh–Ritz and Galerkin method is that in the former some functional (possibly an energy expression for an eigenvalue) is being minimized or maximized, thus the approximate values of the functional represent either upper or lower bounds. In the Galerkin method this information is missing. Exactly the same values are obtained, but we do not know that these are upper or lower bounds. On the other hand, when the functional is of no physical significance the Galerkin method is usually preferred because of its generality. Similar remarks apply to the choice of the Ritz method or the Galerkin procedure when merely a stationary principle exists. For a large class of nonlinear operators Kodnar [115] has demonstrated the equivalence of the Galerkin and Ritz methods. Leipholz [116] has proven the convergence of the Galerkin procedure for quasi-linear boundary-value problems.

3.6 MAXIMUM PRINCIPLES—ORDINARY DIFFERENTIAL EQUATIONS

One of the most useful tools available for the development of approximate solutions is the maximum principle, a subject so well developed that Protter and Weinberger [117] devote an entire book to it. This principle is a generalization of the elementary fact of calculus that any continuous twice differentiable function $f(x)$, which satisfies the *differential inequality* $f'' > 0$ on a closed interval $a \leqslant x \leqslant b$, achieves its maximum value at one of the endpoints of the interval. Solutions of $f'' > 0$ are said to satisfy a *maximum principle.* More generally, functions which satisfy a differential inequality in a domain D and, because of it, achieve their maxima on the boundary of D are said to possess a maximum principle.

The maximum principle permits us to obtain information concerning solutions of differential equations without any explicit knowledge of the solutions themselves. They are also useful in the determination of the bounds for the errors in numerical solutions (see, e.g., Ames [118]). In physical applications, there is usually a natural interpretation of the maximum principle with the consequence that its use helps us apply physical intuition to mathematical models. We begin our discussion with some results for nonlinear one-dimensional operators (see Protter and Weinberger [117] for proofs and a bibliography).

Let $u(x)$ be a solution of the nonlinear equation

$$u'' + H(x, u, u') = 0, \tag{3.85}$$

on $a \leqslant x \leqslant b$. The functions $H(x, y, z)$, $H_y(x, y, z)$, and $H_z(x, y, z)$ are assumed to be continuous functions of x, y, z throughout their domains of definition, and for each x and z

$$H(x, y_1, z) \leqslant H(x, y_2, z), \qquad \text{for} \quad y_1 \geqslant y_2, \tag{3.86}$$

or, equivalently

$$\partial H/\partial y \leqslant 0.$$

Suppose $w(x)$ satisfies the differential inequality

$$w'' + H(x, w, w') \geqslant 0, \tag{3.87}$$

in $a < x < b$ and consider

$$v = w - u.$$

Subtracting Eq. (3.85) from inequality (3.87), we conclude that

$$v'' + H(x, w, w') - H(x, u, u') \geqslant 0.$$

Applying the mean value theorem to the previous equation gives

$$v'' + (\partial H/\partial z)\, v' + (\partial H/\partial y) v \geqslant 0,$$

where H_y and H_z are evaluated at $(x, u + \alpha(w - u), u' + \alpha(w' - u'))$, $0 < \alpha < 1$. Thus v satisfies a linear inequality and a (linear) maximum principle applies.

Theorem 3.6-1. *Suppose* $H'' + H(x, w, w') \geqslant u'' + H(x, u, u')$ *for* $a < x < b$, *where* H, H_y, H_z *are continuous and* $H_y \leqslant 0$. *If* $w(x) - u(x)$ *attains a nonnegative maximum* M *in* $a < x < b$, *then* $w(x) - u(x) \equiv M$.

From these maximum principles, an approximation theorem for boundary- and initial-value problems can be developed.

Theorem 3.6-2. *Let $u(x)$ be a solution of the boundary-value problem*

$$u'' + H(x, u, u') = 0, \qquad a < x < b, \tag{3.88}$$

$$\begin{aligned} -u'(a)\cos\theta + u(a)\sin\theta &= \gamma_1, \\ u'(b)\cos\phi + u(b)\sin\phi &= \gamma_2, \end{aligned} \tag{3.89}$$

where $0 \leqslant \theta \leqslant \pi/2$, $0 \leqslant \phi \leqslant \pi/2$, and θ and ϕ are not both zero. Suppose H, H_y, and H_z are continuous and $H_y \leqslant 0$. If $z_1(x)$ satisfies

$$\begin{aligned} z_1'' + H(x, z_1, z_1') &\leqslant 0, \\ -z_1'(a)\cos\theta + z_1(a)\sin\theta &\geqslant \gamma_1, \\ z_1'(b)\cos\phi + z_1(b)\sin\phi &\geqslant \gamma_2, \end{aligned} \tag{3.90}$$

and if $z_2(x)$ satisfies

$$\begin{aligned} z_2'' + H(x, z_2, z_2') &\geqslant 0, \\ -z_2'(a)\cos\theta + z_2(a)\sin\theta &\leqslant \gamma_1, \\ z_2'(b)\cos\phi + z_2(b)\sin\phi &\leqslant \gamma_2, \end{aligned} \tag{3.91}$$

then the ***upper and lower bounds***

$$z_2(x) \leqslant u(x) \leqslant z_1(x)$$

are valid!

This theorem not only shows how to calculate upper and lower bounds, but it also implies that a solution of Eq. (3.88) satisfying boundary conditions Eq. (3.89) must be unique. If u and $\bar{u}$ are solutions, we can let $z_1 = z_2 = \bar{u}$ to find $u = \bar{u}$.

As a typical application of Theorem 3.6-2, let us consider the development of bounds for the solution $u(x)$ to $u'' - u^3 = 0$, $0 < x < 1$, with boundary conditions $u(0) = 0$, $u(1) = 1$. With $z_1 = x$, Eqs. (3.90) become

$$z_1'' - z_1^3 = -x^3 < 0,$$

$$z_1(0) \geqslant 0 \quad (\text{in fact } z_1 = 0), \quad \text{and} \quad z_1(1) \geqslant 1 \quad [\text{in fact } z_1(1) = 1].$$

With $z_2 = x^\alpha$, Eqs. (3.91) become $\alpha(\alpha - 1)\, x^{\alpha-2} - x^{3\alpha} \geqslant 0$, $z_2(0) = 0$ (if $\alpha > 0$) and $z_2(1) \leqslant 1$, respectively. All three requirements are satisfied if $\alpha = (1 + \sqrt{5})/2$. Thus

$$x^{(1+\sqrt{5})/2} \leqslant u(x) \leqslant x.$$

Theorem 3.6-3. *Suppose $u(x)$ satisfies*

$$u'' + H(x, u, u') = 0, \tag{3.92}$$

with the initial conditions

$$u(a) = \gamma_1, \qquad u'(a) = \gamma_2,$$

and H, H_y, H_z *are continuous and* $H_y \leqslant 0$. *If* $z_1(x)$ *satisfies*

$$z_1'' + H(x, z_1, z_1') \geqslant 0, \qquad z_1(a) \geqslant \gamma_1, \quad z_1'(a) \geqslant \gamma_2,$$

and $z_2(x)$ *satisfies*

$$z_2'' + H(x, z_2, z_2') \leqslant 0, \qquad z_2(a) \leqslant \gamma_1, \quad z_2'(a) \leqslant \gamma_2,$$

then the upper and lower bounds

$$z_2(x) + \gamma_2 - z_2(a) \leqslant u(x) \leqslant z_1(x) + \gamma_1 - z_1(a),$$
$$z_2'(x) \leqslant u'(x) \leqslant z_1'(x),$$

are valid.

Effective utilization of these theorems requires the construction of solutions z_1 and z_2 satisfying the requisite inequalities. The point here is that, generally speaking, there are many such functions. Naturally, we would also like to have available some way to ascertain the "best possible" approximation.

3.7 MAXIMUM PRINCIPLES—PARTIAL DIFFERENTIAL EQUATIONS

Section 6.5 of Volume I discusses the *maximum operation* of Bellman and Kalaba. That analysis provides a systematic way of iteratively calculating and improving upper and lower bounds in theorems obtained from maximum principles for partial differential equations. Theorems providing solution bounds are of considerable importance. Consequently, we present typical results for elliptic and parabolic equations.

Theorem 3.7-1. (*Approximations for elliptic equations.*) *Let* $u(x, y)$ *be a solution of the elliptic*[†] *equation*

$$F(x, y, u, u_x, u_y, u_{xx}, u_{xy}, u_{yy}) = f(x, y), \tag{3.93}$$

[†] We say Eq. (3.93) is *elliptic with respect to a particular function u at a point* (x, y) if for all pairs of real numbers (ξ, η) with $\xi^2 + \eta^2 > 0$, we have

$$\frac{\partial F}{\partial r}\xi^2 + \frac{\partial F}{\partial s}\xi\eta + \frac{\partial F}{\partial t}\eta^2 > 0, \qquad \text{when } p = u_x, \; q = u_y, \; r = u_{xx}, \; s = u_{xy}, \; t = u_{yy},$$

are inserted in F. Equation (3.93) is elliptic in D if it is elliptic at each point of D. A nonlinear equation can be elliptic for some functions u but not for others.

in D and $u = g(x, y)$ on the boundary of D. Let v and V satisfy the inequalities

$$F(x, y, V, V_x, V_y, V_{xx}, V_{xy}, V_{yy}) \leqslant f(x, y) \leqslant F(x, y, v, v_x, v_y, v_{xx}, v_{xy}, v_{yy}), \tag{3.94}$$

in D and

$$v(x, y) \leqslant g(x, y) \leqslant V(x, y), \tag{3.95}$$

on the boundary of D. For each constant θ, $0 \leqslant \theta \leqslant 1$, we assume F is elliptic with respect to $u + \theta(v - u)$ and $u + \theta(V - u)$ in D, and $\partial F/\partial u \leqslant 0$ in D. Then

$$v(x, y) \leqslant u(x, y) \leqslant V(x, y) \qquad \text{in} \quad D.$$

The difficulty in applying this result lies in trying to establish the ellipticity of F within the required class of functions (convexity relation) $\theta v + (1 - \theta)u$, where $\theta = \theta(x, y)$, since u is generally unknown. Nevertheless, some situations of physical significance exist where this difficulty can be resolved.

First let us examine a special case of the Mongé–Amperé equation

$$u_{xx}u_{yy} - u_{xy}^2 = f(x, y). \tag{3.96}$$

Various forms of this equation occur in gas dynamics (Volume I, page 94 and Giese [119]), magnetohydrodynamics (Gunderson [120]), and in wave propagation (Ames and Jones [121]). In abbreviated notation, Eq. (3.96) is $F = rt - s^2 = f(x, y)$, whereupon the ellipticity condition becomes $t\xi^2 - 2s\xi\eta + r\eta^2 > 0$. Thus Eq. (3.96), or the equation obtained from it by multiplying it by -1, is elliptic whenever $rt - s^2 > 0$. If $f(x, y) > 0$ in D, then Eq. (3.96) is elliptic for all solutions in D.

To establish the ellipticity of Eq. (3.96) with respect to a whole family of solutions $u + \theta(v - u)$ is often difficult. For example, the functions

$$V = \tfrac{1}{2} + \tfrac{1}{2}(x^2 + y^2), \qquad v = \tfrac{3}{2} - \tfrac{1}{2}(x^2 + y^2),$$

both satisfy

$$\frac{\partial^2 u}{\partial x^2}\frac{\partial^2 u}{\partial y^2} - \left(\frac{\partial^2 u}{\partial x\, \partial y}\right)^2 = 1$$

in D: $x^2 + y^2 < 1$ and *both* take the boundary value 1 on the boundary of D. Consequently the Dirichlet problem for the Mongé–Amperé equation does not have a unique† solution. Although the equation is elliptic with respect to v and V, it is not elliptic with respect to $\frac{1}{2}(v + V)$, *so the conditions of Theorem* 3.7-1 *are not met*!

As a second example we suppose S is a simple closed curve in space which intersects each line in the z-direction at most once and whose projection on the x, y-plane is convex. Then it is known that there exists a surface of minimum area, which spans the curve S, whose equation $z = u(x, y)$ satisfies the second-order quasi-linear equation

$$(1 + u_y^2)\, u_{xx} - 2u_x u_y u_{xy} + (1 + u_x^2)\, u_{yy} = 0. \tag{3.97}$$

The minimal surface equation has the form

$$F = (1 + q^2)r - 2pqs + (1 + p^2)t = 0,$$

whereupon the condition for ellipticity becomes

$$\begin{aligned} F_r\xi^2 + F_s\xi\eta + F_t\eta^2 &= (1 + q^2)\,\xi^2 - 2pq\xi\eta + (1 + p^2)\,\eta^2 \\ &= \xi^2 + \eta^2 + (q\xi - p\eta)^2 > 0. \end{aligned}$$

Thus Eq. (3.97) is always elliptic.

Clearly $u \equiv$ const satisfies Eq. (3.97) so that Theorem 3.7-1 can be applied by selecting v and V as constants. Noting that $\partial F/\partial u = 0$, we conclude that any minimal surface must take its maximum (and minimum) on the boundary. For additional properties of the minimal surface equation, obtainable by means of the maximum principle, see Bernstein [123], Finn [124], Nitsche [125], Serrin [126], and Bers [127].

A number of studies in fluid mechanics have employed the maximum principle with considerable success. Among these, we find papers involving the comparison of two flows and the use of the maximum principle at infinity by Bers [128], Finn and Gilbarg [129, 130], Gilbarg [131–133], Gilbarg and Shiffman [134], Lavrentiev [135], and Serrin [136–139]. Adams [140] has applied differential inequalities in obtaining estimates on thermal pollution for the Navier–Stokes and energy equations. Survey articles include Ladyzhenskaya and Ural'tzeva [141], and Landis [142, 143].

Maximum principles and approximation theorems for parabolic equations are given by Protter and Weinberger [117] and Friedman [144]. We give an approximation theorem here.

† Courant and Hilbert [122] show that for the general Mongé-Amperé equation $Rr + Ss + Tt + U(rt - s^2) = V$, the Dirichlet problem has at most two solutions when the equation is elliptic.

Let $\bar{x} = (x_1, ..., x_n)$, $\bar{p} = (p_1, ..., p_n)$, where $p_i = \partial u/\partial x_i$, and $\bar{R} = [r_{ij}]$, $i, j = 1, 2, ..., n$, where $r_{ij} = \partial^2 u/\partial x_i \, \partial x_j$. Let $F(\bar{x}, t, u, \bar{p}, \bar{R})$ be a continuously differentiable function of its $n^2 + 2n + 2$ variables.
F is *elliptic with respect to a function u at a given point* $(\bar{x}, t)$ if, for all real vectors $\bar{\xi} = (\xi_1, ..., \xi_n)$, we have

$$\sum_{i,j=1}^{n} \frac{\partial F}{\partial r_{ij}} \xi_i \xi_j > 0 \qquad \text{for} \quad \bar{\xi} \neq 0. \tag{3.98}$$

F is *elliptic in a domain D in* $(\bar{x}, t)$ space if it is elliptic at each point of D. The nonlinear operator

$$L[u] = F\left[\bar{x}, t, u, \frac{\partial u}{\partial x_i}, \frac{\partial^2 u}{\partial x_i \, \partial x_j}\right] - \frac{\partial u}{\partial t}, \tag{3.99}$$

is said to be *parabolic* whenever F is elliptic. Here we have used the notation $F[\bar{x}, t, u, p_i, r_{ij}]$ to denote F with p_i and r_{ij} representing generic arguments of F.

Theorem 3.7-2. (*Approximation for parabolic equations.*) *Let D be a bounded domain in n-dimensional space and $E = Dx(0, T]$.*[†] *Suppose $u(\bar{x}, t)$ is a solution of $L[u] = f(\bar{x}, t)$ in E, with L given by Eq. (3.99), satisfying the initial condition $u(\bar{x}, 0) = g_1(\bar{x})$ in D and the boundary conditions $u(\bar{x}, t) = g_2(\bar{x}, t)$ on $\partial Dx(0, T)$. Let v and V satisfy the inequalities*

$$L[V] \leqslant f(\bar{x}, t) \leqslant L[v] \qquad \text{in} \quad E,$$

where L is parabolic with respect to the functions $\theta u + (1 - \theta)v$ and $\theta u + (1 - \theta)V$ for $0 \leqslant \theta \leqslant 1$. If

$$v(\bar{x}, 0) \leqslant g_1(\bar{x}) \leqslant V(\bar{x}, 0) \qquad \text{in} \quad D,$$

$$v \leqslant g_2 \leqslant V \qquad \text{on} \quad \partial Dx(0, T),$$

then

$$v(\bar{x}, t) \leqslant u(\bar{x}, t) \leqslant V(\bar{x}, t) \qquad \text{in} \quad E.$$

As an example of the application of Theorem 3.7-2, we examine the equation of diffusion in a homogeneous medium,

$$\frac{\partial}{\partial x}\left[k(u) \frac{\partial u}{\partial x}\right] - \frac{\partial u}{\partial t} = 0, \tag{3.100}$$

† Here, $(0, T]$ represents the interval $0 < t \leqslant T$ and $(0, T)$ the open interval $0 < t < T$, and ∂D means the boundary of D.

with positive diffusion coefficient $k(u)$ and $k'(u)$ is bounded. Equation (3.100) is parabolic for all functions u. Any constant satisfies Eq. (3.100). Consequently we can apply Theorem 3.7-2 and conclude that for any solution u the maximum and minimum values must occur either at the initial time or on the boundary.

The maximum principle has been employed to study the boundary-layer flow of a viscous fluid by Nickel [145], Oleinik [146], and Velte [147]. Aronsson [148] treats the parabolic equation

$$(u_x)^2 u_{xx} + 2u_x u_y u_{xy} + (u_y)^2 u_{yy} = 0$$

in some detail. Systems of nonlinear parabolic equations that are coupled only in undifferentiated terms[†] arise in the study of simultaneous diffusion–reaction problems. Maximum principles for such systems have been given by Szarski [149–151], Mlak [152], Besala [153, 154], and McNabb [155]. The paper by McNabb is for multicomponent diffusion systems. Solutions of systems which are coupled in the first derivatives of the unknown functions have been treated by Juberg [156].

In addition to the works of Friedman [144] and Szarski [151] differential and integral inequalities are treated by Walter [157] and Lakshmikantham and Leela [158]. Further application of these results will undoubtedly enhance our understanding of nonlinear equations.

3.8 QUASI LINEARIZATION

At least one very useful procedure, that of quasi linearization, has sprung from maximum principle—differential inequality theories. Introduced by the Bellman and Kalaba school, it is nicely summarized in an introductory book by Bellman and Kalaba [159]. Lee [160] presents a number of interesting applications to a wide variety of engineering problems. Some background material is given in Volume I under the heading "The Maximum Operation." Here, we shall confine our attention to several examples.

Basically the method of quasi linearization leads to a scheme of successive approximations which are quadratically convergent and require moderate amounts of computation at each stage. In some cases the method coincides with the Newton–Raphson–Kantorovich procedure in function space.

---

† Systems that are coupled only in terms which are not differentiated are called *weakly coupled*.

Consider the ordinary differential equation

$$u'' = f(x, u, u'), \tag{3.101}$$

together with the nonlinear boundary conditions

$$g_1[u(0), u'(0)] = 0, \qquad g_2[u(b), u'(b)] = 0, \tag{3.102}$$

or, even more generally,

$$g_i[u(0), u'(0), u(b), u'(b)] = 0, \qquad i = 1, 2.$$

We now apply quasi linearization to both the equation and the boundary conditions. Thus, with some initial approximation $u_0(x)$ we generate the sequence $\{u_n(x)\}$, $n = 1, 2,\ldots$ by means of the equation[†]

$$\begin{aligned}(u_{n+1})'' = f(x, u_n, u_n') &+ (u_{n+1} - u_n)(\partial f/\partial u)(x, u_n, u_n') \\ &+ (u'_{n+1} - u_n')(\partial f/\partial u')(x, u_n, u_n'),\end{aligned} \tag{3.103}$$

with the (linearized) boundary conditions [from Eq. (3.102)]

$$\begin{aligned}&[u_{n+1}(0) - u_n(0)](\partial g_1/\partial u)(u_n(0), u_n'(0)) \\ &\quad + [u'_{n+1}(0) - u_n'(0)](\partial g_1/\partial u')(u_n(0), u_n'(0)) = 0,\end{aligned}$$

and a similar equation from g_2 at $x = b$. Existence, uniqueness, and quadratic convergence for sufficiently small b can be established by a maximum operation, after conversion of the equation to integral form.

As a second example, consider the parabolic equation

$$u_t = u_{xx} + g(u, u_x), \tag{3.104}$$

with initial condition

$$u(x, 0) = h(x), \qquad 0 < x < 1,$$

and boundary conditions

$$u(0, t) = u(1, t) = 0, \qquad x > 0.$$

[†] If a Picard linearization is employed, instead of Eq. (3.103), we find $(u_{n+1})'' = f[x, u_n, u_n']$, which is generally only linearly convergent.

With the initial approximation $u_0(x, t)$, we apply quasi linearization to generate the sequence $\{u_n(x, t)\}$ by means of the equation

$$\begin{aligned}(u_{n+1})_t = (u_{n+1})_{xx} &+ g[u_n, (u_n)_x] \\ &+ [u_{n+1} - u_n](\partial g/\partial u)[u_n, (u_n)_x] \\ &+ [(u_{n+1})_x - (u_n)_x](\partial g/\partial u_x)[u_n, (u_n)_x]. \end{aligned} \tag{3.105}$$

At each step, the linear equation is solved by some numerical scheme such as that of Crank–Nicolson (cf., e.g., Ames [118]).

3.9 REGULAR PERTURBATION AND IRREGULAR DOMAINS

Poincaré's method of perturbation, herein called regular perturbation, was initiated over 70 years ago. Prior to 1950, its use was generally limited to the conventional form set by its founder. While only a few results of perturbation theory are rigorously established, some excellent defenses of its use have been made. Among others we find the Gibbs Lecture of Friedrichs [161]. In the years following 1950, the need for solution of various difficult problems became so great that the perturbation method not only became a common analytical tool but original modifications of the method were created. In this and the next several sections we shall discuss some of these.

The classical procedure modifies only the boundary-value problem itself. However, we can perturb the *boundary* of the integration domain under consideration, the material descriptors, the independent variables, and even the number of space dimensions (e.g., van Dyke [162]). Nowinski [163] has given several applications of a form perturbation technique (modification of the bounding surface) to elastic problems.

Consider a cylindrical bar of arbitrary simply connected cross section with one end fixed in the x,y-plane and the other acted upon by a twisting couple M. The stress function $\phi = \phi(x, y)$ (cf., e.g., Sokolnikoff [164]), for this torsion problem, satisfies Poisson's equation

$$\nabla^2\phi = -2 \tag{3.106}$$

inside the bar and

$$\phi = \text{const}$$

(taken as zero here) on the boundary. If α is the twist per unit length and μ is the Lame constant, the stress components are given by

$$\tau_{zx} = \mu\alpha(\partial\phi/\partial y), \qquad \tau_{zy} = -\mu\alpha(\partial\phi/\partial x). \tag{3.107}$$

The basic, or generating, problem is chosen to be the torsion of a circular cylindrical bar. Deviations from the circle of radius a are conveniently represented by a Fourier series of period 2π,

$$r(\theta) = a + \epsilon \left[(a_0/2) + \sum a_n \cos n\theta + b_n \sin n\theta\right], \tag{3.108}$$

where r, θ are the polar coordinates of a point on the contour and ϵ is the perturbation parameter. Such a representation permits a wide variety of contours, even if one chooses only a single term in Eq. (3.108). In particular

$$r = a(1 + \epsilon \cos 2\theta) \tag{3.109}$$

is the equation for an ellipse-like curve with semiaxes $a(1 - \epsilon)$ and $a(1 + \epsilon)$. We shall describe Nowinski's [163] solution for this boundary curve.

The solution of the torsion problem for a bar of circular cross section is

$$\phi_0(r) = (a^2/2)[1 - (r^2/a^2)]. \tag{3.110}$$

We tentatively select a perturbation expansion, for ϕ in powers of ϵ, in the form

$$\phi(r, \theta; \epsilon) = \phi_0(r) + \epsilon\phi_1(r, \theta) + \cdots . \tag{3.111}$$

Setting this into the full problem, Eqs. (3.106), and with $\phi = 0$ on the boundary, generates a sequence of boundary-value problems. The first two of these are

$$\nabla^2\phi_0 = -2, \tag{3.112}$$

$$\nabla^2\phi_1 = 0, \tag{3.113}$$

with the coupled boundary condition

$$\phi_0(r) + \epsilon\phi_1(r, \theta) + \cdots = 0, \qquad r = a(1 + \epsilon \cos 2\theta). \tag{3.114}$$

Since ϵ appears both explicitly and implicitly in Eq. (3.114), an expansion is necessary to display ϵ explicitly. To this end we assume the functions ϕ_j $(j = 0, 1, 2,...)$ are analytic in r. Upon expanding each term of Eq. (3.114) in a Taylor series about $r = a$, we have

$$\phi_0(a, \theta) + \epsilon a \cos 2\theta(\partial\theta_0/\partial r) + \epsilon\phi_1(a, \theta) + O(\epsilon^2) = 0. \tag{3.115}$$

The boundary condition for the fundamental problem, Eq. (3.112), becomes

$$\phi_0(a, \theta) = 0,$$

which is the only term $O(1)$ in Eq. (3.115). Next we examine all terms which are $O(\epsilon)$ and conclude that the boundary condition for the first approximation is

$$\phi_1(a, \theta) = a^2 \cos 2\theta. \tag{3.116}$$

A solution of Eq. (3.113) satisfying the boundary condition, Eq. (3.116), and remaining bounded at the center of the cross section is

$$\phi_1(r, \theta) = r^2 \cos 2\theta.$$

Consequently, the stress function is

$$\phi = (a^2/2)[1 - (r^2/a^2)] + \epsilon r^2 \cos 2\theta + O(\epsilon^2), \tag{3.117}$$

with

$$\tau_{zx} = -\mu\alpha(1 + 2\epsilon)y,$$

a result that agrees to $O(\epsilon)$ with the classical Saint-Venant solution of the torsion problem for a truly elliptical cross section (cf., e.g., Sokolnikoff [164]).

Nowinski [163] extends his analysis to bars with varying circular cross section along the symmetry axis of the bar. Further generalizations could involve several boundary-shape parameters, parameters arising from nonlinearities in the equation, and parameters describing variability in material descriptors (small changes from homogeniety, etc.).

3.10 CLASSICAL REGULAR PERTURBATION

The classical regular perturbation process is quite straightforward. We shall describe it for the periodic oscillation of systems governed by nonlinear equations.

Let the state of the vibrating system be governed by the equation†

$$[L + \epsilon N + k\, \partial^2/(\partial t^*)^2]\, \bar{u}(\bar{x}, t^*) = 0, \tag{3.118}$$

where ϵ and k are parameters resulting from the material constitutive relations, L is a second-order linear operator, N is a nonlinear operator, $\bar{u}$ is a vector function of the vector $\bar{x}$ describing the state of the vibration

† While our presentation is restricted to this equation, the method has been formally developed by Keller and Ting [165] for the general nonlinear operator equation $F[u] = 0$ which maps u, with values in a Hilbert space H, into a function of the same kind. The Keller–Ting method will be presented in Section 3.11.

at any time t^*. Both L and N depend on the space variables x_i. The operator of Eq. (3.118) is assumed to map the vector $\bar{u}$ into another vector belonging to the same function space as $\bar{u}$. In this space (e.g., a Hilbert space) there is an appropriately defined inner product so that the notion of orthogonality exists. We shall denote by $\langle u, v \rangle$ the *inner product* in the function space.

The problem we consider is that of finding a periodic solution of Eq. (3.118) satisfying the periodicity condition

$$\bar{u}(\bar{x}, t^* + 2\pi/\omega) = \bar{u}(\bar{x}, t^*), \tag{3.119}$$

where ω is the angular frequency of the solution. Before attempting to solve this problem, we stretch the time scale so that the new time variable is

$$t = \omega t^*.$$

Because of this transformation, Eq. (3.118) becomes

$$[L + \epsilon N + k\omega^2 (\partial^2/\partial t^2)]\, \bar{u}(\bar{x}, t) = 0, \tag{3.120}$$

and we search for a solution $\bar{u}(\bar{x}, t)$ which is periodic of period 2π, i.e., $\bar{u}(\bar{x}, t + 2\pi) = \bar{u}(\bar{x}, t)$.

While Keller and Ting [165] present a general development of this method, we shall restrict our discussion in this section to a modification due to Wang [166, 167] for Eq. (3.120).

The basic procedure involves expanding the unknown frequency (Linstedt's method of omitting secular terms) as well as the form of the vibration in powers of the parameter. Once this is done the method leads to a sequence of linear problems. The first of these is homogeneous (the usual linearized problem) while the rest are inhomogeneous. The solvability† conditions for the latter yield the coefficients in the expansion of the frequency in powers of the parameter. The solutions yield the terms in the expansion of the form of the vibration.

Returning now to our problem, we see that the term $[\epsilon N]\bar{u}$ in Eq. (3.120) is negligible if both ϵ and the amplitude of $\bar{u}$ are small compared to one. We suppose that the solutions $\bar{u}$ and ω possess a Taylor series in ϵ about $\epsilon = 0$. Thus‡

$$\omega = \omega_0 + \epsilon\omega_1 + \epsilon^2\omega_2 + \cdots, \qquad \bar{u} = \bar{u}_0 + \epsilon\bar{u}_1 + \epsilon^2 u_2 + \cdots. \tag{3.121}$$

† Lack of secular terms which are either not periodic or become arbitrarily large as $t \to \infty$.

‡ Since ω^2 often appears, the expansion for ω in Eqs. (3.121) can be replaced by one for ω^2.

where ω_i and $\bar{u}_i$ are, respectively, proportional to the ith derivative of ω and $\bar{u}$ with respect to ϵ, evaluated at $\epsilon = 0$.

Setting Eqs. (3.121) into Eq. (3.120) leads, for each power of ϵ, to a sequence of linear equations

$$[L + k\omega_0^2(\partial^2/\partial t^2)]\,\bar{u}_0 = 0, \tag{3.122}$$

$$[L + k\omega_0^2(\partial^2/\partial t^2)]\,\bar{u}_1 = -2k\omega_0\omega_1\bar{u}_{0tt} - [N]\,u_0\,, \tag{3.123}$$

$$[L + k\omega_0^2(\partial^2/\partial t^2)]\,\bar{u}_2 = -2k\{(\omega_0\omega_2 + \tfrac{1}{2}\omega_1^2)\,\bar{u}_{0tt} + \omega_0\omega_1\bar{u}_{1tt}\} - [N]\,\bar{u}_1\,. \tag{3.124}$$

Employing the well-known linear theory for Eq. (3.122), we suppose the eigenvalues and eigenvectors determined from that problem are $\{\lambda_j\}$ and $\{\phi_j\}$, respectively. Then ω_0 will be one of those eigenvalues† and $\bar{u}_0$ can be expressed as a linear combination of the functions $\{\phi_j\}$ with the coefficients determined from the initial conditions by employing the orthogonality conditions. Now Eq. (3.123) will have a periodic solution for $\bar{u}_1$ *if and only if the left-hand side of Eq.* (3.123) *is orthogonal to all solutions of Eq.* (3.122). This requirement eliminates the possibility of secular terms. Taking the inner product of Eq. (3.123) with $\bar{u}_0$, it follows that

$$0 = \langle [L + k\omega_0^2(\partial^2/\partial t^2)]\,\bar{u}_1\,, \bar{u}_0\rangle = -2k\omega_0\omega_1\bar{u}\langle_{0tt}\,, \bar{u}_0\rangle - \langle [N]\,\bar{u}_0\,, \bar{u}_0\rangle,$$

or

$$\omega_1 = \frac{-\langle [N]\,\bar{u}_0\,, \bar{u}_0\rangle}{2k\omega_0\langle \bar{u}_{0tt}\,, \bar{u}_0\rangle}\,. \tag{3.125}$$

Having determined ω_1, Eq. (3.123) can now be solved for $\bar{u}_1$. Then by taking the inner product of Eq. (3.124) with $\bar{u}_0$, ω_2 can be determined as

$$\omega_2 = \frac{-2k\{\omega_0\omega_1\langle \bar{u}_{1tt}\,, \bar{u}_0\rangle + \tfrac{1}{2}\omega_1^2\langle \bar{u}_{0tt}\,, \bar{u}_0\rangle\} - \langle [N]\,\bar{u}_1\,, \bar{u}_0\rangle}{2k\omega_0\langle \bar{u}_{0tt}\,, u_0\rangle}\,.$$

$\bar{u}_2$ is then found by solving Eq. (3.124). Continuing this process will yield as many terms as the analyst desires, but much labor is involved. If $|\,\epsilon\,| < 1$ and $\bar{u}$ has an amplitude of moderate size, then it is reasonable to expect that the first few terms of the series will suffice for this approximate solution.

As an application consider the simple elastic shear vibration of an incompressible hyperelastic slab of thickness h. If $W(\mathrm{I}, \mathrm{II})$ is the strain

† Here, we complete the discussion for simple eigenvalues only. The extension to multiple eigenvalues is found in the work of Keller and Ting [165].

energy function, with I and II the first two invariants of deformation, it follows that

$$\mathrm{I} = \mathrm{II} = 3 + (\partial U/\partial y)^2,$$

for simple shear through a distance $U(y, t)$ in the direction of the horizontal (z-axis). The governing equation of motion for the warping function $U(y, t)$ is

$$(\partial/\partial y)[(\phi + \psi)(\partial U/\partial y)] = \rho(\partial^2 U/\partial t^2), \tag{3.126}$$

where ρ is the constant mass density,

$$\phi = 2\,\partial W/\partial \mathrm{I}, \qquad \psi = 2\,\partial W/\partial \mathrm{II}, \tag{3.127}$$

and a convective coordinate system is employed to describe the motion. The only nonzero shear stress is found to be

$$\tau_{yz} = (\phi + \psi)\,\partial U/\partial y.$$

Specifically, we shall use a form of the strain–energy function known as the IHT–Zahorski type,†

$$W(\mathrm{I}, \mathrm{II}) = C_1(\mathrm{I} - 3) + C_2(\mathrm{II} - 3) + C_3(\mathrm{I}^2 - 9), \tag{3.128}$$

where C_1, C_2, and C_3 are material constants. Upon setting Eqs. (3.128), (3.127) into Eq. (3.126) and nondimensionalizing, we have

$$\frac{\partial^2 u}{\partial \eta^2} + \epsilon \left(\frac{\partial u}{\partial \eta}\right)^2 \frac{\partial^2 u}{\partial \eta^2} - \frac{\partial^2 u}{(\partial \tau^*)^2} = 0, \tag{3.129}$$

$$\frac{\partial u}{\partial \eta}(0, \tau^*) = \frac{\partial u}{\partial \eta}(1, \tau^*) = 0, \tag{3.130}$$

where $u = U/h$, $\eta = y/h$, $\tau^* = Kt$,

$$K^2 = \frac{2(C_1 + C_2 + 6C_3)}{\rho h^2}, \qquad \epsilon = \frac{6C_3}{C_1 + C_2 + 6C_3},$$

and Eq. (3.130) is the boundary condition expressing the vanishing of all surface tractions throughout the motion.

† This type was presented by Zahorski [168] based upon work by Isihara *et al.* [169]. If $C_3 = 0$, the material is known as Mooney–Rivlin type and if $C_1 = C_3 = 0$ it is said to be neo-Hookean.

Comparing Eq. (3.129) with Eq. (3.120), we observe that $L = \partial^2 u/\partial\eta^2$ and $N = (\partial u/\partial\eta)^2\, \partial^2 u/\partial\eta^2$. Proceeding as in the general discussion, we set

$$\begin{aligned} \tau &= \omega\tau^* = (\omega_0 + \epsilon\omega_1 + \epsilon^2\omega_2 + \cdots)\,\tau^*, \\ u(\eta, \tau) &= u_0 + \epsilon u_1 + \epsilon^2 u_2 + \cdots, \end{aligned} \tag{3.131}$$

thereby generating the sequence

$$u_{0\eta\eta} - \omega_0^2 u_{0\tau\tau} = 0, \qquad u_{0\eta}(0, \tau) = u_{0\eta}(1, \tau) = 0, \tag{3.132}$$

$$\begin{aligned} u_{1\eta\eta} - \omega_0^2 u_{1\tau\tau} &= 2\omega_0\omega_1 u_{0\tau\tau} - u_{0\eta}^2 u_{0\eta\eta} \\ u_{1\eta}(0, \tau) &= u_{1\eta}(1, \tau) = 0, \\ &\vdots \end{aligned} \tag{3.133}$$

From the well-known linear theory, we find that Eq. (3.132) is self-adjoint and the general term of the solution is

$$u_0(\eta, \tau) = A_n[\cos\tau + B_n \sin\tau] \cos\omega_0\eta\,,$$

where A_n and B_n are constants and

$$\omega_0 = n\pi, \qquad n = 1, 2, 3, \ldots .$$

We now make the usual assumption that the initial condition associated with the system is satisfied by u_0 alone and that $u_i(\eta, 0) = 0$, $i > 0$. Suppose also, for simplicity, that

$$u_0(\eta, \tau) = A_n \cos\tau \cos\omega_0\eta,$$

where A_n is the dimensionless maximum amplitude. With the inner product

$$\langle u, v\rangle = \int_0^{2\pi}\int_0^1 uv\, d\eta\, d\tau,$$

we find

$$2\omega_0\omega_1 \int_0^{2\pi}\int_0^1 A_n \cos^2\tau \cos^2\omega_0\eta\, d\eta\, d\tau$$
$$- \int_0^{2\pi}\int_0^1 A_n^4\omega_0^4 \sin^2\omega_0\eta \cos^2\omega_0\eta \cos^4\tau\, d\eta\, d\tau = 0.$$

From this, with $\omega_0 = n\pi$, we find

$$\omega_1 = \tfrac{3}{32} n^3\pi^3 A_n^2,$$

and consequently a two-term approximation for the frequency becomes

$$\omega = n\pi[1 + \tfrac{3}{32}\epsilon n^2\pi^2 A_n{}^2].$$

The nonlinear effect becomes significant when the maximum amplitude A_n increases or when $n > 1$, that is, when the system oscillates in its higher modes. Clearly this sytem has hard spring behavior† (frequency increases with amplitude) when $\epsilon > 0$ and soft spring behavior if $\epsilon < 0$.

Wang [170] has used the same approach in studying longitudinal shearing oscillations of thick-walled tubes and torsional oscillations of cylinders both undergoing finite deformations.

Keller and Ting [165] by more general alternative (equivalent in this case) arguments give the details for the following problems.

a. A Nonlinear Wave Equation

$$u_{tt} - u_{xx} = f(u), \qquad f(0) = 0, \quad u(0, t) = u(\pi, t) = 0.$$

A special case of this problem in which $f(u) = \alpha u + \beta u^3$ was analyzed by Stoker [171] using a different method.

b. Longitudinal Vibration of a Bar or String

$$\rho(s)\, x_{tt} = T'(x_s)\, x_{ss}\,, \qquad x(0, t) = 0, \quad x(L, t) = L.$$

c. Transverse Vibration of a String

$$\rho(s)\, \bar{x}_{tt} = [T(|\, \bar{x}_s \,|)\, \bar{x}_s / |\, \bar{x}_s \,|]_s\,,$$

$$\bar{x} = (x^{(1)}, x^{(2)}, x^{(3)}), \qquad \bar{x}(0, t) = (0, 0, 0), \quad \bar{x}(L, t) = (L, 0, 0).$$

A special case of the string has been studied by Carrier [172].

d. Transverse Vibration of a Beam

$$\rho x_{tt} = (N \cos\theta + V \sin\theta)_s\,,$$

$$\rho y_{tt} = (N \sin\theta - V \cos\theta)_s\,,$$

† Even though the initial data are smooth discontinuities may form in the solutions of homogeneous nonlinear hyperbolic equations such as Eq. (3.129). This breakdown is not considered in the perturbation analysis, but the analyst should be aware of its existence (see Chapter 1).

$$V = B\theta_{ss}, \qquad B = \text{constant flexural rigidity},$$

$$x_s = \cos\theta, \qquad y_s = \sin\theta,$$

$$\left.\begin{aligned} N_s + B\theta_s\theta_{ss} &= 0, \quad \text{at} \quad s = 0, \\ N\theta_s - B\theta_{sss} &= 0, \quad \text{at} \quad s = 0, \end{aligned}\right\} \quad (\text{clamped at } s = 0),$$

$$\theta = 0, \qquad \text{at} \quad s = 0,$$

$$N = 0, \qquad \theta_{ss} = 0, \qquad \theta_s = 0, \qquad \text{at} \quad s = L, \quad (\text{free at } s = L).$$

Find $\theta(s, t)$ and $N(s, t)$.

c. Transverse Vibrations of a Circular Membrane (radius a)

$$m\rho u_{tt} = [\rho u_\rho(u_\rho^2 + r_\rho^2)^{-1/2} W_1]_\rho,$$

$$m\rho r_{tt} = [\rho r_\rho(u_\rho^2 + r_\rho^2)^{-1/2} W_1]_\rho - W_2,$$

$$W = W(\epsilon_1, \epsilon_2) \text{ strain energy density}, \qquad W_i = \partial W/\partial\epsilon_i,$$

$$\epsilon_1 = (u_\rho^2 + r_\rho^2)^{1/2} - 1, \qquad \epsilon_2 = r\rho^{-1} - 1; \qquad u(a, t) = 0, \qquad r(a, t) = \sigma a.$$

f. Finite Amplitude Sound Waves in an Enclosure (Irrotational)

$$p_\rho[\rho(p)]\, \nabla^2\phi - \phi_{tt} = 2\phi_{x_i}\phi_{x_i t} + \phi_{x_i}\phi_{x_j}\phi_{x_i x_j},$$

$$\phi_t + \tfrac{1}{2}\phi_{x_i}\phi_{x_i} + \int_{p_0}^{p} dp/\rho(p) = 0,$$

$$\partial\phi/\partial n = 0, \qquad \text{on the surface, } S, \text{ of container.}$$

The method discussed in this section has been applied to partial differential equations for water waves by a number of authors. In particular, Penney and Price [173] examine finite periodic standing gravity waves in a perfect fluid, a topic that also concerns Tadjbakhsh and Keller [174]. Three-dimensional standing surface waves of finite amplitude have been analyzed by the perturbation method by Verma and Keller [175]. Concus [176] discusses standing capillary–gravity waves of finite amplitude in his paper.

The classical Lindstedt method of "casting out the secular terms" (cf., e.g., Ames [177]) has also been extended to partial differential equations by Liu [178] during his study of the nonlinear vibration of beams. Woodall [98] formulates the equations for the finite amplitude, free planar oscillations of a thin elastic beam obtaining the system (neglecting rotatory inertia) of six equations of parabolic type

$$\begin{aligned} \rho A u_{tt} &= (T \cos \theta)_X - (Q \sin \theta)_X , \\ \rho A v_{tt} &= (T \sin \theta)_X + (Q \cos \theta)_X - w, \\ M_X &= -Q, \qquad \theta_X = M/EI, \\ \sin \theta &= v_X = y_X , \qquad \cos \theta = 1 + u_X = x_X , \end{aligned} \tag{3.134}$$

with the six unknowns u, v, θ, T, M, and Q. These are reduced by differentiation and elimination to the two equations

$$\begin{aligned} -\rho A \theta_t^{\,2} &= T_{XX} + 2EI\theta_{XXX}\theta_X + EI\theta_{XX}^2 - T\theta_X^{\,2}, \\ \rho A \theta_{tt} &= 2T_X\theta_X + T\theta_{XX} - EI\theta_{XXXX} + EI\theta_X^{\,2}\theta_{XX} . \end{aligned} \tag{3.135}$$

Before obtaining a perturbation solution, the nondimensional quantities

$$\tau = \omega t, \qquad z = X/L, \qquad N = T/(\pi^2 EI/L^2), \qquad \mu = \rho A L^2/EI, \qquad \lambda = (AL^2)^{-1}$$

are introduced into Eqs. (3.135), thus obtaining

$$\begin{aligned} C_5 \mu\omega^2(\theta_\tau)^2 &= -\pi^2 N_{zz} + C_1\pi^2 N\theta_z^{\,2} - C_3[\theta_{zz}^2 + \theta_z\theta_{zzz}], \\ \mu\omega^2\theta_{\tau\tau} &= C_4[2\pi^2 N_z\theta_z + \pi^2 N\theta_{zz}] - \theta_{zzzz} + C_2\theta_z^{\,2}\theta_{zz} , \end{aligned} \tag{3.136}$$

where $C_1 = 1$, $C_2 = 1$, $C_3 = 1$, $C_4 = 1$, and $C_5 = 1$ are *tracers* introduced to study the influence of various terms on the solution.

This model is solved by Galerkin's method, regular perturbation, and by finite differences, and the results compared for a simply supported beam.† The boundary conditions on θ and T are

$$\left(\int \sin\theta \, dz\right)_{z=0} = 0, \qquad \left(\int \sin\theta \, dz\right)_{z=1} = 0, \qquad \theta_z(0, \tau) = \theta_z(1, \tau) = 0,$$

$$N(0, \tau) = -\pi^{-2}(\partial^2\theta/\partial z^2)(0, \tau), \qquad N(1, \tau) = -\pi^{-2}(\partial^2\theta/\partial z^2)(1, \tau),$$

and the initial conditions are

$$\theta(z, 0) = 0, \qquad \theta_\tau(z, 0) = \lambda P \cos \pi z.$$

† A simply supported beam is a beam whose ends are free of bending moments and free to move in the horizontal direction but restrained from motions in the vertical direction.

The regular perturbation expansion in λ is chosen in the form

$$\begin{aligned} \theta &= \theta_1\lambda + \theta_2\lambda^2 + \theta_3\lambda^3 + \cdots, \\ N &= N_1\lambda + N_2\lambda^2 + N_3\lambda^3 + \cdots, \\ \omega &= \omega_0 + \omega_1\lambda + \omega_2\lambda^2 + \cdots. \end{aligned} \tag{3.137}$$

When a *one-term* Galerkin approximate solution is determined, it is found that it is in closer agreement with the finite difference solution than that due to perturbation. However, to obtain the Galerkin solution the term $T\theta_x{}^2$ had to be neglected in Eq. (3.135).

McQueary [179] and McQueary and Clark [180, 181] develop a regular perturbation method closely related to the previous discussion for the nonautonomous (and autonomous) nonlinear elastic continua governed by the dimensionless equation

$$-a_0L(\psi) + \omega^2\psi_{tt} + \psi + \epsilon N(\psi) = P(x)\cos t, \tag{3.138}$$

where a_0, ω^2, and ϵ are real positive parameters. Their special emphasis is on the recognition and elimination of difficulties created by *small divisors*. The general method is described and details given for $L = \psi_{xx}$ and $N = \psi^3$. In the case of the forced vibration, the fundamental fequency response is known *a priori*, so nothing is gained by expanding ω^2 into a perturbation series. Thus only ψ is expanded as $\psi = \sum_{i=0}^{\infty} \epsilon^i\psi_i$. For certain values of the forcing frequency ω the periodic solution, of period 2π, for ψ_0 takes the form

$$\psi_0 = \cos t \sum_{n=0}^{\infty} P_n\phi_n/(\omega_{n0}^2 - \omega^2), \tag{3.139}$$

where $\omega_{n0}^2 = 1 + \alpha_n a_0$. Here $\{\phi_n\}$ and $\{\alpha_n\}$ represent the eigenfunctions and eigenvalues obtained by solving the homogeneous linearized form of Eq. (3.138), and $P(x) = \sum_{n=0}^{\infty} P_n\phi_n$. In Eq. (3.139) if

$$\omega_{m0}^2 - \omega^2 = O(\epsilon), \tag{3.140}$$

for some specified integer m, with the corresponding $P_m \neq 0$, then we meet the *classic small divisor difficulty* (cf., e.g., Stoker [182] or Rosenberg [183]).The small divisor is encountered because the expanded form of a nonlinear forcing term contains a term that is near a homogeneous solution of the corresponding equation. An apparent resonance condition exists for the particular perturbation equation studied.

The method suggested by McQueary and Clark [179–181] for eliminating the small-divisor difficulty will be discussed under the assumption that in the solution ψ_j of the jth ($j > 0$) perturbation equation we encounter the small divisor. This means that the form of the large term is "near" a homogeneous solution for ψ_j. Since all of the homogeneous equations obtained from the expansion of Eq. (3.138) are of the same form, then the functional form of the term with apparent resonance in the solution of ψ_j is also near a homogeneous solution of ψ_{j-1}. The difficult term in ψ_j is eliminated by including it as an arbitrary homogeneous portion of the solution for ψ_{j-1}. The arbitrariness in the resulting term is then eliminated by choosing its magnitude so that no very large terms exist in the solution of ψ_j. If the small divisor occurs in the solution for ψ_0, we eliminate it by assuming that the part of the forcing function giving rise to the apparent resonance is $O(\epsilon)$. This moves that part of the forcing function to a higher order in the expansion. We then include in ψ_0 a homogeneous solution of the same form as the apparent resonance term and use the arbitrariness to ensure that there are no large terms in ψ_1.

Other applications of regular perturbation include the nonlinear response of an elastic string to a moving load by Yen and Tang [184], who seek steady-state solutions depending on the moving coordinate $\tau = x - Vt$, where V is the (constant) speed of the moving load, with the load intensity as the small parameter. Earlier Steele [185] used the same approach in obtaining the nonlinear effects for a beam on a foundation subjected to a moving load. These analyses are valid for a fixed subcritical or supercritical speed V but not at the critical speed V_c. For a load moving at the critical speed, Steele finds a separate solution by intuitive arguments.

A variation of Liu's method is developed by Thurman and Mote [186, 187] and applied to examine the free periodic nonlinear oscillation of an axially moving strip [186] and to the nonlinear oscillation of a cylinder containing a flowing fluid [187]. As in the other methods, all the dependent and the independent "timelike" variables are expanded in power series in a small parameter. The arbitrariness thus introduced enables one to select the available coefficients to eliminate gradually the secular terms in the subsequent approximations. The secular terms cannot be eliminated identically for all X and t. The authors employ the averaging concept of Krylov–Bogoliubov and select the undetermined coefficients so that the *nonhomogeneous perturbation term is orthogonal, on-the-average over a period*, to the solution of the homogeneous problem. Then the secular terms generated are discarded. The authors find, for their problems, that this method is more convenient than the Keller–Ting [165] procedure.

For nonlinear dispersive waves Luke [188] has developed an asymptotic perturbation method. Kruskal and Zabusky [189] present a method for treating initial-value problems of nonlinear hyperbolic equations. In this restricted procedure the *characteristic variables and the functions of these variables* are expanded in powers of a small parameter ϵ, and the formal solution is uniformly valid over time invervals $O(1/\epsilon)$. The uniform first-order solution is obtained for the equation

$$u_{tt} = (1 + \epsilon u_x)\, u_{xx}\,,$$

subject to the standing-wave initial conditions $u(x, 0) = a \sin \pi x$, $u_t(x, 0) = 0$. Apparently the idea of employing the characteristic variables as the basis for the perturbation expansion is due to Fox [190], who studied a one-dimensional polytropic fluid.

3.11 THE PERTURBATION METHOD OF KELLER *et al.*

The general applicability of the method of Keller, which is described by Keller and Ting [165] and Millman and Keller [191], warrants separate treatment. The paper by Keller and Ting [165] describes applications, listed in Section 3.10, to finite amplitude—free vibrations of undamped continuous systems. The expansion procedure applies equally well to forced and self-excited vibrations as well as to problems possessing parabolic mathematical models.

Before examining a specific problem, we shall describe the method in general terms. Let the state of a system at time t be denoted by u, where the vector u may be a function of one or more variables other than t. Let F denote a nonlinear operator which depends upon a parameter λ, which maps the function u, with values in some unitary vector space (taken here as a Hilbert space H), into itself. For each λ we seek a vector $u(\lambda)$ satisfying

$$F(u, \lambda) = 0.^{\dagger} \tag{3.141}$$

Suppose $u = u_0$ is a solution when $\lambda = \lambda_0$,

$$F(u_0\,, \lambda_0) = 0.$$

† For example, suppose we wish to find a periodic solution of $G(u) = 0$, $u(t + 2\pi/\omega) = u(t)$, where ω is the angular frequency of the solution. Let $t' = \omega t$ and define $u'(t') = u(t)$ so that $u'(t')$ is periodic in t' of period 2π. Setting the new variable into G the parameter ω will appear explicitly. This new operator will be called $F[u'(t'), \omega]$. Dropping all primes, we seek to find a solution of $F[u(t), \omega] = 0$ of period 2π. This is a special case of Eq. (3.141).

To find u for $\lambda \neq \lambda_0$, we attempt an expansion of $u(\lambda)$ in a Taylor series in powers of $\lambda - \lambda_0$. Naturally the coefficients in this series are the derivatives of u with respect to λ evaluated at λ_0. We seek them by differentiating Eq. (3.141) repeatedly with respect to λ and then setting $\lambda = \lambda_0$. The first differentiation yields

$$F_u(u_0, \lambda_0)\, u_\lambda(\lambda_0) + F_\lambda(u_0, \lambda_0) = 0. \tag{3.142}$$

If the linear operator $F_u(u_0, \lambda_0)$ is *nonsingular*, the unique solution of Eq. (3.142) is

$$u_\lambda(\lambda_0) = -[F_u(u_0, \lambda_0)]^{-1} F_\lambda(u_0, \lambda_0). \tag{3.143}$$

The higher derivatives are found in a similar manner, and the resulting series is the perturbation expansion of $u(\lambda)$.

However, in the case that so often occurs, $F_u(u_0, \lambda_0)$ is singular, and Eq. (3.142) does not generally have a solution for $u_\lambda(\lambda_0)$! The straightforward perturbation method fails, in general, as we have already observed. A solution for $u_\lambda(\lambda_0)$ exists if $F_\lambda(u_0, \lambda_0)$ satisfies a *solvability condition* which is developed by introducing a new parameter ϵ and using it in the parametric form

$$u = u(\epsilon), \qquad \lambda = \lambda(\epsilon). \tag{3.144}$$

So that the solution u_0 corresponds to $\epsilon = 0$, we require that $u(0) = u_0$, $\lambda(0) = \lambda_0$. Then we find the derivatives of u and λ with respect to ϵ at $\epsilon = 0$ by successive differentiation of Eq. (3.141) with respect to ϵ. The derivatives of λ with respect to ϵ are determined in order to satisfy the solvability conditions for the derived equations. This is possible under appropriate circumstances such as that of an orthogonality requirement as discussed in the previous section. The resulting Taylor series for $u(\epsilon)$ and $\lambda(\epsilon)$ in powers of ϵ provide the desired expansion.

Millman and Keller [191] give the details for many examples. The list is given below followed by the details for one of them.

(a) Steady-state temperature distribution due to a nonlinear heat source or sink:

$$\nabla^2 T = \lambda S(T), \qquad \text{in the domain } D,$$

$$\partial T/\partial n = \alpha(T - T_0), \qquad \text{on the boundary } B \text{ of } D,$$

$$S(T_0) = 0.$$

(b) Forced vibrations of a "string" with a nonlinear restoring force:

$$u_{tt} - u_{xx} = \epsilon F(u), \qquad 0 < x < \pi,$$

$$u(0, t) = 0, \qquad u(\pi, t) = A \cos \omega t,$$

$$u(x, t + 2\pi/\omega) = u(x, t),$$

$$\frac{1}{2} \int_0^\pi [u_x^2 + u_t^2 - 2\epsilon \int_0^u F(u)\, du]_{t=0}\, dx = E.$$

(c) Superconductivity in a body of arbitrary shape with external magnetic field: The appropriate equations are the Landau–Ginzburg equations (see Millman and Keller [191]).

(d) Comparison of solutions of the Hartree, Fock, and Schrödinger equations for the helium atom: For the equations see Millman and Keller [191].

We now describe the details for a fifth problem, that of self-sustained oscillations of a system with infinitely many degrees of freedom. Here we wish to find periodic solutions of the nonlinear equation

$$u_{tt} - u_{xx} + u = \epsilon f(u_t), \qquad 0 < x < \pi, \tag{3.145}$$

with boundary and periodicity conditions

$$u(0, t) = u(\pi, t) = 0, \qquad u(x, t + 2\pi/\omega) = u(x, t), \tag{3.146}$$

where ω is an undetermined angular frequency. In Eq. (3.145) ϵ is a prescribed small parameter and $f(u_t)$ is a nonlinear function which is of the same sign as u_t when $|u_t|$ is small (real damping) and of the opposite sign when $|u_t|$ is large (negative damping). An example of such a function is†

$$f_0(u_t) = u_t - \tfrac{1}{3}u_t^3. \tag{3.147}$$

The fact that van der Pol's equation has one periodic solution suggests that Eqs. (3.145) and (3.146) will also have at least one solution.

† This is a form obtainable from the van der Pol equation $y'' + \epsilon(y^2 - 1)y' + y = 0$, which is known to have one periodic solution—the self-sustained oscillation. Upon setting $y = u'$ and integrating, we find, after discarding a constant, the equation

$$u'' + \epsilon[\tfrac{1}{3}(u')^2 - 1]\, u' + u = 0.$$

To begin, we first set $t' = \omega t$ and $u'(x, t') = u(x, t)$ in Eqs. (3.145) and (3.146) and then omit the primes to find

$$\omega^2 u_{tt} - u_{xx} + u = \epsilon f(\omega u_t), \tag{3.148}$$

$$u(0, t) = u(\pi, t) = 0, \qquad u(x, t + 2\pi) = u(x, t). \tag{3.149}$$

We seek a solution $u(x, t, \epsilon)$ and an angular frequency $\omega(\epsilon)$ which are differentiable with respect to ϵ at $\epsilon = 0$. To do this the aforementioned method will be employed.

The zero-order term is found by setting $\epsilon = 0$ in Eqs. (3.148). Let us denote $u(x, t, 0) = u_0(x, t)$, $\omega(0) = \omega_0$ so that the (linear) equation for u_0 is

$$\omega_0^2 u_{0tt} - u_{0xx} + u_0 = 0, \tag{3.150}$$

together with

$$u_0(0, t) = u_0(\pi, t) = 0, \qquad u_0(x, t + 2\pi) = u_0(x, t). \tag{3.151}$$

For each positive integer n, Eqs. (3.150) and (3.151) have solutions whose general form is

$$u_0 = A_n \sin nx \cos t, \tag{3.152}$$

where ω_0 is one of the corresponding eigenvalues

$$\omega_0 = (1 + n^2)^{1/2}, \qquad n = 1, 2, \ldots, \tag{3.153}$$

and the amplitude A_n is undetermined to this point.

To obtain the first-order term we differentiate Eqs. (3.148) and (3.149) once with respect to ϵ and set $\epsilon = 0$ to obtain [here $\dot{u} = u_\epsilon(x, t, 0)$, $\dot{\omega} = \omega_\epsilon(0)$]

$$\omega_0^2 \dot{u}_{tt} - \dot{u}_{xx} + \dot{u} = -2\omega_0 \dot{\omega} u_{0tt} + f(\omega_0 u_{0t}), \tag{3.154}$$

and

$$\dot{u}(0, t) = \dot{u}(\pi, t) = 0, \qquad \dot{u}(x, t + 2\pi) = \dot{u}(x, t). \tag{3.155}$$

This is an inhomogeneous form of the problem of Eqs. (3.150), with the same boundary conditions. Equation (3.154) will have a solution only if some solvability condition is satisfied. We show this, and obtain the condition, by multiplying Eq. (3.154) by $w(x, t)$ (a "weighting" function to be discussed) and integrating from 0 to π on x and 0 to 2π on t. After several integrations by parts, this relation becomes

$$\int_0^{2\pi}\int_0^{\pi} \dot{u}(\omega_0{}^2 w_{tt} - w_{xx} + w)\,dx\,dt$$

$$+ \int_0^{\pi} [\dot{u}_t w - \dot{u} w_t]_{t=0}^{2\pi}\,dx + \int_0^{2\pi} [\dot{u}_x w - \dot{u} w_x]_{x=0}^{\pi}\,dt$$

$$= -2\omega_0\dot{\omega}\int_0^{2\pi}\int_0^{\pi} w u_{0tt}\,dx\,dt + \int_0^{2\pi}\int_0^{\pi} w f(\omega_0 u_{0t})\,dx\,dt. \quad (3.156)$$

The first term on the left-hand side of Eq. (3.156) vanishes for *either* of the two functions

$$w = \sin nx \sin t, \qquad w = \sin nx \cos t, \quad (3.157)$$

and for *both* choices of w the second and third terms also vanish because of Eqs. (3.155). Thus Eq. (3.156) yields the two conditions

$$\omega_0\dot{\omega}\pi^2 A_n + \int_0^{2\pi}\int_0^{\pi} \sin nx \cos t f(-\omega_0 A_n \sin nx \sin t)\,dx\,dt = 0, \quad (3.158a)$$

and

$$\int_0^{2\pi}\int_0^{\pi} \sin nx \sin t f(-\omega_0 A_n \sin nx \sin t)\,dx\,dt = 0. \quad (3.158b)$$

Equation (3.158a) determines $\dot{\omega}$ as a function of A_n and Eq. (3.158b) determines A_n. Explicit results are obtained by the authors for $f = f_0$ [Eq. (3.147)].

Clearly, the most difficult part of the preceding computation is the determination of the solvability condition. A useful hint here is to think of this as an orthogonality requirement which reduces the difficulty to that of determining the appropriate weighting function. The extensive collection of examples in the paper of Keller and Ting [165] and Millman and Keller [191] will be of considerable assistance in this effort.

3.12 SINGULAR PERTURBATION

A singular perturbation problem is encountered in seeking a solution for a limiting value (e.g., zero) of a parameter or in seeking a small perturbation solution, when the solution so obtained is *not uniformly valid throughout the domain of definition of the independent variables*. For example, the formal solution may become *infinite* on some locus of the space variables. This occurs near a rounded leading edge in the small perturbation solution of thin-airfoil theory (cf., e.g., Lighthill [192]).

The nonuniformity may be evidenced by a discontinuity of the solution within the region of interest (e.g., the discontinuous pressure across a shock wave in inviscid flow) or by failure of the solution to satisfy a boundary condition (i.e., a discontinuity at the boundary). The well-known example of this type is in the calculation of flow at high Reynolds number over a surface, in which the flow is essentially inviscid except for a thin boundary layer at the surface. The inviscid-flow solution fails to satisfy the no-slip condition at the boundary surface. The Prandtl boundary-layer theory was developed to cope with this mathematical nonuniformity in the flow calculation. The discontinuity or failure to satisfy a boundary condition is often, but not always, due to a reduction in the order of the differential equation, or sets of equations, as the small parameter goes to zero. The discontinuity is an asymptotic representation of a *boundary layer*† (*quick transition region*, *rapid variation*) of one or more dependent variables over a small range of the independent variables.

Friedrichs [161] in his Gibbs Lecture on "Asymptotic Phenomena in Mathematical Physics" discussed the *quick transition regions* which occur in so many problems of mathematical physics. Of course asymptotic expansions were employed prior to the work of Latta [193], Kaplun (collected works, see Lagerstrom *et al.* [194]), and Lagerstrom and Cole [195, 196], but their significant contribution was in the introduction and development of a technique called the method of *matched asymptotic expansions* (inner and outer expansions). This method employs two (or more) distinct asymptotic series expansions of the solution: an *outer* expansion valid away from the nonuniformity and an *inner* expansion valid in the "inner" region of the nonuniformity. Limiting (matching) processes are defined according to which the inner and outer expansions must match in their *overlap* region of common validity. Further refinements and examples are given by van Dyke [197], Cole [198], and in Volume I. More recently (1971) Eckhaus [199] aims at the development of a *rigorous* theory for boundary-layer problems in linear elliptic systems.

Martin [200] presents an admittedly incomplete but useful classification of perturbation problems (one independent variable) according to the type of nonuniformity. Each of these categories is accompanied by an example ordinary differential equation. An understanding of these simple problems can assist in the successful solution of more difficult problems which possess some of the same features.

---

† The term "boundary layer" originated in fluid mechanics but it has become of standard useage, in the above context, by those engaged in singular perturbation problems.

The essential features of the *method of matched asymptotic expansions* are based upon the concept of asymptotics (cf., e.g., Erdelyi [201]). An asymptotic expansion to N terms for a function $f(x, y, \epsilon)$ of the small quantity ϵ, x, and y may have the form

$$f(x, y, \epsilon) = \sum_{n=1}^{N} f_n(x, y)\,\phi_n(\epsilon) + o[\phi_N(\epsilon)], \qquad \text{as} \quad \epsilon \to 0, \quad \text{with} \quad x, y, \text{ fixed,} \tag{3.159}$$

where $\phi_1 = 1$ and $\phi_{n+1}(\epsilon) = o[\phi_n(\epsilon)]$ as $\epsilon \to 0$.† If f can be expanded as in Eq. (3.159) to N terms then for any integer M, $1 \leqslant M < N$, we can write

$$f(x, y, \epsilon) = \sum_{n=1}^{M} f_n(x, y)\,\phi_n(\epsilon) + O[\phi_{M+1}(\epsilon)], \qquad \text{as} \quad \epsilon \to 0, \quad \text{with} \quad x, y, \text{ fixed.}$$

We shall use the symbol $\sim$ to mean "asymptotically equal" and write

$$f(x, y, \epsilon) \sim \sum_{n=1}^{N} f_n(x, y)\,\phi_n(\epsilon), \qquad \text{as} \quad \epsilon \to 0, \quad \text{with} \quad x, y \text{ fixed,} \tag{3.160}$$

without the order symbol.

In using the method of matched asymptotic expansions (MAE) one often starts with a power series expansion in what (intuitively) appears to be the logical form of the small parameter ϵ and then modifies it as necessary when and if difficulties arise. Still, this is not necessary, and the form of the expansion can be left unspecified initially so the expansions can be constructed as the analysis proceeds.

In solving a problem for $f(x, y, \epsilon)$ by MAE, we first represent f by Eq. (3.160), where $\phi_1 = 1$ and $N \to \infty$. Equation (3.160) is then substituted into the differential equation and auxiliary conditions for the problem. The first-order problem, for $f_1(x, y)$, is obtained by setting $\epsilon = 0$. Presumably $f_1(x, y)$ is not uniformly valid for some values of x and y. Equation (3.160) is called the *outer expansion*.

Next *inner variables* are defined by magnifying the physical variables so that the *inner dependent variables remain uniformly valid as* $\epsilon \to 0$ *and as the location of the nonuniformity of the outer solution is approached simultaneously*. The inner variables are required to be $O(1)$ as $\epsilon \to 0$. If the nonuniformity is at $y = 0$, when $\epsilon = 0$, the $O(1)$ requirement on

† We use the "little oh" order symbol $o(\ )$ in the conventional way. That is if $f = o(g)$ as $\epsilon \to 0$, then $\lim_{\epsilon\to 0} f/g = 0$. The "big oh" order symber $O(\ )$ means that if $f = O(g)$ as $\epsilon \to 0$ $(g \neq 0)$, then $\lim_{k\to 0} f/g$ is bounded.

Y is met by magnifying y by a function of ϵ, $\sigma_1(\epsilon)$,

$$y = \sigma_1(\epsilon) Y. \tag{3.161}$$

Similarily the dependent variable f is magnified

$$f(x, y, \epsilon) = \sigma_2(\epsilon) F(x, Y, \epsilon), \tag{3.162}$$

where F is $O(1)$ as $\epsilon \to 0$ with x and Y fixed, and σ_1 and σ_2 are determined from the specific problem, bearing in mind the order requirements on the inner variables.

An *inner expansion* for F, valid in the inner region, is also left unspecified a priori and may be written as

$$F(x, Y, \epsilon) \sim \sum_{n=1}^{N} F_n(x, Y)\, \psi_n(\epsilon), \qquad \text{as} \quad \epsilon \to 0, \quad \text{with} \quad x, Y \text{ fixed}, \tag{3.163}$$

where, again, $\psi_1 = 1$ and $\psi_{n+1} = o(\psi_n)$ as $\epsilon \to 0$.

In the solution process, one of the guides to be kept in mind (van Dyke [197]) is: *When terms are lost or boundary conditions discarded for the outer solution they must be included in the development of the inner solution.* An overlap domain exists between the two solutions. The inner solution must be matched to the outer solution in some fashion. Kaplan and Lagerstrom (cf. Lagerstrom *et al.* [194]) assert that the existence of an overlap domain implies that the *outer expansion of the inner expansion matches the inner expansion of the outer expansion* to appropriate (perturbation parameter) orders for the respective terms. The matching principle can also be used to determine the inner and outer asymptotic sequences of functions of ϵ. While this general matching principle can be bent to the investigator's taste, some systematic forms have been found more useful than others.

The principle found useful by van Dyke [197] is conveniently stated in the following notation: Let f^{mo} be the m-term outer expansion for f, i.e.,

$$f^{mo} = f_1 + \phi_2(\epsilon) f_2 + \cdots + \phi_m(\epsilon) f_m, \tag{3.164}$$

and f^{ni} be the product of $\sigma_2(\epsilon)$ times the n-term inner expansion of F, i.e.,

$$f^{ni} = \sigma_2(\epsilon)[F_1 + \psi_1(\epsilon) F_2 + \cdots + \psi_n(\epsilon) F_n]. \tag{3.165}$$

If we now substitute inner variables, Y and F, into the outer expansion, Eq. (3.164), we have *the m-term outer expansion in inner variables*, denoted by $(f^{mo})^{i}$; if we substitute outer variables, y and f, into the inner expansion, Eq. (3.165), we have the *n-term inner expansion in outer variables*, denoted by $(f^{n\mathrm{i}})^{\mathrm{o}}$. Then the first n terms of $(f^{m\mathrm{o}})^{\mathrm{i}}$ constitute the *n-term inner expansion of the m-term outer expansion*, denoted by $(f^{m\mathrm{o}})^{n\mathrm{i}}$, and the first m terms of $(f^{n\mathrm{i}})^{\mathrm{o}}$ constitute *the m-term outer expansion of the n-term inner expansion*, denoted by $(f^{n\mathrm{i}})^{m\mathrm{o}}$. The *principle* that determines the matching between the inner and outer expansions is then†

$$(f^{m\mathrm{o}})^{n\mathrm{i}} = (f^{n\mathrm{i}})^{m\mathrm{o}}. \tag{3.166}$$

Intermediate expansions and intermediate matching is sometimes necessary or convenient. It is usually helpful to systematize the matching process. One such ordering is given in Fig. 3-4. Beginning with the first

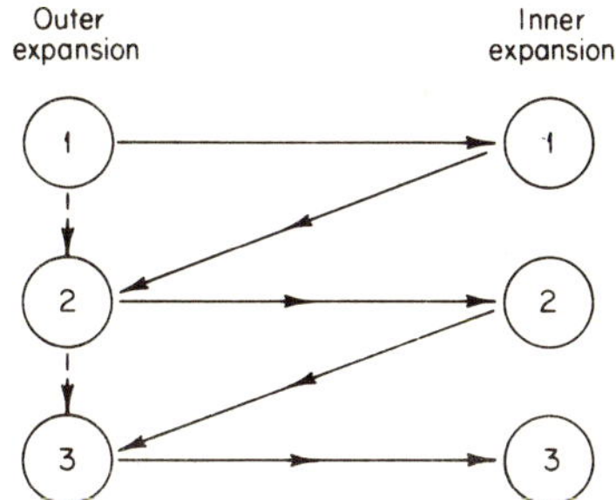

FIG. 3-4. Normal calculation sequence for matched asymptotic expansions.

term of the outer expansion, we proceed as shown, where the numbers in the circles represent the term number in the respective expansion.

Upon completion of the inner and outer expansion (to the desired order in ϵ) it is sometimes desired to combine them into one *composite* solution which is uniformly valid over the entire region of interest—

$$f^{\mathrm{c}} = f^{n\mathrm{i}} + f^{m\mathrm{o}} - (f^{m\mathrm{o}})^{n\mathrm{i}}. \tag{3.167}$$

f^{c} approaches the outer solution in the outer limit and the inner solution in the inner limit.

Martin [200] proposes an asymptotic expansion in *displacement variables*. The influence of the nonuniform region on the outer solution is known as the *displacement effect*. The displacement surface, defined by displacement variables, is a surface that would be the equivalent frictionless body for the outer displaced flow (cf., e.g., Lighthill [202]).

† Here m and n are any two positive integers—usually we choose $m = n$ or $n + 1$.

3.13 LIGHTHILL'S METHOD OF STRAINED COORDINATES

In one class of singular perturbation problems, expanding the unknown solution in powers of a small parameter artificially shifts a singularity to a location within the defined domain of the independent variables. The perturbation solution is then not a valid approximation. Lighthill's [203, 204] *method of strained coordinates* (cf. also van Dyke [197] and Ames [177]) for rendering approximate solutions uniformly valid, is a transformation of variables that moves the singularity back to the proper location of its occurrence (determined from the original exact equation). His transformation, to cure this defect, is a straining of the appropriate independent variable(s) by expanding one (or more) of them as well as the dependent variable in an asymptotic series. The coordinate straining is initially unknown and must be determined term by term as the analysis progresses in accordance with Lighthill's principle:

Higher approximations shall be no more singular than the first. (3.168)

Lighthill's principle is not a definite procedure but a guiding philosophy. Since the *straining is not determined uniquely* by the principle, one can often simplify the analysis by exploiting this nonuniqueness. The application of this principle halts the compounding of singularities that invalidates a straightforward regular expansion in a region of nonuniformity. Uniform validity is restored in a wide class of problems, especially in fluid and gas dynamics, for both subsonic and supersonic flow, where it has been widely used.

Tsien [205] summarizes the Poincaré–Lighthill–Kuo (PLK) method and its application to 1956. Lin [206] generalizes the method to hyperbolic equations in two variables based on the method of characteristics. Pritulo [207] considers some theoretical questions associated with the PLK method. Although Lighthill [208] used the PLK method to cope with the nonuniformity at the leading edge in the airfoil theory governed by an elliptic equation, it has been subsequently presumed that this treatment was a special case and the method could not yield an improved solution beyond the second order. Tsien [205] states that the PLK method fails to produce a solution uniformly valid to all orders. These and other difficulties have lead to the belief that the PLK method was not generally valid for elliptic equations. No way has been found to apply the PLK method on many problems governed by parabolic equations where the nonuniformity is due to an essential singularity which can only be eliminated by stretching the coordinate. Thus it has been widely held that the use of Lighthill's method should be restricted to ordinary and hyperbolic equations.

However, Hoogstraten [209] and Bollheimer and Weissinger [210] independently have been able to use Lighthill's technique in conjunction with a special conformal-mapping technique to find uniformly valid thin-airfoil solutions for round-nosed and sharp-nosed airfoils. The analysis applies only to thin airfoil theory governed by Laplace's equation. Comstock [211], in studying a problem posed by Lin, discovered certain limitations to Lighthill's technique when applied to ordinary differential equations. Presumably these apply to partial differential equations as well.

In that follows, we shall first describe the standard procedure and then discuss Martin's [212, 213] adaptation and extension, using Lagrange's expansion, for ordinary and partial differential equations.

In solving a differential equation with boundary conditions for $f(x, \epsilon)$, the Lighthill† uniformization procedure is to expand both f *and* x in powers of ϵ and in terms of a common independent variable (parameter) s:

$$f(x, \epsilon) = h_1(s) + \epsilon h_2(s) + \epsilon^2 h_3(s) + \cdots, \tag{3.169}$$

$$x = s + \epsilon x_1(s) + \epsilon^2 x_2(s) + \cdots, \tag{3.170}$$

and choose $x_1(s)$ so that $h_2(s)$ is no more singular, say at $s = 0$, than $h_1(s)$;...; in general $x_{n-1}(s)$, $n = 1, 2,...$ is chosen so that $h_n(s)$ *is no more singular than* $h_{n-1}(s)$ (hence h_1), in accordance with Lighthill's principle, Eq. (3.168). The standard method, after noting the singular behavior of $h_n(s)$, is to substitute Eqs. (3.169) and (3.170) back into the problem, collect powers of ϵ, solve the respective differential equations with boundary conditions for the respective terms $h_n(s)$, and then apply the Lighthill rule to obtain the $x_n(s)$. Many examples of this procedure are given in the foregoing references.

Martin [212, 213] claims that the most simple and direct way of applying Lighthill's technique is to use a Lagrange expansion–perturbation scheme and to specify the terms according to Lighthill's principle. The procedure for one variable was introduced by Martin [212]. When more than one independent variable must be strained simultaneously, a higher-dimensional Lagrange expansion is needed. Such generalizations have been developed by Good [214] whose application lay in stochastic processes, Sturrock [215], and Sack [216, 217]. Martin [218] presents two independent derivations for obtaining explicitly the terms, to any desired order, of the N-dimensional Lagrange expansion in simplest form. For his purposes the form Martin [218] develops is most useful.

---

† It is assumed that the straightforward perturbation expansion generates a "solution" which is not uniformly valid, usually because of a singularity.

Lagrange's expansion (e.g., Bellman [219], Sack [216, 217]; for an application to a Mongé–Amperé equation see Ames and Jones [121]) is a generalization of Taylor's expansion where the independent variable is defined by an implicit equation. For a function $f(x)$, the usual form is

$$f(x) = f(s) + \sum_{n=1}^{\infty} (d^{n-1}/ds^{n-1})[g^n(s)\, df(s)/ds](\epsilon^n/n!), \tag{3.171}$$

where

$$x = x(s, \epsilon) = s + \epsilon g(x). \tag{3.172}$$

A parameter α may be included in the above, with the obvious generalization

$$f(x, \alpha) = f(s, \alpha) + \sum_{n=1}^{\infty} (\partial^{n-1}/\partial s^{n-1})[g^n(s, \alpha)\, \partial f(s, \alpha)/\partial s](\epsilon^n/n!), \tag{3.173}$$

where

$$x = x(s, \alpha, \epsilon) = s + \epsilon g(x, \alpha). \tag{3.174}$$

Clearly Eqs. (3.173) and (3.174) may be specialized to that case where $\alpha = \epsilon$. Thus we have

$$f(x, \epsilon) = f(s, \epsilon) + \sum_{n=1}^{\infty} (\partial^{n-1}/\partial s^{n-1})[g^n(s, \epsilon)\, \partial f(s, \epsilon)/\partial s](\epsilon^n/n!), \tag{3.175}$$

where

$$x = x(s, \epsilon, \epsilon) = s + \epsilon g(x, \epsilon). \tag{3.176}$$

Equation (3.176) has the expansion

$$x = s + \sum_{n=1}^{\infty} (\partial^{n-1}/\partial s^{n-1})[g^n(s, \epsilon)](\epsilon^n/n!). \tag{3.177}$$

If now we assume classical expansions for f and g,

$$f(x, \epsilon) = f_1(x) + \epsilon f_2(x) + \epsilon^2 f_3(x) + \cdots, \tag{3.178}$$

and

$$g(x, \epsilon) = g_1(x) + \epsilon g_2(x) + \epsilon^2 g_3(x) + \cdots, \tag{3.179}$$

then Eqs. (3.175) and (3.177) become

$$\begin{aligned} f(x, \epsilon) = f_1(s) &+ \epsilon[f_2(s) + g_1(s) f_1'(s)] \\ &+ \epsilon^2[f_3(s) + g_2(s) f_1'(s) + g_1(s) f_2'(s) \\ &+ g_1(s) g_1'(s) f_1'(s) + \tfrac{1}{2} g_1^2(s) f_1''(s)] \\ &+ O(\epsilon^3), \end{aligned} \tag{3.180}$$

where

$$x = s + \epsilon g_1(s) + \epsilon^2[g_2(s) + g_1(s)\, g_1'(s)] + O(\epsilon^3). \tag{3.181}$$

Equations (3.180) and (3.181) are valid, at least asymptotically as $\epsilon \to 0$, for arbitrary f and g that can be expanded as above.

An alternative to the standard process in applying Lighthill's method is now available. Comparing Eqs. (3.169) and (3.170) with Eqs. (3.180) and (3.181), respectively, and assuming the expansions are unique it follows that

$$\begin{aligned} h_1(s) &= f_1(s), \\ h_2(s) &= f_2(s) + g_1(s) f_1'(s), \\ h_3(s) &= f_3(s) + g_2(s) f_1'(s) + g_1(s) f_2'(s) \\ &\quad + g_1(s)\, g_1'(s) f_1'(s) + \tfrac{1}{2} g_1^2(s) f_1''(s), \\ &\;\;\vdots \end{aligned} \tag{3.182}$$

and

$$\begin{aligned} x_1(s) &= g_1(s), \\ x_2(s) &= g_2(s) + g_1(s)\, g_1'(s), \\ &\;\;\vdots \end{aligned} \tag{3.183}$$

Since the f_n's are already determined, we need only specify the $g_n(s)$ in Eqs. (3.182) so that Lighthill's principle [Eq. (3.168)] holds. Thus the tedious procedure of substituting Eqs. (3.169) and (3.170) into the differential equation and boundary conditions and again solving the problem in the new form to determine the uniformly valid solution has been circumvented, yet Lighthill's uniformization has been accomplished.

The specification of $g_n(s)$ in Eqs. (3.182) is well known to be nonunique (cf., e.g., van Dyke [197]). But by making certain predetermined specifications of $g_n(s)$ that satisfy Lighthill's principle, Eq. (3.168), we can establish formulas for a completely determined uniformly valid solution for all problems in the class. One such solution is obtained by Martin [212] by requiring that all $h_n(s) = 0$, for $n \geqslant 2$.

For partial differential equations, where the need arises for straining several variables simultaneously, Martin [213] has shown the great utility of his method.

3.14 MISCELLANEOUS ASYMPTOTIC PROCEDURES

The *two-time method* has been applied to obtain asymptotic expansions of a variety of physical problems governed by ordinary differential equations. Basic references include Kevorkian [220] and Cole [198], who devotes his Chapter 3 to this subject. Frieman, Sandri, and others (cf., e.g., Frieman [221]) apply the method to problems in statistical mechanics. Reiss [222] generalizes to an *N-time method* and notes that the indeterminacy created by the introduction of these new "auxiliary" variables frequently requires appeal to intuitive arguments in order to evaluate the expansion coefficients. He demonstrates, for a linear ordinary equation, how to eliminate intuitive arguments and presents a systematic procedure to determine the expansion coefficients and prove the uniform asymptotic convergence of the expansion.

In partial differential equations at least two applications to nonlinear problems exist. One of these by Keller and Kogelman [223] will be briefly described in the sequel. The other by Davy [224] concerns wave propagation in rate dependent materials governed by the dimensionless equation

$$u_{ttt} - u_{xxt} + \alpha u_{tt} + \beta[f(u_x)]_x = 0.$$

Keller and Kogelman [223] wish to determine the solution of the initial boundary value problem,

$$u_{tt} - u_{xx} + u = \epsilon f(u, u_t, \epsilon), \qquad 0 < x < \pi, \quad t > 0, \tag{3.184}$$

$$\begin{aligned} u(0, t) &= u(\pi, t) = 0, \\ u(x, 0) &= g(x, \epsilon), \\ u_t(x, 0) &= h(x, \epsilon), \end{aligned} \tag{3.185}$$

and its *asymptotic behavior when ϵ is small and t is large*. Details are given for the problem specified by Eq. (3.184) with $f = f_0(u_t) = u_t - \frac{1}{3}u_t^3$, $u(0, t) = u(\pi, t) = 0$, $u(x, t + 2\pi/\omega) = u(x, t)$, in which case it is known (cf. Millman and Keller [191]) that the problem possesses the infinitely many periodic solutions (limit cycles)

$$u_n(x, t, \epsilon) = [4/3^{1/2}(1 + n^2)^{1/2}] \sin nx \cos[\omega_n(\epsilon)t] + \epsilon \dot{u}_n(x, t, 0) + O(\epsilon^2),$$

and

$$\omega_n(\epsilon) = (1 + n^2)^{1/2} + \tfrac{1}{2}\ddot{\omega}_n(0)\, \epsilon^2 + O(\epsilon^3).$$

u_n and $\ddot{\omega}_n(0)$ were also determined in the Millman and Keller work.

To solve Eqs. (3.184) and (3.185) for small ϵ, we introduce *two-time variables*,† σ and τ, which are obtained by stretching t as follows:

$$\sigma = \omega(\epsilon)t = [1 + O(\epsilon^2)]t, \qquad \tau = \epsilon t. \tag{3.186}$$

† The introduction of these auxiliary variables gives the method its flexibility but it also creates a certain amount of indeterminacy.

The angular frequency $\omega(\epsilon)$ is at our disposal, so we have set $\omega(0) = 1$ and $\omega_\epsilon(0) = 0$. Upon writing

$$u(x, t, \epsilon) = v(x, \sigma, \tau, \epsilon), \tag{3.187}$$

we use Eqs. (3.186) and (3.187) in Eqs. (3.184) and (3.185) to find

$$\begin{aligned}
\omega^2 v_{\sigma\sigma} - v_{xx} + v &= -2\epsilon\omega v_{\sigma\tau} - \epsilon^2 v_{\tau\tau} + \epsilon f(v, \omega v_\sigma + \epsilon v_\tau, \epsilon),\\
v(0, \sigma, \tau, \epsilon) &= v(\pi, \sigma, \tau, \epsilon) = 0,\\
v(x, 0, 0, \epsilon) &= g(x, \epsilon),\\
\omega v_\sigma(x, 0, 0, \epsilon) + \epsilon v_\tau(x, 0, 0, \epsilon) &= h(x, \epsilon).
\end{aligned}$$

The solution of this problem is then obtained by the perturbation method previously discussed, wherein v and ω are expanded in Taylor series in ϵ about $\epsilon = 0$.

Closely associated processes include an extension of the Krylov–Bogoliubov–Mitropolsky perturbation method (cf. e.g., Bogoliubov and Mitropolsky [225]) due to Montgomery and Tidman [226]. That method is used by Tidman and Stainer [227] to treat the nonlinear longitudinal traveling oscillations in a hot plasma whose equations are taken to be

$$\begin{aligned}
n_t + (nu)_x &= 0,\\
u_t + uu_x + (mn)^{-1} p_x &= -eE/m,\\
p_t + 3(pu)_x - 2up_x &= 0,\\
E_t &= 4\pi enu,
\end{aligned} \tag{3.188}$$

where n is the electron density, u the velocity, p the pressure, and E the electric field.

Nayfeh [228] proposes an alternative pertubation procedure which he calls the *derivative-expansion method*. Each dependent quantity Q is assumed to have an asymptotic representation of the form

$$Q(x, t, \epsilon) = \sum_{n=0}^{N} \epsilon^n Q_n(x_0, \ldots, x_N, t_0, \ldots, t_N) + O(\epsilon^{N+1}),$$

where $x_n = \epsilon^n x$, $t_n = \epsilon^n t$, and Q_0, Q_n/Q_{n-1} are bounded for each x_n and t_n.

The parameter ϵ is introduced for convenience and is set equal to one in the final solution. Thus

$$\frac{\partial Q}{\partial t} = \sum_{n=0}^{N} \epsilon^n \frac{\partial Q}{\partial t_n}, \qquad \frac{\partial Q}{\partial x} = \sum_{n=0}^{N} \epsilon^n \frac{\partial Q}{\partial x_n},$$

and the dependent variables Q in Eqs. (3.188) are p, E, n, and u. The details for Eqs. (3.188) are given by Nayfeh [228].

References

1. Finlayson, B. A., and Scriven, L. E., *Chem. Eng. Sci.* **20**, 395 (1965).
2. Finlayson, B. A., and Scriven, L. E., *Appl. Mech. Rev.* **19**, 735 (1966).
3. Mikhlin, S. G., "Variational Methods in Mathematical Physics." Pergamon/Macmillan, London and New York, 1964.
4. Goodman, T. R., *Advan. Heat Transfer* **1**, 51. Academic Press, New York, 1964.
5. Schuleshko, P., *Aust. J. Appl. Sci.* **10**, 1 (1959).
6. Snyder, L. J., Spriggs, T. W., and Stewart, W. E., *AIChE J.* **10**, 535 (1964).
7. Bolotin, V. V., "Nonconservative Problems of the Theory of Elastic Stability," p. 60. Pergamon/Macmillan, London and New York, 1963.
8. Finlayson, B. A., Approximate Solutions of the Equations of Change. Convective Instability by Active Stress, Ph.D. Dissertation, Univ. of Minnesota, 1965.
9. Hildebrand, F. B., "Methods of Applied Mathematics." Prentice-Hall, Englewood Cliffs, New Jersey, 1952.
10. Duncan, W. F., Galerkin's Method in Mechanics and Differential Equations. *Gt. Brit. Min. Aviat. Aeronaut. Res. Coun. Curr. Pap.* **1**, 484 (1937).
11. Hugelman, R. D., and Haworth, D. R., *AIAA J.* **3**, 1367 (1965).
12. Hugelman, R. D., *AIAA J.* **3**, 2158 (1965).
13. Collatz, L., "The Numerical Treatment of Differential Equations." Springer-Verlag, Berlin and New York, 1960.
14. Kaplan, S., *Nucl. Sci. Eng.* **13**, 22 (1962).
15. Kaplan, S., and Bewick, J. A., "Space and Time Synthesis by the Variational Method." Bettis Tech. Rev., Reactor Tech. 27-44, WAPD-BT-28 (1963).
16. Kaplan, S., Marlowe, O. J., and Bewick, J. A., *Nucl. Sci. Eng.* **18**, 163 (1964).
17. Oseen, C. W., *Ark. Mat. Astron. Fys.* **6**, 29 (1910).
18. Oseen, C. W., "Hydrodynamik." Leipzig, Germany, 1927.
19. Lewis, J. A., and Carrier, G. F., *Quart. Appl. Math.* **7**, 228 (1949).
20. Carrier, G. F., *J. Soc. Ind. Appl. Math.* **13**, 68 (1965).
21. Schetz, J. A., *J. Appl. Mech.* **30**, 263 (1963).
22. Schetz, J. A., *J. Appl. Mech.* **33**, 425 (1966).
23. Schetz, J. A., and Jannone, J., *J. Appl. Mech.* **32**, 757 (1965).
24. Schetz, J. A., and Jannone, J., *J. Heat Transfer* **87C**, 157 (1965).
25. Schetz, J. A., "Approximate Solution of Boundary Layer Problems with Mass Transfer," Dept. Aerospace Eng. Rep. #65-2, Univ. of Maryland, November (1965).
26. Hamielec, A. E., Hoffman, T. W., and Ross, L. L., *AIChE J.* **13**, 212 (1967).
27. Kantorovich, L. V., and Krylov, V. L., "Approximate Methods in Higher Analysis." Wiley (Interscience), New York, 1958.

28. Snyder, L. J., and Stewart, W. E., *AIChE J.* **12**, 167 (1966).
29. Kawaguti, M., *J. Phys. Soc. Jap.* **10**, 694 (1955).
30. Hays, D. F., *Acad. Roy. Belg. Bull. Cl. Sci.* **49**, 576 (1963).
31. Biot, M. A., *J. Aeronaut. Sci.* **24**, 857 (1957).
32. Biot, M. A., and Agrawal, H. C., *J. Heat Transfer* **86**, 437 (1964).
33. Richardson, P. D., *J. Heat Transfer* **86**, 298 (1964).
34. Donnelly, R. J., Herman, R., and Prigogine, I., (eds.), "Nonequilibrium Thermodynamics, Variational Techniques and Stability." Univ. of Chicago Press, Chicago, Illinois, 1966.
35. Hays, D. F., pp. 17–43 of Reference [34].
36. Collings, W. F., The Method of Undetermined Functions as Applied to Nonlinear Diffusion Problems. M.M.E. thesis, Univ. of Delaware, Newark, Delaware, 1962.
37. Finlayson, B. A., *Brit. Chem. Eng.* **14**, 53–57, 179–182 (1969).
38. Grotch, S. L., *AIChE J.* **15**, 463 (1969).
39. Hamielec, A. E., and Johnson, A. I., *Can. J. Chem. Eng.* **40**, 41 (1962).
40. Hamielec, A. E., Storey, S. H., and Whitehead, S. M., *Can. J. Chem. Eng.* **41**, 246 (1963).
41. Dorodnitsyn, A. A., "On a Method of Numerical Solution of Certain Non-linear Problems of Aero-Hydromechanics," Proc. Third All-Soviet Math. Congress 1956, Vol. III, *Izv. Akad. Nauk SSSR*, Moscow (1958) [Translation: AEC Transl. UCRL-Trans. 993 (L)]. *See also* Proc. Ninth Int. Congr. Appl. Mech. Vol. 1, 485 (1957).
42. Dorodnitsyn, A. A., *Prikl. Mat. Tekh.-Fiz.* **1**, 111 (1960). See also *Advan. Aeronaut. Sci. Proc. Int. Congr. A, 2nd.* **3**, Pergamon, Oxford, 1962.
43. Dorodnitsyn, A. A., *Arch. Mech. Stosowanej.* **14**, 343 (1962). (Translation: Foreign Tech. Div. Tech. Translation FTD-TT-65-67). See also "Exact Numerical Methods in the Boundary Layer Theory," W. Fiszdon (ed.), Vol. 1, *Fluid Dyn. Trans.* Pergamon, Oxford, 1964.
44. Golubev, V. V., "Laminar Boundary Layers" Loytsyanskiy (ed.), Chapter 3. Gos. en. Iz dat. Moscow (1962). (Translation: Foreign Tech. Div. Tech. Translation FTD-TT-63-749).
45. Pavlovskii, Y. N., *Z. Vycisl. Mat. i Mat. Fiz.* **2**, 884 (1962). (Translation available from M. Holt, Univ. of Calif. Berkeley).
46. Pavlovskii, Y. N., *Z. Vycisl. Mat. i Mat. Fiz.* **4**, 178 (1964).
47. Liu, Shen-Tsuan, *Z. Vycisl. Mat. i Mat. Fiz.* **2**, 638 (1962).
48. Liu, Shen-Tsuan, *Z. Vycisl. Mat. i Mat. Fiz.* **2**, 868 (1962).
49. Abbott, D. E., and Bethel, H. E., *Ing. Arch.* **37**, 110 (1968). (See also H. E. Bethel and D. E. Abbott, "On the Convergence and Exactness of Solutions of the Laminar Boundary-Layer Equations using the N-Parameter Integral Formulation of Galerkin-Kantorovich-Dorodnitsyn" Tech. Rept. FMTR-66-2, School of Mech. Eng., Purdue Univ. 1966. Also published under the same title as ARL 66-0090, 1966 Office of Aerospace Research, U.S.A.F., Wright Patterson Air Forse Base, Ohio).
50. Goertler, H., *J. Math. Mech.* **6**, 1 (1957).
51. Babenko, K. I., "Numerical Methods for Solving Problems in Gasdynamics," Proc. All-Union Conf. on Theo. Appl. Mech. 27 Jan.–3 Feb. (1960) *Izv. Akad. Nauk SSSR*, Moscow (1962). (Translation as Foreign Tech. Div. Tech. Transl. FTD-TT-63-471).
52. Truckenbrodt, E., *Ing. Arch.* **20**, 211 (1952).
53. Pallone, A., *J. Aeronaut. Sci.* **28**, 449 (1961).
54. Medvedev, V. A., *Prikl. Mat. Meh.* **27**, 1148 (1963).

55. Hanson, F. B., and Richardson, P. D., *J. Mech. Eng. Sci.* **7**, 131 (1965).
56. Nielsen, J. N., Lynes, L. L., and Goodwin, F. K., *Tech. Rep.* AFFDL-TR-65-107, Itek Corp. (1965).
57. Koob, S. J., and Abbott, D. E., *J. Basic Eng.* **90D**, 563 (1968).
58. Koob, S. J., and Abbott, D. E., *Tech. Rep.* FMTR-68-2, Purdue Univ., Lafayette, Indiana (1968).
59. Deiwert, G. S., and Abbott, D. E., *Tech. Rep.* FMTR67-2, Purdue Univ., Lafayette, Indiana (1967).
60. Deiwert, G. S., and Abbott, D. E., *AIAA J.* Paper 69-397 (1969).
61. Abbott, D. E., Deiwert, G. S., Forsnes, V. G., and Deboy, G. R., "Application of the Method of Weighted Residuals to the Turbulent Boundary Layer Equations," presented at AFOSR-IFP-Stanford 1968 Conference on Computation of Turbulent Boundary Layers, Stanford Univ., Stanford, California.
62. Smith, A. M. O., and Clutter, D. W., *AIAA J.* **1**, 2062 (1963).
63. Hicks, J., and Wei, J., *ACM J.* **14**, 549 (1967).
64. Acrivos, A., Shah, M. J., and Peterson, E. E., *Chem. Eng. Sci.* **20**, 101 (1965).
65. Kapur, J. N., and Gupta, R. C., *Z. Angew. Math. Mech.* **43**, 135 (1963); **44**, 277 (1964).
66. Rajeshwari, G. K., and Rathna, S. L., *Z. Angew. Math. Phys.* **13**, 43 (1962).
67. Lemlich, R., and Steinkamp, J. S., *AIChE J.* **10**, 445 (1964).
68. Siegel, R., and Sparrow, E. M., *AIChE J.* **5**, 73 (1959).
69. Belotserkovskii, O. M., and Chuskin, P. I., "The Numerical Method of Integral Relations," *NASA Tech. Transl.* F-8356 (1964).
70. Carter, L. F., and Gill, W. N., *AIChE J.* **10**, 330 (1964).
71. Launder, B. E., *J. Heat Transfer* **86**, 360 (1964).
72. Conway, H. D., *Trans. ASME* **82**, 275 (1960).
73. Jeffreys, H., *Quart. J. Mech. Appl. Math.* **4**, 283 (1951).
74. Kadner, H., *Z. Angew. Math. Mech.* **40**, 99 (1960).
75. Villadsen, J. V., and Stewart, W. E., "Solution of Boundary-Value Problems by Orthogonal Collocation." Chemical Eng. Dept., Univ. of Wisconsin, Madison, Wisconsin, 1967.
76. Becker, M., The Principles and Applications of Variational Methods," Research Monograph No. 27. MIT Press, Cambridge, Massachusetts, 1964.
77. Yamada, H., "A Method of Approximate Integration of the Laminar Boundary Layer Equations." *Rep. Res. Inst. Fluid Eng., Kyushu Univ.* **6**, 87 (1950).
78. Prager, S., *Chem. Eng. Progr., Symp. Ser.* **55**, #25, 11 (1959).
79. Poots, G., *Int. J. Heat Mass Transfer* **5**, 339 (1962).
80. Lourie, A., and Tchekmarev, A., *Prikl. Mat. Meh.* **1**, 324 (1937).
81. Panov, D. J., *Prikl. Mat. Meh.* **3**, 139 (1939).
82. Schwesinger, G., *J. Appl. Mech.* **17**, 202 (1950).
83. Eringen, A. C., *J. Appl. Mech.* **20**, 461 (1953).
84. Fung, Y. C., *J. Aerosp. Sci.* **25**, 145 (1958).
85. Tanner, R. I., *Int. J. Mech. Sci.* **3**, 13 (1961).
86. Nowinski, J. L., *AIAA J.* **1**, 617 (1963).
87. Nowinski, J. L., and Woodall, S. R., *J. Acoust. Soc.* **36**, 2113 (1964).
88. Nowinski, J. L., *J. Appl. Mech.* **31**, 72 (1964).
89. Chu, H -n., and Herrmann, G., *J. Appl. Mech.* **23**, 532 (1956).
90. Chu, H -n., *J. Aerosp. Sci.* **28**, 602 (1961).
91. Kuo, C -p., Nonlinear Vibration of Thin Plates. Ph.D. Dissertation, Univ. of Iowa, Iowa City, Iowa, 1971.

92. Prescott, J., "Applied Elasticity." Dover, New York, 1961.
93. Evensen, D. A., Nonlinear Flexural Vibrations of Thin Circular Rings. Ph.D. Dissertation, California Inst. of Technology, Pasadena, California, 1964; see also "A Theoretical and Experimental Study of the Nonlinear Flexural Vibrations of Thin Circular Rings," NASA TR R-227 (1965).
94. Dowell, E. H., *AIAA J.* **5**, 1508 (1967).
95. Advani, S. H., Non-linear Vibrations of Spinning Dics and Wave Propagation in an Infinite Plate. Ph. D. Dissertation, Stanford University, Stanford, California, 1965.
96. Advani, S. H., *J. Appl. Mech.* **34**, 1044 (1967).
97. Advani, S. H., *Int. J. Mech. Sci.* **9**, 307 (1967).
98. Woodall, S. R., *Internat. J. Non-linear Mech.* **1**, 217 (1966).
99. Repman, Yu. V., *Prikl. Mat. Meh.* **4**, 3 (1940).
100. Petrov, G. I., *Prikl. Mat. Meh.* **4**, 31 (1940).
101. Keldysh, M. V., *Izv. Akad. Nauk SSSR Ser. Mat.* **6**, 309 (1942); Also *NASA Tech. Transl.* F-195 (1964).
102. Sani, R. L., "Convergence of a Generalized Galerkin Method for Certain Fluid Stability Problems." *Rensselaer Polytech. Inst. Math. Rep.* No. 63, Troy, New York, 1964.
103. Di Prima, R. C., and Sani, R. L., *Quart. Appl. Math.* **23**, 183 (1965).
104. Faedo, S., *Ann. Scuola Norm. Sup. Pisa* **1**, 1 (1949).
105. Green, J. W., *J. Res. Nat. Bur. Stand.* **51**, 127 (1953).
106. Hopf, E., *Math. Nachr.* **4**, 213 (1950/1951).
107. Il'in, A. M., Kalashnikov, A. S., and Oleinik, O. A., *Russian Math. Surveys* **17**, 1 (1962).
108. Meister, B., *Arch. Rational Mech. Anal.* **14**, 81 (1963).
109. Finlayson, B. A., and Scriven, L. E., *Proc. Roy. Soc. Ser. A* **A310**, 183 (1969).
110. Kadner, H., *Z. Angew. Math. Mech.* **40**, 99 (1960).
111. Kravchuk, M. F., Zh. In-ta Mat., *Akad. Nauk UkSSR* **1**, 23 (1936).
112. Krasnosel'ski, M. A., "Topological Methods in the Theory of Nonlinear Integral Equations." Pergamon/Macmillan, Oxford and New York, 1964.
113. Glansdorff, P., "Mechanical Flow Processes and Variational Methods Based on Fluctuation Theory," *in* "Non-equilibrium Thermodynamics, Variational Techniques and Stability" (Donnelly, R. J., Herman, R., and Prigogine, I., eds.), p. 45. Univ. of Chicago Press, Chicago, Illinois, 1966.
114. Finlayson, B. A., and Scriven, L. E., "On the Search for Variational Principles." Chem. Eng. Dept., Univ. of Minnesota, Minneapolis, Minnesota, 1967.
115. Kodnar, R., *Z. Angew. Math. Mech.* **44**, 579 (1964).
116. Leipholz, H., *Acta Mech.* **1**, 339 (1966).
117. Protter, M. H., and Weinberger, H. F., "Maximum Principles in Differential Equations." Prentice-Hall, Englewood Cliffs, New Jersey, 1967.
118. Ames, W. F., "Numerical Methods for Partial Differential Equations." Nelson & Sons, London; Barnes & Noble, New York, 1969.
119. Giese, J., *J. Math. Anal. Appl.* **15**, 397 (1966).
120. Gunderson, R. M., *J. Math. Mech.* **17**, 491 (1967).
121. Ames, W. F., and Jones, S. E., *J. Math. Anal. Appl.* **21**, 479 (1968).
122. Courant, R., and Hilbert, D., "Methods of Mathematical Physics," Vol. 2. Wiley (Interscience), New York, 1962.
123. Bernstein, S. N., *Math. Ann.* **69**, 82 (1910).
124. Finn, R., *Arch. Rational Mech. Anal.* **14**, 337 (1963).

125. Nitsche, J. C. C., *Ann. of Math.* **66**, 543 (1957).
126. Serrin, J. B., *Arch. Rational Mech. Anal.* **14**, 337 (1963).
127. Bers, L., *J. Res. Nat. Bur. Standards, Sect. B* **51**, 229 (1953).
128. Bers, L., *Comm. Pure Appl. Math.* **7**, 441 (1954).
129. Finn, R., and Gilbarg, D., *Acta Math.* **98**, 265 (1957).
130. Finn, R., and Gilbarg, D., *Comm. Pure Appl. Math.* **10**, 23 (1957).
131. Gilbarg, D., *J. Rational Mech. Anal.* **1**, 309 (1952).
132. Gilbarg, D., *J. Rational Mech. Anal.* **2**, 233 (1953).
133. Gilbarg, D., *Math. Z.* **72**, 165 (1959).
134. Gilbarg, D., and Shiffman, M., *J. Rational Mech. Anal.* 3, 209 (1954).
135. Lavrentiev, M., *Mat. Sb.* **46**, 391 (1938).
136. Serrin, J. B., *Amer. J. Math.* **74**, 492 (1952).
137. Serrin, J. B., *J. Rational Mech. Anal.* **1**, 563 (1952).
138. Serrin, J. B., *J. Rational Mech. Anal.* **2**, 563 (1953).
139. Serrin, J. B., *J. Math. and Phys.* **33**, 27 (1954).
140. Adams, E., "Theoretical Bounds of Temperature in a River, Lake or Bay with Thermal Discharge." Wissen. Rat. Rep., Univ. Karlsruhe, Karlsruhe, West Germany, 1971; Abstract p. 539 Proc. CANCAM 1971, Univ. Calgary, Calgary, Canada.
141. Ladyzhenskaya, O. A., and Ural'tzeva, N. N., *Uspehi Mat. Nauk* **16**, 19 (1960); translated in *Russian Math. Surveys* **16**, 17 (1961).
142. Landis, E. M., *Uspehi Mat. Nauk* **14**, 21 (1959); translated as *Amer. Mat. Soc. Transl.*, ser 2, #20.
143. Landis, E. M., *Uspehi Mat. Nauk.* **18**, 3 (1963); translated in *Russian Math. Surveys* **18**, 1 (1963).
144. Friedman, A., "Partial Differential Equations of Parabolic Type." Prentice-Hall, Englewood Cliffs, New Jersey, 1964.
145. Nickel, K., *Arch. Rational Mech. Anal.* **2**, 1 (1958).
146. Oleinik, O. A., "Seminari 1962-1963 di Analise, Algebra, Geometria, e Topologia." Istituto Nazionale di Alta Matematica, Edizioni Cremonese, **1**, 372 (1965).
147. Velte, W., *Arch. Rational Mech. Anal.* **5**, 420 (1960).
148. Aronsson, G., *Ark. Mat.* **7**, 395 (1968).
149. Szarski, J., *Ann. Polon. Math.* **2**, 237 (1955).
150. Szarski, J., *Ann. Polon. Math.* **15**, 15 (1964).
151. Szarski, J., "Differential Inequalities." Polish Scientific Publishers, Warsaw, 1965.
152. Mlak, W., *Ann. Polon. Math.* **3**, 349 (1957).
153. Besala, P., *Ann. Polon. Math.* **13**, 247 (1963).
154. Besala, P., *Ann. Polon. Math.* **14**, 289 (1964).
155. McNabb, A., *J. Math. Anal. Appl.* **3**, 133 (1961).
156. Juberg, R. K., "Several Observations Concerning a Maximum Principle for Parabolic Systems." *Tech. Rep.*, Univ. of Calif. Irvine, California, 1967.
157. Walter, W., "Differential und Integral-Ungleichungen." Springer-Verlag, Berlin and New York, 1964. (English Translation, 1970.)
158. Lakshmikantham, V., and Leela, S., "Differential and Integral Inequalities," Vol. II. Academic Press, New York, 1969.
159. Bellman, R. E., and Kalaba, R. E., "Quasilinearization and Nonlinear Boundary Value Problems." Elsevier, Amsterdam, 1965.
160. Lee, E. S., "Quasilinearization and Invariant Imbedding." Academic Press, New York, 1968.
161. Friedrichs, K. O., *Bull. Amer. Math. Soc.* **61**, 485 (1955).

162. van Dyke, M., "Perturbation Methods in Fluid Mechanics." Academic Press, New York, 1964.
163. Nowinski, J. L., *Acta Mech.* **4**, 79 (1967).
164. Sokolnikoff, I. S., "Mathematical Theory of Elasticity." McGraw-Hill, New York, 1956.
165. Keller, J. B., and Ting, L., *Comm. Pure Appl. Math.* **19**, 371 (1966).
166. Wang, A. S. D., *Internat. J. Engrg. Sci.* **7**, 1199 (1969).
167. Wang, A. S. D., *J. Franklin Inst.* **289**, 361 (1970).
168. Zahorski, S., *Arch. Mech. Stos.* **11**, 613 (1959).
169. Isihara, A., Hashitrume, N., and Tatibana, M., *J. Chem. Phys.* **19**, 1508 (1951).
170. Wang, A. S. D., Large Amplitude Oscillations of Thick and Thick-Walled Elastic Incompressible Bodies. Ph.D. Thesis, Univ. of Delaware, Newark, Delaware, 1967.
171. Stoker, J. J., "Periodic Motions in Non-Linear Systems Having Infinitely Many Degrees of Freedom," Actes du Colloque Int. des Vibrations Non Lineaires Pub. Sci. et Tech. du Ministeres de L'air, No. 281, Paris p. 61 (1953).
172. Carrier, G., *Quart. Appl. Math.* **3**, 157 (1945); **7**, 97 (1949).
173. Penney, W. G., and Price, A. T., *Philos. Trans. Roy. Soc. London Ser. A* **A244**, 254 (1952).
174. Tadjbakhsh, I., and Keller, J. B., *J. Fluid Mech.* **8**, 442 (1960).
175. Verma, G. R., and Keller, J. B., *Phys. Fluids* **5**, 52 (1962).
176. Concus, P., *J. Fluid Mech.* **14**, 568 (1962).
177. Ames, W. F., "Nonlinear Ordinary Differential Equations in Transport Processes." Academic Press, New York, 1968.
178. Liu, F. C., "Nonlinear Vibration of Beams." *Aero-Astro. Res. Dev. Rev.* **1**, 74 (1964).
179. McQueary, C. E., On Periodic Solutions of Autonomous and Nonautonomous Nonlinear Partial Differential Equations of Motion of Certain Elastic Continua. Ph.D. Dissertation, Univ. of Texas, Austin, Texas, 1966.
180. McQueary, C. E., and Clark, L. G., *Proc. 10th Midwest. Mech. Conf.* (1967).
181. McQueary, C. E., and Clark, L. G., *Internat. J. Non-linear Mech.* **2**, 331 (1967).
182. Stoker, J. J., "Nonlinear Vibrations." Wiley (Interscience), New York, 1950.
183. Rosenberg, R. M., *Appl. Mech. Rev.* **14**, 837 (1961).
184. Yen, D. H. Y., and Tang, S-c., "On the Nonlinear Response of an Elastic String to a Moving Load," Preprint, Dept. of Metallurgy, Mechanics and Materials Science, Michigan State Univ., East Lansing, Michigan, 1970.
185. Steele, C. R., *Int. J. Solids Struct.* **3**, 565 (1967).
186. Thurman, A. L., and Mote, C. D., Jr., *J. Appl. Mech.* **36**, 83 (1969); see also *J. Appl. Mech.* **38**, 279 (1971).
187. Thurman, A. L., and Mote, C. D., Jr., *J. Eng. Ind.* (to be published).
188. Luke, J. C., *Proc. Roy. Soc. Ser. A* **A292**, 403 (1966).
189. Kruskal, M. D., and Zabusky, N. J., *J. Math. Phys.* **5**, 231 (1964).
190. Fox, P. A., *J. Math. and Phys.* **34**, 133 (1955).
191. Millman, M. H., and Keller, J. B., *J. Math. Phys.* **10**, 342 (1969).
192. Lighthill, M. J., *Aerosp. Quart.* **3**, 193 (1951).
193. Latta, G. E., Singular Perturbation Problems, Ph.D. Dissertation, California Inst. Tech., Pasadena, California, 1951.
194. Lagerström, P. A., Howard, L. N., and Liu, C-s., (eds.), "Fluid Mechanics and Singular Perturbations"—A Collection of Papers by Saul Kaplun. Academic Press, New York, 1967.
195. Lagerström, P. A., and Cole, J. D., *J. Rational Mech. Anal.* **4**, 817 (1955).
196. Lagerström, P. A., *J. Math. Mech.* **6**, 605 (1957).

197. van Dyke, M., "Perturbation Methods in Fluid Mechanics." Academic Press, New York, 1964.
198. Cole, J. D., "Perturbation Methods in Applied Mathematics." Ginn (Blaisdell), Boston, Massachusetts, 1968.
199. Eckhaus, W., "Boundary Layers in Linear Elliptic Singular Perturbation Problems." Rept. Dept. of Mathematics, Technische Hogeschool Delft, Netherlands, 1971.
200. Martin, E. D., "A Method of Asymptotic Expansions for Singular Perturbation Problems with Application in Viscous Flow." NASA-TN-D-3899 (1967).
201. Erdelyi, A., "Asymptotic Expansions." Dover, New York, 1956.
202. Lighthill, M. J., *J. Fluid Mech.* **4**, 383 (1958).
203. Lighthill, M. J., *Phil. Mag.* **40**, 1179 (1949).
204. Lighthill, M. J., *Z. Flugwiss.* **9**, 267 (1961).
205. Tsien, H. S., "The Poincare-Lighthill-Kuo Method." *Advan. Appl. Mech.* **4**, 281. Academic Press, New York, 1956.
206. Lin, C. C., *J. Math. and Phys.* **23**, 117 (1954).
207. Pritulo, M. F., *J. Appl. Math. Mech.* **26**, 661 (1962).
208. Lighthill, M. J., *Aerosp. Quart.* **3**, 193 (1951).
209. Hoogstraten, H. W., *J. Engrg. Math.* **1**, 51 (1967).
210. Bollheimer, L., and Weissinger, J., *Z. Flugwiss.* **16**, 6 (1968).
211. Comstock, C., *SIAM J. Appl. Math.* **16**, 596 (1968).
212. Martin, E. D., *J. Inst. Math. Appl.* **3**, 16 (1967).
213. Martin, E. D., "Generalizations of Lagrange's Expansion Combined with Lighthill's Technique for Uniformizing Solutions of Partial Differential Equations." NASA-TN-D-6245 (1971).
214. Good, I. J., *Proc. Cambridge Phil. Soc.* **56**, 367 (1960).
215. Sturrock, P. A., *J. Math. Phys.* **1**, 405 (1960).
216. Sack, R. A., *SIAM J. Appl. Math.* **13**, 913 (1965).
217. Sack, R. A., *SIAM J. Appl. Math.* **14**, 1 (1966).
218. Martin, E. D., Study of Directional-Mean-Free-Path Method for Gas Flow in Translational Nonequilibrium, Ph.D. Dissertation, Stanford Univ., Stanford, California, 1968; also NASA-TN-D5762 (1970).
219. Bellman, R. E., "Perturbation Techniques in Mathematics, Physics and Engineering." Holt, New York, 1964.
220. Kevorkian, J., "The Two Variable Expansion Procedure for the Approximate Solution of Certain Nonlinear Differential Equations." Lectures in Appl. Math., *Amer. Math. Soc.*, Providence, Rhode Island **7**, 206 (1966).
221. Frieman, E. A., *J. Math. Phys.* **4**, 410 (1963).
222. Reiss, E. L., *SIAM Rev.* **13**, 189 (1971).
223. Keller, J. B., and Kogelman, S., *SIAM J. Appl. Math.* **18**, 748 (1970).
224. Davy, D. T., On Wave Propagation in Nonlinear Rate Dependent Materials, Ph.D. Dissertation, Univ. of Iowa, Iowa City, Iowa, 1971.
225. Bogoliubov, N., and Mitropolsky, Y. A., "Asymptotic Methods in the Theory of Nonlinear Oscillations." Gordon & Breach, New York, 1961.
226. Montgomery, D., and Tidman, D. A., *Phys. Fluids* **7**, 242 (1964).
227. Tidman, D. A., and Stainer, H. M., *Phys. Fluids* **8**, 345 (1965).
228. Nayfeh, A. H., *Phys. Fluids* **8**, 1896 (1965).

CHAPTER **4**

Numerical Methods

4.0 INTRODUCTION

Volume I contains an extensive chapter (Chapter 7) on numerical methods, which is primarily concerned with finite-difference procedures and the related method of characteristics. Of all the methods available, only that of finite differences seemed to be of universal applicability to both linear and nonlinear problems. However, the approximation of large systems of equations by finite differences very often leads to severe computational complexities. The impatient engineer or physicist, in immediate need of numbers to describe his design process, often formulates crude *ad-hoc* models or methods which, having served their purpose, are later discarded. Included are Heaviside's operator methods, Southwell's demonstration of the practical use of finite differences in his relaxation schemes, and, more recently, the *finite-element* method. The use of finite elements is an alternative approach to that of finite differences and its considerable advantages and relatively simple logic make it ideally suited for digital computation. Indeed, its popularity among engineers is already assured, even though its recognized "birth date" of 1955 is little more than one and one-half decades distant.

Since the formulation of the Navier–Stokes equations; one of the most challanging problems has been that of extracting solutions from them for the flow of viscous fluids. It is only through the essential nonlinearity of the equations that many of the interesting observed phenomena of fluid mechanics can be explained. Those necessary nonlinearities render the equations analytically difficult, except in special cases. Consequently, one can understand the accelerating use of finite difference and finite element methods in solving problems of fluid mechanics, plasma physics, and related phenomena. A great variety of techniques have been proposed and successfully applied on one or more problems,[†] in fact, there seem to be as many numerical methods as problems. Procedures for multidimensional problems are in existence, and some are very versatile.

This chapter will consist of three parts. In the first, we shall present the basic concepts of *finite elements* together with examples and a collection of references for nonlinear problems. In the second part, we discuss the numerical solution of various nonlinear problems of *viscous flow* and present a literature review. In the last section, we shall discuss some of the *new directions* being pursued in the numerical solution of nonlinear equations.

A. FINITE ELEMENTS

4.1 INTRODUCTION TO FINITE ELEMENTS

Methods for the analysis of discrete structural, hydraulic, and electrical networks are well understood by the practicing engineer. In treating the continuum, it was quite natural to attempt to reduce it, in an essentially physical manner, to an assembly of discrete, equivalent elements. The first attempts employed a decomposition of elastic continua into an assembly of rod or beams which were standard engineering structures. The early works of Hrenikoff [1], in which elasticity problems were solved by a "framework" method, and McHenry [2], who introduced a lattice analogy for plane stress problems, bear special mention. Courant [3], in a now classic work, formulated the essence of a triangular finite element. We should also note the paper of Newmark [4] and the parallel work of Prager and Synge [5] which led to the hypercircle procedure (cf., e.g., Synge [6]).

Credit for approximating directly to a continuum by an element with mutliple connecting points and the use of the term *finite element* must go to Turner *et al.* [7], Argyris [8], and Clough [9]. These pioneering efforts developed the basic properties by physical arguments related to the stress or displacement distribution within the subregion. One rational way of deriving stiffness properties of elements is to assume a *displacement*

[†] Some assert that any solution procedure which solves *more than one* nontrivial nonlinear problem should be called a "general" method.

pattern defined in terms of nodal displacements, element by element, so that continuity is observed and internal forces are obtained by virtual work. This process is essentially equivalent to the approximate minimization of total potential energy (e.g., Rayleigh–Ritz processes) as discussed by Clough [10]. There are two advantages in this procedure over the standard approximation. First, the piecewise continuous field definition permits irregular boundaries to be simply fitted. The second advantage lies in the fact that this piecewise process generates, from the minimization equations, banded (or sparse) matrices unlike the Ritz process. Solutions by direct or iterative matrix methods can be conveniently obtained.

The "finite-element" method, arising as it did in structural mechanics, has now become ubiquitous in that field. A rapid extension into nonstructural fields was promoted by Zienkiewicz and Cheung [11, 12] in general field problems, Wilson and Nickell [13] in heat conduction, Herrman [14] in elastic and torsional analysis of irregular shapes, Zienkiewicz *et al.* [15] in three-dimensional field problems, and Winslow [16] in certain fluid flow problems governed by a quasi-linear Poisson equation.

Reduction of a continuum to a discrete formulation can be accomplished in a variety of ways. We shall begin with the popular *displacement formulation* so often used in structural mechanics (cf., e.g., Zienkiewicz and Cheung [12]). The steps are as follows.

(1) Separate the continuum by imaginary lines or surfaces into a number of *finite elements* (triangles, rectangles, cubes, tetrahedrons).

(2) The elements are assumed to be interconnected at a discrete number of nodal points on their boundaries. The displacements of these nodal points will be the basic unknown parameters.†

(3) A function, or functions, is chosen to uniquely define the state of displacement within each finite element, in terms of its nodal displacements.

(4) The displacement functions of (3) uniquely define the state of strain within an element in terms of the nodal displacements. These strains, together with any initial strains and the elastic properties of the material, will define the state of stress throughout the element and on its boundaries.

(5) A system of forces concentrated at the nodes and equilibrating

† Pian [17], in the context of elasticity, shows how nodeless parameters can be introduced. But the retention of some of the nodal parameters is always essential to preserve required interelement continuity. Zienkiewicz [18] discusses the introduction of parameters other than nodal ones also.

the boundary stresses and any distributed loads is determined resulting in a stiffness relation,[†]

$$\{F\} = [k]\{\delta\} + \{F\}_p + \{F\}_{\epsilon 0}\,, \tag{4.1}$$

where $\{F\}$ is the vector of forces acting on all the nodes, $\{\delta\}$ is the vector of corresponding nodal displacements, $[k]$ is the element stiffness matrix, $\{F\}_p$ is the vector of nodal forces required to balance any distributed loads acting on the element, and $\{F\}_{\epsilon 0}$ is the vector of nodal forces required to balance any initial strains such as may be due to thermal changes, if the nodes are not subject to any displacement. The first of the terms represents the forces induced by nodal displacements. In a similar way we obtain a corresponding unique definition of stresses at any specified point or points of the element in terms of the nodal displacements;

$$\{\sigma\} = [s]\{\delta\} + \{\sigma\}_p + \{\sigma\}_{\epsilon 0}\,, \tag{4.2}$$

where $\{\sigma\}$ are the stresses and $[s]$ is the element stress matrix.

The approximations employed in the procedure outlined above lead to some difficulties which need discussion. It is sometimes difficult, with the chosen displacement functions, to satisfy the displacement continuity requirement between adjacent elements. Thus the *compatibility condition on such lines may be violated* though within each element it is satisfied due to the uniqueness of displacements implied in its definition. *Second, by concentrating the equivalent forces at the nodes, equilibrium conditions are satisfied in the overall sense only. Thus we can expect local violation of equilibrium conditions within each element and on its boundaries. Third, since the choice of element shape and the form of the displacement function, for specific cases, is not specified the degree of approximation which can be achieved will depend on the ingenuity, skill, and physical knowledge of the investigator.*

4.2 FORMULATION OF FINITE ELEMENT CHARACTERISTICS

We wish to develop results in a general form applicable to any situation. However, to motivate the development, the relations will be illustrated with the example of plane stress analysis of a region[‡] with triangular elements shown in Fig. 4-1. A typical finite element e is defined by the nodes i, j, m, etc., and the straight line boundaries. We define the

[†] We denote by $\{F\}$ a *vector* and by $[k]$ a *matrix*.

[‡] Note how the elements can be chosen to closely approximate the boundary, yet the number of interior and boundary nodal points is usually less than by other methods.

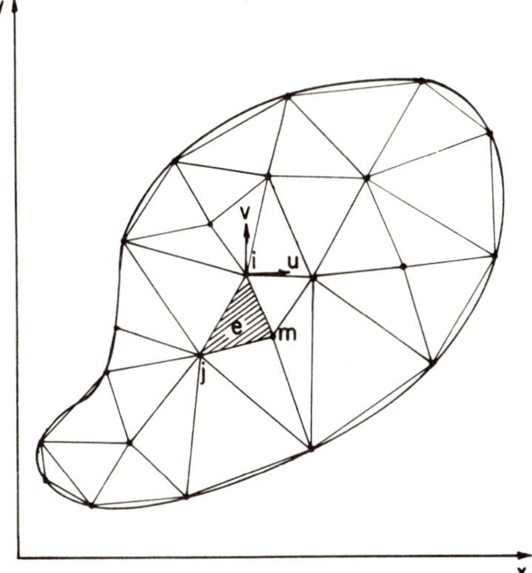

FIG. 4-1. Plane stress region divided into triangular finite elements.

"*displacements*"† at any point within the element, as a column vector $\{f(x, y)\}$, by means of the general relation

$$\{f\} = [N]\{\delta\}^e = [N_i N_j N_m \cdots] \begin{Bmatrix} \delta_i \\ \delta_j \\ \delta_m \\ \vdots \end{Bmatrix}, \tag{4.3}$$

where the components of $[N]$ are functions of position and $\{\delta\}^e$ represents a listing of nodal displacements for a particular element.

In the *simple example of plane stress* (Fig. 4-1)

$$\{f\} = \begin{Bmatrix} u(x, y) \\ v(x, y) \end{Bmatrix}$$

represents horizontal and vertical displacements of a typical point in e and

$$\{\delta_i\}^e = \begin{Bmatrix} u_i \\ v_i \end{Bmatrix}$$

stands for the displacements of a node i.

† We use the terms "displacements", "strains", and "stresses" in a general way. In the example of plane stress, the meaning is obvious. For other cases, as in plate elements, the "displacements" may be given by the lateral deflection and slopes of the plate, the "strains" by curvatures of the middle surface, and the "stresses" by internal bending moments (see Zienkiewicz and Cheung [12]). All the expressions obtained here are generally valid, provided the external work and the total internal work are as given by the general terms.

The functions N_i, N_j, N_m, etc., have to be selected to give appropriate nodal displacements when the coordinates of the appropriate nodes are substituted into Eq. (4.3). If the nodes are denoted by (x_i, y_i), (x_j, y_j), etc., then this requirement for our simple case becomes

$$\begin{aligned} N_i(x_i, y_i) &= I \quad \text{(identity matrix)}, \\ N_i(x_j, y_j) &= N_i(x_m, y_m) = 0, \end{aligned} \tag{4.4}$$

with similar relationships for N_j and N_m. These are satisfied by suitable linear functions of x and y. In this case, we note that each of the nodal displacements of

$$\{\delta\}^e = \begin{Bmatrix} \delta_i \\ \delta_j \\ \delta_m \end{Bmatrix} \tag{4.5}$$

has two components

$$\{\delta_i\} = \begin{Bmatrix} u_i \\ v_i \end{Bmatrix}. \tag{4.6}$$

Consequently, the displacements within an element must be uniquely defined by these six values. The simplest representation is provided by two linear relations

$$u = \alpha_1 + \alpha_2 x + \alpha_3 y, \qquad v = \beta_1 + \beta_2 x + \beta_3 y, \tag{4.7}$$

containing six constants α_n, β_n. These can be evaluated by solving the two sets of three simultaneous arising when the nodal coordinates are inserted and the displacements are equated to the nodal displacements. Thus we find

$$u = 2\Delta^{-1}\{(a_i + b_i x + c_i y)\, u_i + (a_j + b_j x + c_j y)\, u_j + (a_m + b_m x + c_m y)\, u_m\}, \tag{4.8}$$

with

$$\begin{aligned} a_i &= x_j y_m - x_m y_j, \\ b_i &= y_j - y_m, \\ c_i &= x_m - x_j. \end{aligned} \tag{4.9}$$

The other coefficients are obtained by a cyclic permutation of the subscripts. Thus

$$\begin{matrix} i, & j, & m \\ \downarrow & \downarrow & \downarrow \\ j, & m, & i \\ \downarrow & \downarrow & \downarrow \\ m, & i, & j, \end{matrix} \tag{4.10}$$

so that, in particular, $a_j = x_m y_i - x_i y_m$, etc. The symbol

$$2\Delta = \det \begin{vmatrix} 1 & x_i & y_i \\ 1 & x_j & y_j \\ 1 & x_m & y_m \end{vmatrix} = 2(\text{area of triangle } ijm).^{\dagger}$$

In an entirely similar manner, we obtain the equation for the displacement component v within the element as

$$v = 2\Delta^{-1}\{(a_i + b_i x + c_i y)\, v_i + (a_j + b_j x + c_j y)\, v_j + (a_m + b_m x + c_m y)\, v_m\}. \tag{4.11}$$

Combining Eqs. (4.8) and (4.9), we have

$$\{f\} = \begin{Bmatrix} u \\ v \end{Bmatrix} = \begin{bmatrix} N_i' & 0 & N_j' & 0 & N_m' & 0 \\ 0 & N_i' & 0 & N_j' & 0 & N_m' \end{bmatrix} \begin{Bmatrix} u_i \\ v_i \\ u_j \\ v_j \\ u_m \\ v_m \end{Bmatrix}, \tag{4.12}$$

where $N_i' = (a_i + b_i x + c_i y)/2\Delta$, etc. In the standard form of Eq. (4.3), the preceding equation becomes

$$\{f\} = \begin{Bmatrix} u \\ v \end{Bmatrix} = [N_i'I \quad N_j'I, \quad N_m'I]\{\delta\}^e, \tag{4.13}$$

where I is a 2×2 identity matrix.

We especially note that the *matrix* $[N]$ *of Eq.* (4.3) *is not generally a square matrix*. The chosen displacement function, Eqs. (4.7), automatically guarantees continuity of displacements with adjacent elements. This follows because the displacements vary linearly along any side of the triangle, and with identical displacements imposed at each of the nodes the same displacements will exist at each point of the interface.

Since the displacements are now available [Eq. (4.3)] within the element, the "strains" at any point can be determined by means of a relation of the general form

$$\{\epsilon\} = [B]\{\delta\}^e. \tag{4.14}$$

For our simple plane-stress problem (using a linear strain-displacement relation) the strains in the plane are defined by the well-known

† We shall customarily number the nodes in a counterclockwise direction.

relations

$$\{\epsilon\} = \begin{Bmatrix} \epsilon_x \\ \epsilon_y \\ \gamma_{xy} \end{Bmatrix} = \begin{Bmatrix} u_x \\ v_y \\ u_y + v_x \end{Bmatrix}. \tag{4.15}$$

Thus, from Eq. (4.12), the matrix $[B]$ can be computed, and Eq. (4.14) becomes

$$\begin{Bmatrix} \epsilon_x \\ \epsilon_y \\ \gamma_{xy} \end{Bmatrix} = \frac{1}{2\Delta} \begin{bmatrix} b_i & 0 & b_j & 0 & b_m & 0 \\ 0 & c_i & 0 & c_j & 0 & c_m \\ c_i & b_i & c_j & b_j & c_m & b_m \end{bmatrix} \begin{Bmatrix} u_i \\ v_i \\ u_j \\ v_j \\ u_m \\ v_m \end{Bmatrix}. \tag{4.16}$$

Consequently, the strains will be constant throughout the element.

Of course the material within the element boundaries may be subjected to initial strains resulting from thermal changes, crystal growth, piezo-effects, etc. With these initial strains denoted by $\{\epsilon_0\}$ and assuming linear elastic behavior, the stress–strain relationship will be given by

$$\{\sigma\} = [D](\{\epsilon\} - \{\epsilon_0\}), \tag{4.17}$$

where the matrix $[D]$ contains the appropriate material constants.

For our particular case of plane stress, the three components of stress

$$\{\sigma\} = \begin{Bmatrix} \sigma_x \\ \sigma_y \\ \tau_{xy} \end{Bmatrix},$$

corresponding to the strains already defined, must be determined. From the usual isotropic stress–strain relationship,

$$\epsilon_x - (\epsilon_x)_0 = E^{-1}[\sigma_x - \nu\sigma_y],$$

$$\epsilon_y - (\epsilon_y)_0 = E^{-1}[\sigma_y - \nu\sigma_x],$$

$$\gamma_{xy} - (\gamma_{xy})_0 = [2(1+\nu)/E]\,\tau_{xy},$$

we find

$$\begin{Bmatrix} \sigma_x \\ \sigma_y \\ \tau_{xy} \end{Bmatrix} = [E/(1-\nu^2)] \begin{bmatrix} 1 & \nu & 0 \\ \nu & 0 & 0 \\ 0 & 0 & (1-\nu)/2 \end{bmatrix} \begin{Bmatrix} \epsilon_x - (\epsilon_x)_0 \\ \epsilon_y - (\epsilon_y)_0 \\ \gamma_{xy} - (\gamma_{xy})_0 \end{Bmatrix}. \tag{4.18}$$

Next, we need to discuss the development of nodal forces which are statically equivalent to the boundary stresses and distributed loads on the element. We denote these by

$$\{F\}^e = \begin{Bmatrix} F_i \\ F_j \\ F_m \\ \vdots \end{Bmatrix}, \tag{4.19}$$

and note that each of these, say $\{F_i\}$, must contain the same number of components as the corresponding nodal displacement $\{\delta_i\}$ and be ordered in the same way. The distributed loads $\{p\}$ are defined as those acting on a unit volume of material within the element with directions corresponding to those of the displacements $\{f\}$ at that point.

For the particular case of plane stress, the nodal forces, with components U and V corresponding to the directions of u and v displacements, are

$$\{F_i\} = \begin{Bmatrix} U_i \\ V_i \end{Bmatrix},$$

and the distributed load is

$$\{p\} = \begin{Bmatrix} X \\ Y \end{Bmatrix},$$

where X and Y are the "body force" components.

The simplest strategy, to make the nodal forces statically equivalent to the actual boundary stresses and distributed loads, is to impose an arbitrary (virtual) nodal displacement and equate the external and internal work done by the various forces and stresses during that displacement. With a virtual displacement $\{\delta^*\}^e$ at the elemental nodes, the displacements and strains within the element [Eqs. (4.3) and (4.14)] become, respectively,

$$\{f^*\} = [N]\{\delta^*\}^e \qquad \text{and} \qquad \{\epsilon^*\} = [B]\{\delta^*\}^e, \tag{4.20}$$

The *work done by the nodal forces* is equal to the sum of the products of the individual force components and corresponding displacements

$$(\{\delta^*\}^e)^{\mathrm{T}}\{F\}^e, \tag{4.21}$$

where the notation $(\ )^{\mathrm{T}}$ indicates the transpose operation of the matrix. The *internal work per unit volume* done by the stresses and distributed forces is

$$\{\epsilon^*\}^{\mathrm{T}}\{\sigma\} - \{f^*\}^{\mathrm{T}}\{p\} = (\{\delta^*\}^e)^{\mathrm{T}}([B]^{\mathrm{T}}\{\sigma\} - [N]^{\mathrm{T}}\{p\}. \tag{4.22}$$

Upon equating the external work, Eq. (4.21), with the *total* internal work, obtained by integrating Eq. (4.22) over the volume of the element, we have

$$(\{\delta^*\}^e)^{\mathrm{T}}\{F\}^e = (\{\delta^*\}^e)^{\mathrm{T}} \left(\int [B]^{\mathrm{T}}\{\sigma\}\, d(\mathrm{vol}) - \int [N]^{\mathrm{T}}\{p\}\, d(\mathrm{vol})\right). \tag{4.23}$$

Since Eq. (4.23) is valid for any virtual displacement, the multipliers must be equal. Upon substitution of Eqs. (4.14) and (4.17) into the equal multipliers of Eq. (4.23), we find

$$\{F\}^e = \left(\int [B]^{\mathrm{T}}[D][B]\, d(\mathrm{vol})\right) \{\delta\}^e - \int [B]^{\mathrm{T}}[D]\{\epsilon_0\}\, d(\mathrm{vol})$$

$$- \int [N]^{\mathrm{T}}\{p\}\, d(\mathrm{vol}). \tag{4.24}$$

By examining the characteristics of the general equation (4.1) for any structural element, we see that Eq. (4.24) is typical if the *stiffness matrix* is

$$[k]^e = \int [B]^{\mathrm{T}}[D][B]\, d(\mathrm{vol}), \tag{4.25}$$

the *nodal forces due to distributed loads* are

$$\{F\}_p{}^e = - \int [N]^{\mathrm{T}}\{p\}\, d(\mathrm{vol}), \tag{4.26}$$

and *those due to initial strain* are

$$\{F\}_{\epsilon 0}^e = - \int [B]^{\mathrm{T}}[D]\{\epsilon_0\}\, d(\mathrm{vol}). \tag{4.27}$$

In the simple case of the plane stress triangular element, both [*B*] [Eq. (4.16)] and [*D*] [Eq. (4.18)] are independent of the coordinates, so the indicated integrations are very simple.

Lastly, the interconnection and solution of the entire assembly follows standard procedures which we sketch here for completeness. To obtain a complete solution, the two conditions of *displacement continuity* and *equilibrium* must be satisfied throughout the body. Assuming that the displacement functions have been chosen to satisfy continuity, we turn our attention to equilibrium. Overall equilibrium conditions have already been satisfied *within* an element. Therefore all that is necessary is to establish equilibrium conditions at the nodes. The resulting equations will contain the nodal displacements as unknowns, and once these have been solved, the problem is completed.

Suppose the structure is loaded by external concentrated forces

$$\{R\} = \begin{Bmatrix} R_1 \\ \vdots \\ R_n \end{Bmatrix},$$

applied at the nodes *in addition* to the distributed loads applied to the individual elements. Each one of the forces $\{R_i\}$ must have the same number of components as that of the elements considered. For equilibrium to be established at a typical node i, each component of R_i has, in turn, to be equated to the sum of the component forces contributed by all the elements meeting at the node. Thus considering all the four components

$$\{R_i\} = \sum \{F_i\}, \tag{4.28}$$

with summation over all the elements. Using the characteristics of Eq. (4.1), for each element, and taking note only of the appropriate forces F_i, by using the submatrices of the square matrix

$$[k] = \begin{bmatrix} k_{ii} & k_{ij} & k_{im} & \cdots \\ k_{mi} & k_{mj} & k_{mm} & \cdots \end{bmatrix},$$

(here k_{ii} are square submatrices of size $l \times l$, where l is the number of force components at each node) Eq. (4.28) becomes

$$\{R_i\} = \sum_{m=1} \sum_{i} [k_{im}]^e \{\delta_m\} + \sum_{i} \{F_i\}_p{}^e + \sum \{F_i\}_{\epsilon 0}^e, \tag{4.29}$$

where the sum on i is over all elements. This linear system is of the form

$$[K]\{\delta\} = \{R\} - \{F\}_p - \{F\}_{\epsilon 0},$$

$$[K_{im}] = \sum [k_{im}]^e, \qquad \{F_i\}_p = \sum \{F_i\}_p^e, \qquad \{F_i\}_{\epsilon 0} = \sum \{F_i\}_{\epsilon 0}^e. \tag{4.30}$$

Equation (4.29) or (4.30) must be solved by direct or iterative methods for the δ's. Once the nodal displacements have been determined, the stresses at any point of an element can be found via Eq. (4.17):

$$\{\sigma\} = [D][B]\{\delta\}^e - [D]\{\epsilon_0\}. \tag{4.31}$$

Zienkiewicz and Cheung [12] show that if the system of displacements is defined throughout the structure by the element displacement functions, with nodal displacements as the undetermined parameters, then the procedure of minimizing the total potential energy generates the same formulation.

4.3 THEORETICAL COMMENTS ON DISPLACEMENT FUNCTIONS

Since the assumed displacement (shape) functions limit the infinite number of degrees of freedom the true minimum will probably not be achieved no matter the fineness of the subdivision. To ensure *convergence* to the correct result some simple requirements have to be satisfied (Zienkiewicz and Cheung [12]).

(a) The selected displacement function should not permit straining of an element to occur when the nodal displacements are caused by a rigid body rotation [special case of (b)];
(b) The displacement function has to be such that if nodal displacements are compatible with a constant strain condition such constant strain will be obtained.

Condition (b) arises because as elements become smaller, nearly constant strain conditions will prevail in them. If constant strain conditions do, in fact, exist, it is desirable for accuracy that a finite size element is able to reproduce these exactly.

We note that the discontinuity of displacement will cause *infinite strains* at the interfaces, a factor ignored in the formulation because the energy contribution is limited to the elements themselves. However, if *in the limit*, as the size of subdivision decreases, continuity is restored, then the formulation given will still converge to the correct answer. Zienkiewicz and Cheung [12] deal with *discontinuous displacement functions* which have the properties that:

(i) A constant strain condition automatically ensures displacement continuity.
(ii) The constant strain criterion of condition (b) is satisfied.

If the chosen displacement functions *do not* give rise to discontinuities between displacements of adjacent elements, then at any stage of the solution the *total energy* is above that of the minimum, but the *strain energy is always below* that of the exact solution (see de Veubeke [19]) when the displacements are prescribed. Thus a *lower bound* on the overall product sum of displacements and loadings is available to assist the engineer in assessing his results.

4.4 ADDITIONAL ELEMENTS IN TWO AND THREE DIMENSIONS

Two-dimensional problems can be treated by using more nodes, or by using elements of other polygonal shapes. If a suitable displacement function can be found for such elements, then presumably the accuracy can be improved for a given number of nodes, since the increase in degrees of freedom permits a closer approximation to the displacements within an element.

In two dimensions we shall briefly describe the *triangular element with six nodes* and *quadrilaterals*.

The triangular element with six nodes (Fig. 4-2), additional nodes being placed along the sides, is really in the spirit of a refinement into

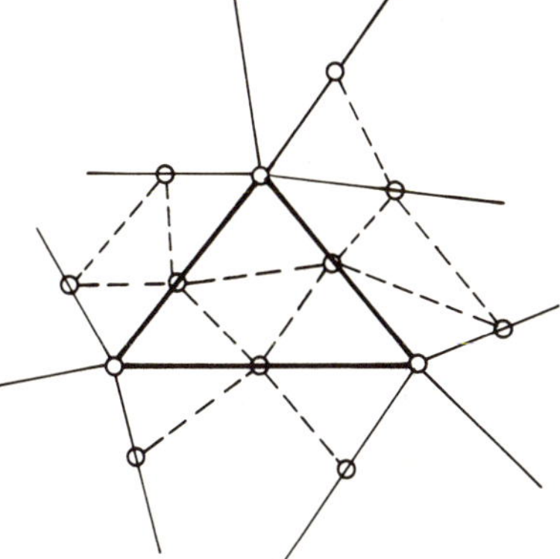

FIG. 4-2. Triangular element with six nodes.

more triangular elements as shown by the dashed lines in Fig. 4-2. But an element with six nodes permits the use of a full quadratic form containing six constants. These can be uniquely evaluated in terms of the nodal values of the function. Thus, for example, the u-displacement can be written

$$u = \alpha_1 + \alpha_2 x + \alpha_3 y + \alpha_4 x^2 + \alpha_5 xy + \alpha_6 y^2, \tag{4.32}$$

where the six nodal displacements u_γ at the nodes (x_γ, y_γ), $\gamma = 1,..., 6$, determine six linear equations for the α_i, $i = 1,..., 6$ (cf. de Veubeke [19] and Argyris [20]).

The use of a full quadratic for the displacements permits all strains and stresses to take any prescribed linear variation throughout the element and the constant stress criterion of Section 4.3 is still applicable.

With Eq. (4.32) for the u-displacement, the variation of u along any side of the element is now parabolic, that is,

$$u(s) = a + bs + cs^2,$$

where s is the parameter describing distance along the particular side. Thus the three values of u at the three nodes uniquely determine the parabola. Consequently, continuity of displacements at these nodes with those of the adjacent element is ensured automatically, guaranteeing continuity of the displacement along the entire interface. For a given number of nodes, a *better* representation of true stress and displacement is obtained than would be obtained, with the same number of nodes using a much finer subdivision (Fig. 4-2) and a linear displacement function in the four times larger number of element (cf. Zienkiewicz and Cheung [12]). For practical use, it is often convenient to place the additional nodes at the midpoints of the sides.

Pertinent equations are listed below for the triangular element with six nodes. The nodes are listed at (x_γ, y_γ), $\gamma = 1, \ldots, 6$, for convenience. Thus

$$\{\delta\}^e = \begin{Bmatrix} u_1 \\ v_1 \\ \vdots \\ u_6 \\ v_6 \end{Bmatrix} = \begin{bmatrix} 1, & x_1, & y_1, & x_1^2, & x_1y_1, & y_1^2, & 0, & 0, & 0, & 0, & 0, & 0 \\ 0, & 0, & 0, & 0, & 0, & 0, & 1, & x_1, & y_1, & x_1^2, & x_1y_1, & y_1^2 \\ 1, & x_2, & y_2, & & \cdots & & & & & & & \\ & \vdots & & & & & & & & & & \end{bmatrix} \begin{Bmatrix} \alpha_1 \\ \alpha_2 \\ \vdots \\ \alpha_6 \\ \beta_1 \\ \beta_2 \\ \vdots \\ \beta_6 \end{Bmatrix} \tag{4.33}$$

or

$$\{\delta\}^e = [C]\{\alpha\}. \tag{4.34}$$

Since at any interior point of the element

$$\{f\} = \begin{Bmatrix} u \\ v \end{Bmatrix} = \begin{bmatrix} 1, & x, & y, & x^2, & xy, & y^2, & 0, & 0, & 0, & 0, & 0, & 0 \\ 0, & 0, & 0, & 0, & 0, & 0, & 1, & x, & y, & x^2, & xy, & y^2 \end{bmatrix} \{\alpha\} = [P]\{\alpha\},$$

then

$$\{f\} = [P][C]^{-1}\{\delta\}^e = [N]\{\delta\}^e,$$

and

$$\{\epsilon\} = [Q]\{\alpha\} = [Q][C]^{-1}\{\delta\}^e = [B]\{\delta\}^e.$$

The final calculations are then made as in Section 4.1. It is convenient to evaluate $[C]^{-1}$ numerically, but only one 6×6 matrix needs inversion, since the two sets of equations for u and v are identical.

Another element, that of the quadrilateral, is an obvious extension of the triangular element. Of course, rectangular elements are but a special case. If suitable displacement functions can be found for the quadrilateral element, then the accuracy should be improved because of the increased degree of freedom. Quadrilaterals of arbitrary shape will also permit close adherence to any boundary shape thus retaining this advantage of the triangular shape.

Quadrilaterals can be obtained by the obvious *combination of adjacent triangles*. There is some computational advantage because of the reduction in input data, only half as many elements are present. There is also considerable advantage when stress averaging is considered. Zienkiewicz and Cheung [12] discusses the advantages and convergence of this combination.

For an arbitrary quadrilateral, Irons [21] has shown how to obtain a displacement (shape) function in terms of special coordinates[†] which take constant values along the sides of any arbitrary quadrilateral. The new coordinates ξ and η are such that lines of constant ξ and η are *straight* and take values of ± 1 at the sides of the quadrilateral. Both values increase along a *linear distance scale* as shown in Fig. 4-3. Along

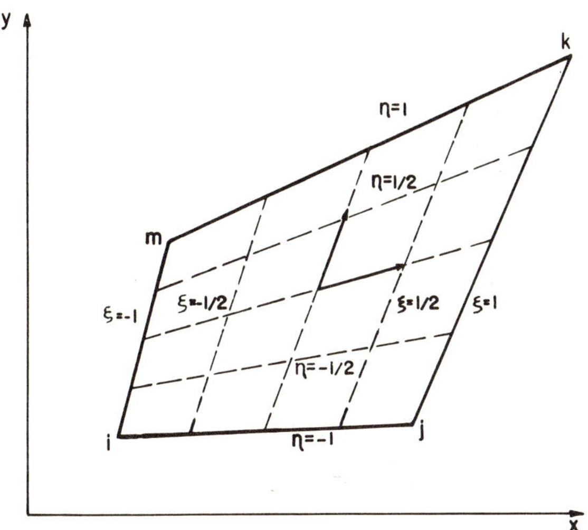

FIG. 4-3. General quadrilateral element.

[†] A *skew* coordinate system is introduced.

any $\xi =$ constant line, x and y vary linearly with η, and along any $\eta =$ constant line, x and y vary linearly with ξ. Thus we find the interpolation formula

$$\begin{aligned} x &= \tfrac{1}{4}\{(1-\xi)(1-\eta)\, x_i + (1+\xi)(1-\eta)\, x_j \\ &\quad + (1+\xi)(1+\eta)\, x_k + (1-\xi)(1+\eta)\, x_m\}, \\ y &= \tfrac{1}{4}\{(1-\xi)(1-\eta)\, y_i + (1+\xi)(1-\eta)\, y_j \\ &\quad + (1+\xi)(1+\eta)\, y_k + (1-\xi)(1+\eta)\, y_m\}, \end{aligned} \tag{4.35}$$

between the two coordinate systems.

If the expression[‡]

$$\begin{aligned} u &= \tfrac{1}{4}\{(1-\xi)(1-\eta)\, u_i + (1+\xi)(1-\eta)\, u_j \\ &\quad + (1+\xi)(1+\eta)\, u_k + (1-\xi)(1+\eta)\, u_m\}, \end{aligned} \tag{4.36}$$

is adopted, then the displacement u will vary linearly along the element sides, thus ensuring continuity. While it is complicated to express Eq. (4.36) directly in terms of x and y coordinates, it is simple to derive the strain matrices, explicitly, by noting that

$$\begin{Bmatrix} \partial/\partial\xi \\ \partial/\partial\eta \end{Bmatrix} = [J] \begin{Bmatrix} \partial/\partial x \\ \partial/\partial y \end{Bmatrix}, \qquad \begin{Bmatrix} \partial/\partial x \\ \partial/\partial y \end{Bmatrix} = [J]^{-1} \begin{Bmatrix} \partial/\partial\xi \\ \partial/\partial\eta \end{Bmatrix}, \tag{4.37}$$

where

$$[J] = \begin{bmatrix} x_\xi & y_\xi \\ x_\eta & y_\eta \end{bmatrix} = \frac{1}{4} \begin{bmatrix} -(1-\eta), & (1-\eta), & (1+\eta), & -(1+\eta) \\ -(1-\xi), & -(1+\xi), & (1+\xi), & (1-\xi) \end{bmatrix} \begin{bmatrix} x_i, & y_i \\ x_j, & y_j \\ x_k, & y_k \\ x_m, & y_m \end{bmatrix} \tag{4.38}$$

Thus the strain matrices can be given explicitly in terms of η and ξ.

Integration with respect to $dx\, dy$ becomes integration with respect to $d\xi\, d\eta$, over the intervals $-1 \leqslant \xi \leqslant 1$, $-1 \leqslant \eta \leqslant 1$, where

$$dx\, dy = |\det[J]|\, d\xi\, d\eta.$$

Three-dimensional problems will require many more elements to achieve a reasonable approximation. Consequently even the largest

---

‡ For the rectangle of sides a and b, and dimensionless coordinates $x' = x/a$, $y' = y/b$, the shape function, Eq. (4.36) becomes

$$u = \tfrac{1}{4}\{(1-x')(1-y')u_i + (1+x')(1-y')u_j + (1+x')(1+y')u_k + (1-x')(1+y')u_m\}.$$

computers may have their storage capacity and speed taxed. This may tip the economics to the side of complex elements, with a limited number of nodal connections, as opposed to the simplest elements. In three dimensions, the simplest element is a tetrahedron—an element with four nodes. Adopting a counterclockwise ordering in the plane representation (Fig. 4-4a), we can generalize the two-dimensional displacement equation.

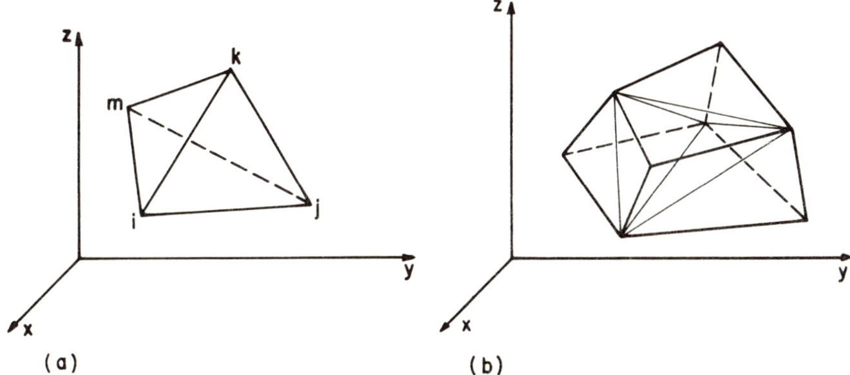

FIG. 4-4. Three-dimensional element: (a) Tetrahedral; (b) Composite element with 8 nodes (subdivision into five tetrahedra by one method).

Since the state of displacement has three components *u*, *v*, *w*, in the three coordinate-directions, we can uniquely determine these at interior points, and ensure displacement continuity on the boundary, with

$$u = \alpha_1 + \alpha_2 x + \alpha_3 y + \alpha_4 z.$$

All other expressions carry over as was first shown by Gallagher *et al.* [22], Melosh [23], and Argyris [24, 25].

More complex elements, one of which is shown in Fig. 4-4b, have been suggested and shape functions developed by Irons [26], Argyris [27], and Zienkiewicz and Cheung [12].

4.5 FINITE ELEMENTS AND FIELD PROBLEMS

In Section 4.2 we noted that the basic ideas of the finite element formulation of elasticity problems could be obtained by minimizing a functional (the total potential energy of the system) without reference

† Included are diffusion, conduction, flow through porous media, torsion of shafts, bending of beams, irrotational flow of ideal fluids, etc.

to detailed equilibrium conditions. Thus the method, *as a variational procedure*, can be applied to a great variety of problems[†] where the minimization of some functional, subject to boundary conditions, gives the exact solution. This functional (often some integral) may represent a physically recognizable quantity, but that is not necessary in order to apply the finite element formulation.

From the calculus of variations, if u minimizes[†] (or maximizes) the integral (functional)

$$I(u) = \iiint_R f(x, y, z, u, u_x, u_y, u_z)\, dx\, dy\, dz, \tag{4.39}$$

then u must satisfy the Euler equation

$$\frac{\partial}{\partial x}\left\{\frac{\partial f}{\partial u_x}\right\} + \frac{\partial}{\partial y}\left\{\frac{\partial f}{\partial u_y}\right\} + \frac{\partial}{\partial z}\left\{\frac{\partial f}{\partial u_z}\right\} - \frac{\partial f}{\partial u} = 0, \tag{4.40}$$

within the same region R, provided u satisfies the same boundary conditions. Thus the equation

$$\frac{\partial}{\partial x}\left(k_x \frac{\partial \phi}{\partial x}\right) + \frac{\partial}{\partial y}\left(k_y \frac{\partial \phi}{\partial y}\right) + Q = 0, \tag{4.41}$$

is the Euler equation of the functional

$$E(\phi) = \iint_S \{\tfrac{1}{2}[k_x(\partial\phi/\partial x)^2 + k_y(\partial\phi/\partial y)^2] - Q\phi\}\, dx\, dy, \tag{4.42}$$

where k_x, k_y, and Q are known specific functions. The functions k_x and k_y may represent anisotropic conduction coefficients, while Q is the heat generation and ϕ a temperature, for example.

To treat the aforementioned field problem, with the functional $E(\phi)$, Eq. (4.42), the plane region S is divided into finite elements as shown in Fig. 4-5 (triangles displayed). The nodal values of ϕ are employed to define the function within each element. For the typical element ijm, we have,

$$\{\phi\} = [N_i, N_j, N_m]\{\phi\}^e, \qquad \{\phi\}^e = \begin{Bmatrix} \phi_i \\ \phi_j \\ \phi_m \end{Bmatrix}, \tag{4.43}$$

$$N_i = (a_i + b_i x + c_i y)/2\Delta, \qquad \Delta = \text{area of element},$$

as in Eq. (4.8) et seq.

[†] A complementary formulation may require a maximization. We usually subsume both under the more general stationary requirement.

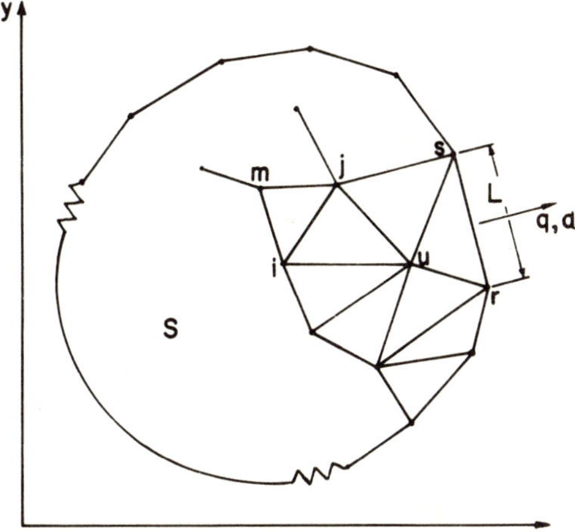

Fig. 4-5. Two-dimensional domain for field problems.

The nodal values now define the function ϕ uniquely and continuously throughout the region, so we can now move to the minimization of the function† $E(\phi)$, Eq. (4.42). This is probably best done by evaluating the contributions of an element to the differential $\partial E/\partial \phi_i$, then adding all such contributions and equating the result to zero. As in the nodal equilibrium equations of plane elasticity only the elements adjacent to node i will contribute to $\partial E/\partial \phi_i$. With E^e designating an integration limited to the area of the element, we find

$$\frac{\partial E^e}{\partial \phi_i} = \iint \left\{ k_x \frac{\partial \phi}{\partial x} \frac{\partial}{\partial \phi_i} \left(\frac{\partial \phi}{\partial x} \right) + k_y \frac{\partial \phi}{\partial y} \frac{\partial}{\partial \phi_i} \left(\frac{\partial \phi}{\partial y} \right) - Q \frac{\partial \phi}{\partial \phi_i} \right\} dx\, dy. \tag{4.44}$$

With ϕ defined by Eqs. (4.43), this becomes

$$\partial E^e/\partial \phi_i = (2\Delta)^{-2} \iint \{k_x[b_i, b_j, b_m]\{\phi\}^e + k_y[c_i, c_j, c_m]\{\phi\}^e\}\, dx\, dy$$
$$- 2\Delta^{-1} \iint Q(a_i + b_i x + c_i y)\, dx\, dy. \tag{4.45}$$

† Different variational forms can be employed. De Veubeke [19, 28] demonstrates the use of equilibrating stress distributions and of complementary potential energy as the functional to be optimized. Herrmann [29] uses the Reissner functional with success, and Pian and Tong [30] show how other hybrid formulations are possible. Hybrid formulations based on physical intuition were introduced and used successfully by Pian [31] and Severn and Taylor [32].

Clearly, any element contributes to only three of the differentials associated with its nodes. These are

$$\{\partial E^e/\partial\phi\} = \begin{Bmatrix} \partial E^e/\partial\phi_i \\ \partial E^e/\partial\phi_j \\ \partial E^e/\partial\phi_m \end{Bmatrix},$$

whereupon Eq. (4.45) becomes

$$\{\partial E^e/\partial\phi\} = [h]\{\phi\}^e + \{F^e.\} \tag{4.46}$$

Equation (4.46) is of the standard "stiffness" form [compare Eq. (4.1)].

If k_x, k_y are taken as constant within the element, and we note that $\iint_e dx\, dy = \Delta$, then $[h]$ becomes

$$[h] = \frac{k_x}{4\Delta}\begin{bmatrix} b_ib_i & b_jb_i & b_mb_i \\ b_ib_j & b_jb_j & b_mb_j \\ b_ib_m & b_jb_m & b_mb_m \end{bmatrix} + \frac{k_y}{4\Delta}\begin{bmatrix} c_ic_i & c_jc_i & c_mc_i \\ c_ic_j & c_jc_j & c_mc_j \\ c_ic_m & c_jc_m & c_mc_m \end{bmatrix}, \tag{4.47}$$

a symmetric matrix.

If Q is assumed to be constant within an element, then $\{F\}^e$ can be calculated. In particular

$$\begin{aligned} F_i &= -(Q/2\Delta)\iint (a_i + b_ix + c_iy)\, dx\, dy \\ &= -Q\Delta/3, \end{aligned}$$

whereupon

$$\{F\}^e = -Q\Delta/3 \begin{Bmatrix} 1 \\ 1 \\ 1 \end{Bmatrix}.$$

The assembly consists of summing all differentials of E and equating these to zero. The boundary conditions must be incorporated.

Some field problems are given in Zienkiewicz and Cheung [12] who also discuss eigenvalue problems and an introduction to nonlinear problems. Other papers on conduction–diffusion problems include the work of Visser [33] on thermal elasticity, Zienkiewicz *et al.* [34] on seepage problems in anisotropic media, and Adler and Gallagher [35] in conduction. An extensive review paper by Zienkiewicz [36] lists over 100 applications from many fields.

4.6 FINITE ELEMENTS AND NONLINEAR PROBLEMS

Nonlinear problems of fluid mechanics, solved by the finite element formulation, will be discussed in Part B. Problems of solid mechanics with geometric or material nonlinearities can be handled by means of some iterative approach. Indeed, the simple direct physical formulation of finite elements and consequent close visualization of the phenomena sometimes suggests iterative processes which might not be obvious mathematically.

A great variety of nonlinear problems have been published in the *Proceedings of the First and Second Air Force Conferences on Matrix Methods in Structural Analysis* [37, 38] and in the *Proceedings of the U.S.–Japan Seminar (Tokyo) on Recent Advances in Matrix Methods of Structural Analysis and Design* [39]. We shall briefly describe the details for the finite element analysis of problems containing *geometric nonlinearities* and for those with *material nonlinearities.* A bibliography for each class will be included as the discussion progresses.

Geometric nonlinearity results in two classes of problems—the large deflection problem† and the problem of structural stability. The initial paper by Turner *et al.* [40] concerned the large deflection of structures subjected to heating and external loads. The treatment by finite elements showed how each of the effects could be introduced into the analysis. First, with respect to the equilibrium equation, an *incremental* or *piecewise linear calculation procedure* was adopted. Moving from the initial to the final equilibrium state in a series of steps, the changing geometry could be included in forming the stiffness (equilibrium) equations. Martin (see Gallagher *et al.* [39, p. 343]) has presented an extensive review of finite element treatments of problems with geometric nonlinearity. Gallagher [41] has reviewed the status of finite elements in elastic instability analyses.

Marcal (see Gallagher *et al.* [39, p. 257]) has reviewed the literature of finite element analysis with material nonlinearities. Treatments here are an outgrowth of the method of *initial strains* (or thermal strains) by Mendelson and Manson [42]. This method is based on the idea of modifying the elastic equations of equilibrium to compensate for the

† The term "large deflection" is misleading, for problems lying in this area need not have actual deflections which are in any sense large. Indeed, they can be and often are as small as those arising in the linear problem. What is important is that the *deformed configuration* must be used when writing the equilibrium equations and the strain-displacement equations must include appropriate higher-order nonlinear terms. Each effect introduces nonlinear complications.

fact that the inelastic strains do not cause any change in stress. However, lack of *a priori* knowledge of the inelastic strains forces the method to be iterative. Originally formulated for finite differences, various instabilities kept the method from being too successful in its initial form. Gallagher *et al.* [22], Padlog *et al.* [43], and Argyris [44] adapted the method of initial strains to the finite element method by calculating the so-called initial force vector. Combined problems of creep and plasticity were considered, and two approaches, constant stress and constant strain, to the iterative solution were examined. The *constant strain* option was found to be the more stable iterative technique. Subsequent applications to plane solids were made by Percy *et al.* [45], Argyris *et al.* [46], and Jensen *et al.* [47]. Marcal and Mallett [48] have applied the method to plates, and Witmer and Kotanchik (see [38]) use it in a shell study. The latter paper contains several references on studies of shells in the elastic–plastic range. Greenbaum and Rubinstein [49] investigate the application of the method in creep analysis.

Under the *incremental-initial-strain method*, the load is considered to be applied in small increments. Although the process of plastic deformation is not time dependent, it is convenient to assume here that each step is associated with a "time" interval. This immediately permits incorporation of the effects of time-variable temperature changes and paves the way for subsequent creep investigations.

In any one increment of time, the total strain increment $\Delta\{\epsilon\}$ is assumed to be expressible as

$$\begin{aligned}\Delta\{\epsilon\} &= \Delta\{\epsilon_e\} + (\Delta\{\epsilon_p\} + \Delta\{\epsilon_\theta\}) \\ &= \Delta\{\epsilon_e\} + \Delta\{\epsilon_0\} = [D]^{-1}\{\sigma\} + \Delta\{\epsilon_0\},\end{aligned} \tag{4.48}$$

where ϵ_e, ϵ_p, and ϵ_θ represent the elastic, plastic, and thermal strain components, respectively. If $\Delta\{\epsilon_p\}$ and $\Delta\{\epsilon_\theta\}$ can be considered to be known, then the change of stress can be calculated in the standard way. If the state of stress or strain is known at the start of the interval, the state can be ascertained at the end by addition of the increments. The change of thermal strain creates no problem, but the increment of plastic strain depends upon *both* the initial and final stress levels and generally cannot be directly determined. Thus we need usually carry out an iteration:

(i) Time (load) intervals are made sufficiently small so that the plastic strain increments of the *previous interval* can be used to calculate stress and consequently increments in the present interval.

(ii) The increments of stress and strain of (i) are employed to obtain a new estimate of $\Delta\{\epsilon_p\}$, and the process is repeated until convergence is obtained.

The computation of plastic strain increments can be as complicated as desired in the foregoing iterative process. Any material laws can be used (see the exceptional case of ideally or nearly ideally plastic below). In particular most typical situations permit

$$\Delta\{\epsilon_p\} = C[D_0]^{-1}\Delta\{\sigma\}, \tag{4.49}$$

where $[D_0]^{-1}$ is that part of the elasticity matrix depending only on Poisson's ratio. The Prager–Mises (Drucker [50]) theory for strain hardening material, has C as a function of the second stress invariant

$$\bar{\sigma} = \{(\sigma_1 - \sigma_2)^2 + (\sigma_2 - \sigma_3)^2 + (\sigma_3 - \sigma_1)^2\}^{1/2}/3,$$

where σ_i are the principal stress components. Generally, C is a function of $\bar{\sigma}$ and the temperature θ. Sometimes the explicit relation

$$\bar{\epsilon}_p = K\bar{\sigma}^n$$

is used for the second strain invariant. With K and n constant,

$$C = d\bar{\epsilon}_p/d\bar{\sigma} = nK\bar{\sigma}^{n-1}.$$

When the material is ideally plastic, C becomes arbitrarily large, and the alternative *tangent modulus method*, also called the *incremental-variable elasticity* procedure, must be used.

In the tangent modulus method, the load is again considered to be applied in an incremental way but the *total strain* occurring during the increment is treated by employing a suitably modified modulus, as if the material were elastic.

The elastic strain increment is found in the usual way by

$$\Delta\{\epsilon_e\} = [D]^{-1}\Delta\{\sigma\},$$

and the plastic strain increment (from the "total strain" theory of Hencky and Sokolovsky, e.g., Drucker [50])

$$\Delta\{\epsilon_p\} = C[D_0]^{-1}\Delta\{\sigma\}.$$

Thus,

$$\Delta\{\sigma\} = ([D]^{-1} + C[D_0]^{-1})^{-1}\,\Delta(\{\epsilon_e\} + \{\epsilon_p\}),$$

so that with only the thermal strain as the initial strain in an increment the change of stress can be found, in an elastic manner with the elasto-plastic elasticity matrix

$$[D_{ep}] = ([D]^{-1} + C[D_0]^{-1})^{-1}. \tag{4.50}$$

Returning now to a typical situation in which *geometric nonlinearities* are present, we sketch the treatment by Martin (in [37, p. 697]) of a thin triangular element initially in the x, y-plane, in plane stress. The nodal locations are chosen as $1(0, 0)$, $2(x_2, 0)$, $3(x_3, y_3)$. In this initial position, the element carries initial stresses $\sigma_x{}^0$, $\sigma_y{}^0$, τ_{xy}^0. While these do not have influence on subsequent displacements of the element within the initial plane, they do have an effect on displacements of the element *out* of its initial plane—thus a w-component of displacement must be introduced. Starting from the initial situation, subsequent deformation is of the form

$$\epsilon_x = \epsilon_x{}^0 + \epsilon_x{}^a, \qquad \epsilon_y = \epsilon_y{}^0 + \epsilon_y{}^a, \qquad \gamma_{xy} = \gamma_{xy}^0 + \gamma_{xy}^a,$$

where

$$\begin{aligned} \epsilon_x{}^a &= \frac{\partial u}{\partial x} + \frac{1}{2}\left(\frac{\partial w}{\partial x}\right)^2, \\ \epsilon_y{}^a &= \frac{\partial v}{\partial y} + \frac{1}{2}\left(\frac{\partial w}{\partial y}\right)^2, \\ \gamma_{xy}^a &= \frac{\partial u}{\partial y} + \frac{\partial v}{\partial x} + \frac{\partial w}{\partial x}\frac{\partial w}{\partial y}. \end{aligned} \tag{4.51}$$

The strain energy functional is given by†

$$\begin{aligned} U &= \frac{1}{2}\iiint (\sigma_x\epsilon_x + \sigma_y\epsilon_y + \tau_{xy}\gamma_{xy})\,dx\,dy\,dz \\ &= U_0 + U_1 + U_2, \end{aligned}$$

where

$$U_0 = [Et/2(1-\nu^2)] \iint [(\epsilon_x{}^0)^2 + 2\nu\epsilon_x{}^0\epsilon_y{}^0 + (\epsilon_y{}^0)^2 + \lambda_1(\gamma_{xy}^0)^2]\,dx\,dy,$$

$$U_1 = [Et/(1-\nu^2)] \iint [\epsilon_x{}^0\epsilon_x{}^a + \nu(\epsilon_x{}^0\epsilon_y{}^a + \epsilon_y{}^0\epsilon_y{}^a) + \epsilon_y{}^0\epsilon_y{}^a + \lambda_1\gamma_{xy}^0\gamma_{xy}^a]\,dx\,dy,$$

$$U_2 = [Et/(1-\nu^2)] \iint [(\epsilon_x{}^a)^2 + 2\nu\epsilon_x{}^a\epsilon_y{}^a + (\epsilon_y{}^a)^2 + \lambda_1(\gamma_{xy}^a)^2]\,dx\,dy, \tag{4.52}$$

and t is the plate thickness. Since U_0 contains only the initial strain, we discuss it no further.

---

† We have used Hooke's law in what follows, i.e.,

$$\begin{Bmatrix} \sigma_x \\ \sigma_y \\ \tau_{xy} \end{Bmatrix} = \frac{E}{1-\nu^2}\begin{bmatrix} 1 & \nu & 0 \\ \nu & 1 & 0 \\ 0 & 0 & \lambda_1 \end{bmatrix}\begin{Bmatrix} \epsilon_x \\ \epsilon_y \\ \gamma_{xy} \end{Bmatrix}, \qquad \lambda_1 = \frac{1-\nu}{2}.$$

Employing the definition of subsequent strains and Hooke's law, Eq. (4.52) becomes

$$U_1 = t \iint \left[\sigma_x{}^0 \frac{\partial u}{\partial x} + \sigma_y{}^0 \frac{\partial v}{\partial y} + \tau_{xy}^0 \left(\frac{\partial u}{\partial y} + \frac{\partial v}{\partial x}\right)\right] dx\, dy \qquad \text{(linear part)}$$

$$+ \frac{1}{2} t \iint \{\partial w/\partial x, \partial w/\partial y\} \begin{bmatrix} \sigma_x{}^0 & \tau_{xy}^0 \\ \tau_{xy}^0 & \sigma_y{}^0 \end{bmatrix} \begin{Bmatrix} \partial w/\partial x \\ \partial w/\partial y \end{Bmatrix} dx\, dy \qquad \text{(nonlinear part).} \tag{4.53}$$

Using linear displacement functions,

$$u = a_0 + a_1 x + a_2 y, \qquad v = b_0 + b_1 x + b_2 y, \qquad w = c_0 + c_1 x + c_2 y,$$

where

$$\begin{aligned} a_0 &= u_1, & a_1 &= (u_2 - u_1)/x_2, & a_2 &= (x_{32} u_1 - x_3 u_2 + x_2 u_3)/x_2 y_3, \\ b_0 &= v_1, & b_1 &= (v_2 - v_1)/x_2, & b_2 &= (x_{32} v_1 - x_3 v_2 + x_2 v_3)/x_2 y_3, \\ c_0 &= w_1, & c_1 &= (w_2 - w_1)/x_2, & c_2 &= (x_{32} w_1 - x_3 w_2 + x_2 w_3)/x_2 y_3, \end{aligned}$$

$$x_{32} = x_3 - x_2, \qquad x_2 y_3 = 2\ (\text{area of triangle}) = 2A,$$

it follows that the first integral in U_1 contains only linear terms in the nodal displacements. The second term becomes

$$\begin{aligned} U_1' &= \frac{1}{2} t \iint \{c_1, c_2\} \begin{bmatrix} \sigma_x{}^0 & \tau_{xy}^0 \\ \tau_{xy}^0 & \sigma_y{}^0 \end{bmatrix} \begin{Bmatrix} c_1 \\ c_2 \end{Bmatrix} dx dy \\ &= \frac{1}{2} \frac{t}{4A} \{w\}^{\mathrm{T}} [x, y]^{\mathrm{T}} [\sigma^0, \tau^0][x, y]\{w\}, \end{aligned}$$

where

$$\{w\} = \begin{Bmatrix} w_1 \\ w_2 \\ w_3 \end{Bmatrix}, \qquad [x, y] = \begin{bmatrix} -y_3 & y_3 & 0 \\ x_{32} & -x_3 & x_3 \end{bmatrix}, \qquad [\sigma_0, \tau_0] = \begin{bmatrix} \sigma_x{}^0 & \tau_{xy}^0 \\ \tau_{xy}^0 & \sigma_y{}^0 \end{bmatrix}.$$

Thus the *initial stress matrix* $[K^1]$ is

$$\begin{aligned} [K^1] &= (t/4A)[x, y]^{\mathrm{T}} [\sigma^0, \tau^0][x, y] \\ &= [K^1](\sigma_x{}^0) + [K^1](\sigma_y{}^0) + [K^1](\tau_{xy}^0), \end{aligned}$$

where

$$[K^1](\sigma_x^0) = \frac{\sigma_x^0 t}{4A} \begin{bmatrix} y_3^2 & -y_3^2 & 0 \\ -y_3^2 & y_3^2 & 0 \\ 0 & 0 & 0 \end{bmatrix}, \quad \text{etc.}$$

Similarily, U_2 can be evaluated and written out. This initial stress formulation has been generalized by Purdy and Przemieniecki [51] (see also Martin, in Gallagher *et al.* [39, p. 374]) by retaining all terms in the strain energy expression.

We now list some references for various solid mechanics problems treated by finite elements, in which one or more nonlinearity is present.

1. Geometric Nonlinearity Only. Oden [52] on strings; Mallett and Berke [53] on columns with nonlinear lateral support; Bogner *et al.* [54] and Murray and Wilson [55] on large deflections of thin plates; Gallagher [41] on stability analysis; shells treated by conical elements by Grafton and Strome [56], and buckling by Navaratna [57] and Navaratna *et al.* [58]. A considerable generalization is due to Stricklin *et al.* (see [38]). Review of geometric nonlinearities by Martin (see Gallagher *et al.* [39, p. 343]; [37, p. 697]), general discussion by Kuwai (see Gallagher *et al.* [39, p. 383]), and elasticity by Hartz and Nathan (see Gallagher *et al.* [39, p. 415]).

2. Material Nonlinearity. Review article by Marcal (see Gallagher *et al.* [39, p. 257]); elastic–plastic behavior by Pope [59], Swedlow and Yang [60], Marcal and King [61], Yamada *et al.* [62], Levy [63], Marcal [64], and Yamada (see Gallagher *et al.* [39, p. 283]). The limit load problem in plasticity is treated by Hayes and Marcal [65].

3. Combined Material and Geometric Nonlinearity. Large deflection of membranes by Oden and Kubitza [66]; arches by Armin *et al.* (in [38]); plates by Murray and Wilson [55]; plates and axisymmetric shells of revolution by Marcal [67–69], and by Popov and Yaghmai [70].

While the original derivation of element matrices was primarily based on energy minimization principles, more recently it has been realized that the formulation can be based *directly* on the governing equation by means of weighted residual methods. Szabo and Lee [71, 72] use Galerkin's method to obtain the stiffness matrices for plates and plane elastic problems, Langhaar and Chu [73] employ piecewise polynomials and partitioning for ordinary equations while Leonard and Bramlette [74] consider n coupled linear equations. With the problem thus freed of a variational formulation penetration of finite element concepts into other areas will proceed at a rapid pace.

B. NUMERICAL SOLUTIONS IN FLUID MECHANICS

4.7 PRELIMINARY REMARKS

Before the era of the third generation computing machines (early 1960s), numerical computations in fluid mechanics, wherein the nonlinearities are retained, are very limited in number. One due to Thom [75] in 1933 studied the wake associated with steady laminar flow past a circular cylinder. Solutions were obtained on a desk calculator for Reynolds numbers ten and twenty. Flugge-Lotz and her students (cf. Volume I, page 349) have carried on extensive calculations of boundary-layer flow since the early 1950s. That period also saw advances in our ability to compute gas flows. To be particularly noted is the work of Lax and his students (cf. Volume I, page 445ff). Here, we shall not repeat the discussions available in Volume I and other references (e.g., Richtmyer and Morton [76], Ames [77]), but will confine our attention to the numerical methods developed in the last several years (1965–1971) for solving the nonlinear equations of fluid mechanics.

Strictly numerical methods† for integrating these nonlinear models for fluid flow fall into four general categories. These are the methods of:

(i) finite elements;
(ii) stream function–vorticity;
(iii) primitive variables;
(iv) vector potential.

Our discussion will be in the above order.

4.8 FINITE ELEMENTS AND UNSTEADY FLOW

As we have already observed in Part A, the finite element method, originally developed for structural mechanics, has been applied to field problems of many types. Applications in fluid mechanics were inevitable and, while limited (1971), are being rapidly developed. Specific computations have been made in the areas of potential flow by Martin (in [38]), Argyris *et al.* (two and three dimensions) [78], Doctors [79], Argyris [80], and deVries and Norrie [81]. Flow in porous media has been studied

† Procedures discussed do not include the very versatile approximate methods, such as that of Galerkin, etc., which are presented in earlier chapters and in Volume I.

by finite element computation by Zienkiewicz *et al.* [34], Javandel and Witherspoon [82], Taylor and Brown [83], Sandhu and Wilson [84], and Volker [85]. Fluid motion in a container has occupied the attention of Tong and Fung [86], Tong [87], Luk [88], and Archer and Rubin (in [37]). Argyris and Scharpf [89], Reddi [90], and Reddi and Chu [91] demonstrate the applicability of finite element computation to lubrication problems.

Studies in compressible flow by finite elements are due to Argyris *et al.* [78] and Argyris [80], while Skiba [92] examines natural convection in rectangular cavities. One of the earliest studies (1964) by Oden and Somoggi [93] concerned low Reynolds number flows, a topic also investigated by Tong (see Gallagher *et al.* [39]; [94]), and Atkinson *et al.* [95].

Studies involving fluid mechanics and structural vibration include the vibration of submerged structures by Zienkiewicz and Newton [96] and Zienkiewicz *et al.* [97] and an application to supersonic panel flutter by Olson [98]. In these applications for shell-fluid coupled motions, the solid wall displacements are assumed to be small. Other references are provided in the review paper by Zienkiewicz [36]. A book on the application of finite elements to fluid mechanics is in preparation by Norrie and deVries [99].

As a typical example, we describe herein a finite element procedure for steady inviscid two-dimensional compressible flow. The governing equations are those of momentum

$$-\rho^{-1}(\partial p/\partial x) = uu_x + vu_y, \qquad -\rho^{-1}(\partial p/\partial y) = uv_x + vv_y, \tag{4.54}$$

continuity

$$(\rho u)_x + (\rho v)_y = 0, \tag{4.55}$$

and the constitutive relation

$$p = p(\rho), \tag{4.56}$$

where u and v are the velocity components in the x- and y-directions, p is pressure, and ρ is density. From these we may eliminate the pressure, and with a potential function $\phi(u = -\phi_x, v = -\phi_y)$ obtain the equation (cf., e.g., Kuethe and Schetzer [100] or Volume I)

$$(\phi_x^2 - a^2)\phi_{xx} + 2\phi_x\phi_y\phi_{xy} + (\phi_y^2 - a^2)\phi_{yy} = 0, \tag{4.57a}$$

where

$$a^2 = A + b(\phi_x^2 + \phi_y^2). \tag{4.57b}$$

In the case of a polytropic gas, $p = k\rho^\gamma$, $b = (\gamma - 1)/2$, and

$$A = a_\infty^2 + bq_\infty^2 = \gamma RT + [(\gamma - 1)/2](u_\infty^2 + v_\infty^2).$$

We shall be concerned with a finite-element solution of Eqs. (4.57) in the interior of a plane region D subject to *either* ϕ being specified or $d\phi/dn + Q + \alpha\phi = 0$ on the boundary C of D. Here Q and α are prescribed functions of x and y along C. The boundary curve C is assumed to be sufficiently regular so that the divergence theorem is satisfied. As discussed in Part A a variational or weighted residual formulation can be employed in the finite element development. We shall adopt the former, after the work of Norrie and deVries [99].

If a, G, H, Q, and α are functions only of x and y, then a necessary condition[†] for the functional

$$\Gamma(\theta) = \iint_D \{(12a^2)^{-1}[\theta_x^4 + \theta_y^4] - \tfrac{1}{2}[\theta_x^2 + \theta_y^2] + (G/a^2)\,\theta_x\theta_y\}\,dx\,dy - \int_C [H\theta + Q\theta + \tfrac{1}{2}\alpha\theta^2]\,ds, \tag{4.58}$$

to be stationary is that there exist ϕ such that

$$(\phi_x^2 - a^2)\,\phi_{xx} + 2G\phi_{xy} + (\phi_y^2 - a^2)\,\phi_{yy} = 0 \tag{4.59}$$

in the interior of D. A comparison of Eq. (4.59) with Eq. (4.57a) shows that they are of the same general form, but a and G are known functions of x and y. If the variation is taken as $h(x, y)$, then

$$\int_C h[(3a^2)^{-1}(\phi_x^3 n_x + \phi_y^3 n_y) - (\phi_x n_x + \phi_y n_y) + (G/a^2)(\phi_x n_y + \phi_y n_x) - H - Q - \alpha\phi]\,ds = 0. \tag{4.60}$$

To satisfy Eq. (4.60), the following choices of h are of interest:

(a) $h = 0$ on C but otherwise arbitrary and nonzero in D. This is the case where the boundary condition is $\phi = g(x, y)$ on C.

(b) h is arbitrary and nonzero on C and in D. This requires

$$(3a^2)^{-1}(\phi_x^3 n_x + \phi_y^3 n_y) - (\phi_x n_x + \phi_y n_y) + (G/a^2)(\phi_x n_y + \phi_y n_x) - H - Q - \alpha\phi = 0 \tag{4.61}$$

[†] Equation (4.59) is the Euler equation of the functional Eq. (4.58). This is found by standard variational techniques. That is, take $\theta(x, y) = \phi(x, y) + \epsilon h(x, y)$ with the variation h arbitrary. n_x and n_y are x and y components of the unit outward normal to the boundary curve C.

on C. This is the so-called natural boundary condition associated with the functional, Eq. (4.58).

The variational procedure has an Euler equation resembling Eq. (4.57) except for the terms a and G. Further, the natural boundary condition, Eq. (4.61), differs from $d\phi/dn + Q + \alpha\phi = 0$ on C, originally specified. Thus it is clear that a direct application of this variational procedure to the functional does not yield a solution to the required boundary-value problem. An *iterative scheme*, described below, will overcome this difficulty.

Beginning with an initial value $\phi^0(x, y)$, we calculate

$$G^0 = \phi_x{}^0\phi_y{}^0,$$
$$(a^0)^2 = A + b[(\phi_x{}^0)^2 + (\phi_y{}^0)^2], \tag{4.62}$$

and

$$H^0 = [3(a^0)^2]^{-1}[(\phi_x{}^0)^3\, n_x + (\phi_y{}^0)^3\, n_y] + [G^0/(a^0)^2][\phi_x{}^0 n_y + \phi_y{}^0 n_x]. \tag{4.63}$$

From the geometry, the unit outward normal to the boundary curve n is known, and therefore, so are its x- and y-components, n_x and n_y. The value H is chosen via Eq. (4.63), so that the remaining condition on C [Eq. (4.61)] becomes

$$(d\phi^0/dn) + Q + \alpha\phi^0 = 0. \tag{4.64}$$

Upon substituting ϕ^0, Eqs. (4.62) and (4.63) into the functional, Eq. (4.58), and minimizing we obtain a solution surface ϕ^1 which is also a solution to the following boundary-value problem:

$$[(\phi_x{}^1)^2 - (a^0)^2]\,\phi_{xx}^1 + 2\phi_x{}^0\phi_y{}^0\phi_{xy}^1 + [(\phi_y{}^1)^2 - (a^0)^2]\,\phi_{yy}^1 = 0, \tag{4.65}$$

in D, subject to

$$\begin{aligned}(d\phi^1/dn) + Q + \alpha\phi^1 = {} & [3(a^0)^2]^{-1}\{n_x[(\phi_x{}^1)^3 - (\phi_x{}^0)^3] + n_y[(\phi_y{}^1)^3 - (\phi_y{}^0)^3] \\ & + [(a^0)^2]^{-1}\{(\phi_x{}^1\phi_y{}^1)(\phi_x{}^1 n_y + \phi_y{}^1 n_x) \\ & - (\phi_x{}^0\phi_y{}^0)(\phi_x{}^0 n_y + \phi_y{}^0 n_x)\}.\end{aligned} \tag{4.66}$$

Using the solution surface ϕ^1, the same procedure as that used for ϕ^0 is used to generate a new surface ϕ^2. More generally, the iteration technique is obtained from Eqs. (4.65) and (4.66) by replacing 0 by n and 1 by $n + 1$. If the solution converges, that is to say,

$$\lim_{n\to\infty} | \phi^{n+1} - \phi^n | = 0,$$

and there exists a unique limit function ϕ, $\lim \phi^n = \phi$, then ϕ is a solution to Eq. (4.57) subject to the boundary condition $(d\phi/dn) + Q + \alpha\phi = 0$ on C.

We should note that the functional given by Eq. (4.58) is not unique. Indeed any functional which upon minimization and iteration satisfies Eq. (4.57) together with the appropriate boundary conditions constitutes a *proper functional* for this problem.

The finite element procedure will now be applied to the iteration method just developed. *We remark again that over the region in which the functional, Eq.* (4.58), *is applied the functions* a^2, G, H, *and* Q *are assumed to be functions of* x *and* y *only*. They do not depend upon the particular ϕ^{n+1} whose solution we seek.

To apply the technique, the domain D is discretized into finite elements. A particular element, denoted by e, is shown in Fig. 4-1 with nodal points labeled i, j, and m in a counterclockwise manner. Next a *shape function* is chosen for ϕ† as an explicit relation for the element whose form is uniquely determined by the nodal values of ϕ. Choice of the linear expression

$$\phi = \alpha_1 + \alpha_2 x + \alpha_3 y \tag{4.67}$$

has the advantage that the derivatives of ϕ are constant, whereupon the functions a^2, G, and H appearing in the iteration are constants over each element. The values for the α_p, $p = 1, 2, 3$ are exactly those specified in Section 4.1 for u [Eqs. (4.7) et seq.].

We now complete the formulation for that boundary-value problem in which $d\phi/dn + Q + \alpha\phi = 0$ must hold on C. We suppose that there are m nodal points in D and n elements. Then it follows that the functional

$$\Gamma(\phi) = F[\phi_1, \phi_2, \ldots, \phi_m] \tag{4.68}$$

and the set of equations to be solved in the minimization‡ process reduces to

$$\partial\Gamma/\partial\phi_p = 0, \qquad p = 1, 2, \ldots, m.$$

† There is only one nodal value for this field problem. At the node i, we shall label the potential value ϕ_i. Equation (4.67) is the simplest possible shape function corresponding to an approximation of the solution surface by elemental triangular planes.

‡ If some of the nodal values are prescribed, then these are omitted in the further discussion and the remaining values are renumbered.

It will usually be convenient to treat the interior and boundary nodes separately. Thus we write

$$\Gamma(\phi) = \sum_{e=1}^{n} (X^e + Y^e), \tag{4.69}$$

where n is the total number of elements in D,

$$X^e = \iint_{D_e} \{(12a^2)^{-1}(\phi_x^{\,4} + \phi_y^{\,4}) - \tfrac{1}{2}(\phi_x^{\,2} + \phi_y^{\,2}) + (G/a^2)\,\phi_x\phi_y\}\,dx\,dy,$$

$$Y^e = -\int_{C_e} (H\phi + G\phi + \tfrac{1}{2}\alpha\phi^2)\,ds,$$

D_e is the domain given by triangle e, and $\int_{C_e}$ is the contribution of the triangle e, to the functional, only over those sides of such triangles which constitute part of the bounding curve C.

Only those elements which have a vertex at node p contribute to $\partial\Gamma/\partial\phi_p$. When calculating $\partial\Gamma/\partial\phi_p$, for an interior point of D, only those terms $\partial X^e/\partial\phi_p$ contribute, since the terms $\partial Y^e/\partial\phi_p$ are functions only of the boundary points. Thus for *interior* points

$$\partial\Gamma/\partial\phi_p = \sum_{e=1}^{n} \partial X^e/\partial\phi_p = 0, \qquad p = 1, 2,\ldots, m, \tag{4.70}$$

where the p's are only interior points and n is the total number of elements in D. After some elementary calculations, we find

$$\begin{aligned}\frac{\partial X^e}{\partial\phi_p} = \frac{1}{3a^2}&\left(\left[\frac{1}{2\Delta_e}\{b_i, b_j, b_m\}\begin{Bmatrix}\phi_i\\ \phi_j\\ \phi_m\end{Bmatrix}\right]^3 \frac{b_i}{2} + \left[\frac{1}{2\Delta_e}\{c_i, c_j, c_m\}\begin{Bmatrix}\phi_i\\ \phi_j\\ \phi_m\end{Bmatrix}\right]^3 \frac{c_i}{2}\right)\\ &- \left[\frac{b_i}{4\Delta_e}\{b_i, b_j, b_m\} + \frac{c_i}{4\Delta_e}\{c_i, c_j, c_m\}\right]\begin{Bmatrix}\phi_i\\ \phi_j\\ \phi_m\end{Bmatrix}\\ &+ \frac{G}{a^2}\left[\frac{c_i}{4\Delta_e}\{b_i, b_j, b_m\} + \frac{b_i}{4\Delta_e}\{c_i, c_j, c_m\}\right]\begin{Bmatrix}\phi_i\\ \phi_j\\ \phi_m\end{Bmatrix},\end{aligned} \tag{4.71}$$

where $\Delta_e = \iint_{D_e} dxdy$ and $p = 1, 2,\ldots, m$. Equations (4.71) are nonlinear in the ϕ's.

To overcome the difficulty introduced by the nonlinearities in Eqs. (4.71), we modify the iteration previously developed to account for a^2 and G. Following that procedure, it is assumed that the $(n-1)$st solution,

ϕ^{n-1} is known, and we wish to calculate ϕ^n. From† Eqs. (4.62) and (4.63), we find

$$(a^{n-1})^2 = A + \frac{1}{4\Delta^2}\left(\left[\{b_i, b_j, b_m\}\begin{Bmatrix}\phi_i^{n-1}\\ \phi_j^{n-1}\\ \phi_m^{n-1}\end{Bmatrix}\right]^2 + \left[\{c_i, c_j, c_m\}\begin{Bmatrix}\phi_i^{n-1}\\ \phi_j^{n-1}\\ \phi_m^{n-1}\end{Bmatrix}\right]^2\right),$$

and (4.72)

$$G^{n-1} = \frac{1}{4\Delta^2}\left[\{b_i, b_j, b_m\}\begin{Bmatrix}\phi_i^{n-1}\\ \phi_j^{n-1}\\ \phi_m^{n-1}\end{Bmatrix}\right]\left[\{c_i, c_j, c_m\}\begin{Bmatrix}\phi_i^{n-1}\\ \phi_j^{n-1}\\ \phi_m^{n-1}\end{Bmatrix}\right].$$

Using these same equations, the nonlinear terms in Eq. (4.71) are replaced by

$$\frac{1}{8\Delta^3}\left[\{b_i, b_j, b_m\}\begin{Bmatrix}\phi_i^{n-1}\\ \phi_j^{n-1}\\ \phi_m^{n-1}\end{Bmatrix}\right]^2\left[\{b_i, b_j, b_m\}\begin{Bmatrix}\phi_i^{n}\\ \phi_j^{n}\\ \phi_m^{n}\end{Bmatrix}\right],$$

and (4.73)

$$\frac{1}{8\Delta^3}\left[\{c_i, c_j, c_m\}\begin{Bmatrix}\phi_i^{n-1}\\ \phi_j^{n-1}\\ \phi_m^{n-1}\end{Bmatrix}\right]^2\left[\{c_i, c_j, c_m\}\begin{Bmatrix}\phi_i^{n}\\ \phi_j^{n}\\ \phi_m^{n}\end{Bmatrix}\right],$$

which are now in terms of the known solution ϕ^{n-1} and the required one, ϕ^n.

With the notation,

$$\alpha_2^{n-1} = \frac{1}{2\Delta}\{b_i, b_j, b_m\}\begin{Bmatrix}\phi_i^{n-1}\\ \phi_j^{n-1}\\ \phi_m^{n-1}\end{Bmatrix}, \qquad \alpha_3^{n-1} = \frac{1}{2\Delta}\{c_i, c_j, c_m\}\begin{Bmatrix}\phi_i^{n-1}\\ \phi_j^{n-1}\\ \phi_m^{n-1}\end{Bmatrix},$$

Eqs. (4.72) become

$$(a^{n-1})^2 = A + b[(\alpha_2^{n-1})^2 + (\alpha_3^{n-1})^2],$$

$$G^{n-1} = (\alpha_2^{n-1})(\alpha_3^{n-1}).$$

Similarly, the nonlinear terms in Eq. (4.71) become

$$\frac{(\alpha_2^{n-1})^2}{2\Delta}\{b_i, b_j, b_m\}\begin{Bmatrix}\phi_i^{n}\\ \phi_j^{n}\\ \phi_m^{n}\end{Bmatrix}, \qquad \frac{(\alpha_3^{n-1})^2}{2\Delta}\{c_i, c_j, c_m\}\begin{Bmatrix}\phi_i^{n}\\ \phi_j^{n}\\ \phi_m^{n}\end{Bmatrix}.$$

† We drop the subscript e on all subsequent terms.

Combining the above results, Eq. (4.70), for the (ϕ_i^n)'s becomes

$$\frac{\partial \Gamma}{\partial \phi_p} = \sum_{\substack{\text{sum over} \\ \text{all elements}}} \frac{1}{4\Delta} \Big[\frac{1}{\{A + b[(\alpha_2^{n-1})^2 + (\alpha_3^{n-1})^2]\}}$$

$$\cdot \Big(\frac{(\alpha_2^{n-1})^2}{3} b_i\{b_i, b_j, b_m\} + \frac{(\alpha_3^{n-1})^2}{3} c_i\{c_i, c_j, c_m\}$$

$$+ (\alpha_2^{n-1})(\alpha_3^{n-1}) c_i\{b_i, b_j, b_m\} + (\alpha_2^{n-1})(\alpha_3^{n-1}) b_i\{c_i, c_j, c_m\} \Big)$$

$$- b_i\{b_i, b_j, b_m\} - c_i\{c_i, c_j, c_m\} \Big] \begin{Bmatrix} \phi_i^n \\ \phi_j^n \\ \phi_m^n \end{Bmatrix} = 0. \tag{4.74}$$

Although the summation is carried out over all elements, it is recalled that if, in fact, a certain element does not contribute to the nodal point i, then the right-hand side of Eq. (4.74) does not contain a submatrix with an i suffix for that particular element.

For each interior point, an equation of the type (4.74) is generated. The set of linear equations of the (ϕ_i^n)'s thus obtained completely characterizes the solution surface at the interior points only. For points on the boundary C, where the relation $d\phi/dn + Q + \alpha\phi = 0$ holds, Eq. (4.74) must be modified to include effects of $\partial Y^e/\partial \phi_p$. That discussion is available in the work of Zienkiewicz and Cheung [12] and Norrie and deVries [99].

4.9 STREAM FUNCTION—VORTICITY TECHNIQUES

This second group of numerical methods has the common feature that the stream function and vorticity are used as the dependent variables. The equations modeling the time dependent flow of a two-dimensional viscous incompressible Newtonian fluid in cartesian coordinates are those of momentum (Navier–Stokes)

$$u_t + uu_x + vu_y = -\rho^{-1}p_x + \nu(u_{xx} + u_{yy}), \tag{4.75a}$$

$$v_t + uv_x + vv_y = -\rho^{-1}p_y + \nu(v_{xx} + v_{yy}), \tag{4.75b}$$

and continuity

$$u_x + v_y = 0. \tag{4.76}$$

The dependent variables are the velocity components u and v in the

x- and y-directions, respectively, and the pressure p. The kinematic viscosity ν and the density ρ are material constants. A stream function ψ and vorticity ω are defined by means of the relations

$$u = \psi_y, \qquad v = -\psi_x, \qquad \omega = -u_y + v_x, \tag{4.77}$$

whereupon we find one form of the vorticity equation†

$$\omega_t + (u\omega)_x + (v\omega)_y = \nu \nabla^2 \omega. \tag{4.78}$$

An alternative form of Eq. (4.78) is

$$\omega_t + \psi_y \omega_x - \psi_x \omega_y = \nu \nabla^2 \omega. \tag{4.79}$$

We also note that the definition of ω [see Eq. (4.77)] may be written in terms of ψ as

$$\nabla^2 \psi = -\omega. \tag{4.80}$$

A knowledge of the pressure is often useful as an aid in understanding features of the flow. A suitable pressure equation is found by computing the x derivative of Eq. (4.75a), the y-derivative of Eq. (4.75b), and summing the results. This generates

$$\nabla^2 p = -\rho[(u^2)_{xx} + 2(uv)_{xy} + (v^2)_{yy}]. \tag{4.81}$$

Probably Emmons [101] was the first author to use the stream function–vorticity method in a digital computer calculation. He was concerned with the numerical solution of Eqs. (4.79) and (4.80), at a Reynolds number of 4000, with the goal of understanding turbulence. His explicit finite difference discretization employs a forward difference in time and a standard five-point molecule for each Laplacian. Later Payne [102] employs essentially the same ideas in his calculation of nonsteady flow. In particular Payne [103] applies his method to the calculation of wake structure behind a circular cylinder.

Fromm [104, 105], building on the ideas of the preceding investigators prefers to discretize $u = \psi_y$, $v = -\psi_x$ and Eqs. (4.78) and (4.80). He maintains the structure of Eq. (4.78), since the velocity components are always of interest and Eq. (4.78) is simpler than Eq. (4.79). With $\Delta x = \Delta y = a$, we use the notation $\omega_{i,j}^n = \omega(ia, ja, n\,\Delta t)$. The vorticity equation (4.78) is discretized with a time-centered scheme wherein

† We shall use ∇^2 to denote the Laplace operator in the appropriate coordinate systems.

the diffusion terms $\omega_t = \nu\nabla^2\omega$ are approximated by a DuFort and Frankel [106] molecule

$$\omega_{i,j}^{n+1} - \omega_{i,j}^{n-1} = \frac{2\nu\Delta t}{a^2}\{\omega_{i+1,j}^{n} + \omega_{i-1,j}^{n} + \omega_{i,j+1}^{n} + \omega_{i,j-1}^{n} - 2\omega_{i,j}^{n-1} - 2\omega_{i,j}^{n+1}\}. \tag{4.82}$$

Equation (4.82) is known to be unconditionally stable for the diffusion equation but does not always satisfy the consistency condition (cf., e.g., Richtmyer and Morton [76] or Ames [77]). The final algorithm for advancing the vorticity to a new time is the explicit scheme

$$\begin{aligned}\omega_{i,j}^{n+1} = [1 + (4\nu\, \Delta t/a^2)]^{-1}\{&\omega_{i,j}^{n-1} \\ &+ (2\,\Delta t/a)[(u\omega)_{i-1/2,j}^{n} - (u\omega)_{i+1/2,j}^{n} + (v\omega)_{i,j-1/2}^{n} - (v\omega)_{i,j+1/2}^{n}] \\ &+ (2\nu\,\Delta t/a^2)[\omega_{i+1,j}^{n} + \omega_{i-1,j}^{n} + \omega_{i,j+1}^{n} + \omega_{i,j-1}^{n} - 2\omega_{i,j}^{n-1}]\}.\end{aligned} \tag{4.83}$$

Velocities and vorticities required by the calculations are obtained by averaging over the values specified at the nearest points in accordance with the relations

$$u_{i-1/2,j} = \tfrac{1}{4}(u_{i-1,j+1/2} + u_{i-1,j-1/2} + u_{i,j+1/2} + u_{i,j-1/2}), \tag{4.84}$$

$$v_{i,j-1/2} = \tfrac{1}{4}(v_{i-1/2,j} + v_{i+1/2,j} + v_{i-1/2,j-1} + v_{i+1/2,j-1}), \tag{4.85}$$

$$\omega_{i-1/2,j} = \tfrac{1}{2}(\omega_{i-1,j} + \omega_{i,j}), \tag{4.86}$$

and the remaining quantities in Eq. (4.83) are obtained by a suitable permutation of the indices. For the first time advancement, Fromm takes $\omega^1 = \omega^0$. While vorticities on an obstacle are not changed at this stage special consideration is given to the boundary values, particularly in the case of containing walls. For computation at the wall, with fluid below the wall, we use

$$\begin{aligned}\omega_{i,j_0}^{n+1} = [1 + (4\nu\,\Delta t/a^2)]^{-1} \\ \cdot\{\omega_{i,j}^{n-1} + (2\,\Delta t/a^2)[(u\omega)_{i-1/2,j_0}^{n} - (u\omega)_{i+1/2,j_0}^{n} + (v\omega)_{i,j_0-1/2}] \\ + (2\nu\,\Delta t/a^2)[\omega_{i+1,j_0}^{n} + \omega_{i-1,j_0}^{n} + \omega_{i,j_0}^{n} + \omega_{i,j_0-1}^{n} - 2\omega_{i,j_0}^{n-1}]\},\end{aligned}$$

where j_0 is the y index for the upper wall. If the wall velocity is u_0, then Eq. (4.84) becomes

$$u_{i-1/2,j_0} = \tfrac{1}{4}[2u_0 + u_{i-1,j_0-1/2} + u_{i,j_0-1/2}].$$

Similar boundary treatment occurs for fluid above a wall.

Without viscosity and when a time-centered difference is used, a restriction must be imposed on Δt for stability of Eq. (4.83). However, with diffusion only, the difference form requires no restriction on Δt. Experience (Harlow [107]) indicates that conditions for achieving accuracy of solution of stable equations are very nearly the same as stability criteria for corresponding (possibly) unstable equations. Consequently the two conditions

$$\Delta t\,(|\,u_0\,| + |\,v_0\,|)/a \leqslant 1, \qquad \nu\,\Delta t/a^2 \leqslant \tfrac{1}{4}, \tag{4.87}$$

were imposed.

Once the vorticity field is advanced to the $(n+1)$st time value, the advanced field values for ψ are obtained by a Gauss–Seidel iteration through the application[†] of

$$\psi_{i,j}^{(p+1)} = \tfrac{1}{4}\{\psi_{i+1,j}^{(p)} + \psi_{i-1,j}^{(p+1)} + \psi_{i,j+1}^{(p)} + \psi_{i,j-1}^{(p+1)} + a^2\omega_{i,j}\}. \tag{4.88}$$

While Fromm did not use an acceleration technique, it seems clear that one such as successive overrelaxation (cf., e.g., Ames [77]) should be used. Values of ψ on an obstacle are taken as the reference value ψ_r, while wall values are obtained by averaging the individual values. For fluid above a wall, since the wall must contain a streamline,

$$\psi_0 = \psi_1 - u_0 a + a^2\omega_0\,. \tag{4.89}$$

When the ψ field has converged sufficiently, as determined by some criteria, the new velocities are calculated by means of

$$u_{i,j-1/2}^{n+1} = a^{-1}\{\psi_{i,j}^{n+1} - \psi_{i,j-1}^{n+1}\},$$

$$v_{i-1/2,j}^{n+1} = a^{-1}\{\psi_{i-1,j}^{n+1} - \psi_{i,j}^{n+1}\},$$

and the vorticity values are brought up to the $(n+1)$st time by applying the definition [Eq. (4.77)]

$$\omega^{n+1} = a^{-1}\{u_{i,j-1/2} - u_{i,j+1/2} + v_{i+1/2,j} - v_{i-1/2,j}\}.$$

At walls the vorticity values are modified to conform to the ψ field locally. At the lower wall we simply solve Eq. (4.89) for ω_0. The process is then repeated.

Harlow and Fromm [108] have used essentially the same method for

† The p superscript indicates the iteration step at the $(n+1)$st time. It should not be confused with the time index. The ω's are the advanced values.

the two-dimensional problem of heat transfer from a rectangular cylinder into a surrounding fluid. Wilkes [109] solved the thermal convection problem in a two-dimensional cell essentially using an alternating direction implicit procedure (e.g., Ames [77]). Though the method is unconditionally stable for the linear diffusion equation, Wilkes found it to be unstable for his nonlinear problem at a Grashof number of 200,000.† Pearson [110] solved the unsteady, axisymmetric viscous flow generated by the differential rotation of two infinite parallel disks using an adaptation of the Crank–Nicolson centered scheme (cf., e.g., Ames [77] or Volume I) for the vorticity transport equation, Eq. (4.79). The resulting algebraic problem was solved by successive overrelaxation when the flow was almost linear or by a modified alternating direction implicit method (ADI) of Peaceman and Rachford when the nonlinear terms are not small. Additional stability is achieved by employing a smoothing process at boundary points.

Keller and Takami [111] study the steady viscous incompressible flow about circular cylinders. They introduce new cartesian coordinates, $\xi + i\eta = (1/\pi)\ln(x + iy)$, which map the exterior of the unit circle in the (x, y)-plane onto a semi-infinite strip in the (ξ, η)-plane. The transformed vorticity-stream function equations are discretized and an "*extrapolated line Liebmann*" *method* (Keller [112]) is used, resulting in coupled tridiagonal linear systems which are easily solved by the tridiagonal algorithm.

A large number of additional modifications of the basic stream function vorticity concept are given in the 1969 IUTAM Symposium on High Speed Computing in Fluid Mechanics [113]. We shall have occasion to refer to several articles in the proceedings of that conference.

Jenson [114], Hamielec *et al.* [115], and Rimon and Cheng [116] have solved the axisymmetric stream function–vorticity formulation for viscous incompressible flow in logarithmically contracted spherical polar coordinates ($z = \ln r$, θ). Different numerical procedures of second-order accuracy are employed. The novel feature here is the use of a varying (exponential) step size in the radial direction with the small step size near the boundary of the sphere. Cheng (in [113, p. 34]) discusses the accuracy of these results and questions the claimed accuracy.

† This instability may have been due to the manner in which boundary conditions on the vorticity were handled. Vorticity at the boundary always lagged one time-step behind the rest of the field. The extension of the ADI method used by Wilkes to three dimensions is not unconditionally stable for even the linear diffusion equation (see Ames [77] and references therein). Difficulty in proper boundary condition treatment has been reported in a number of studies.

Alonso [117] also employs a *graded mesh* structure in his solutions of the time dependent confined rotating flow of an incompressible viscous fluid. Small cells are employed in regions of high velocity and vorticity gradients, intermediate cells in regions of intermediate gradients, and large cells in regions of small gradients. Like Fromm he finds a circular cylindrical form of Eq. (4.78) to be more convenient. The tangential momentum and vorticity transport equations are discretized by a Crank–Nicolson centered scheme for the z (axial)-derivatives and an explicit one for the r-derivatives. The discretized linear equations are subjected to ADI resulting in two tridiagonal systems. The stream function is also obtained by ADI.

Fromm (in [113, pp. 1, 113]) amplifies his earlier work and herein stresses numerical dispersion effects. Fourth-order methods are shown to have better phase properties so that distortions resulting from dispersion are reduced.

4.10 PRIMITIVE VARIABLE METHODS

The use of the Navier–Stokes and continuity equations in the primitive variables u, v, w, and p, instead of the vorticity and stream function equations, is attractive in several ways. First the primitive equations can be applied to three-dimensional flow problems, whereas no three-dimensional counterpart of the vorticity transport equation is known. As we shall see in one of the methods, due to Chorin, the treatment of boundary conditions does not show any tendency to numerical instability so common in schemes based on the vorticity transport equation. Finally, if the pressure varies slowly with time, as occurs in many incompressible flows, the rate of convergence of the Chorin scheme is greatly improved.

Perhaps the earliest of these primitive variable procedures is the "marker and cell" method of Harlow and Welch [118] and Welch [119]. With the velocity field at the old time level given, these authors determine the pressure by means of an iterative scheme on the divergence of the Navier–Stokes equations. Boundary conditions for these calculations are obtained by application of the Navier–Stokes equations. The continuity equation is satisfied at the boundaries by the introduction of an *artificial reflection principle*. Finally, the velocity field is predicted at the new time level using an explicit discretization of the Navier–Stokes equations. Chorin [120] notest hat no reflection principle is known to hold for the boundary conditions appended to the Navier–Stokes equations. Therefore the application of one means, in effect, that the assigned boundary data are not imposed on the continuity equation.

Chorin [120, 121] set about designing what Ames [122] has called a semi-implicit procedure which would be free of the deficiencies of the marker and cell (MAC) technique of Harlow and Welch. In the first work, Chorin [121], studying the thermal convection of a fluid heated from below (Bénard problem), introduced an artificial compressibility into the continuity equation, and the pressure became a function of an artificial equation of state. The velocity field was advanced in time using an ADI scheme on the complete Navier–Stokes equations, with the pressure term expressed by means of the artificial equation of state. At the new time level the corresponding artificial density was obtained using a DuFort–Frankel molecule on the perturbed continuity equation. The paper includes a rigorous treatment of boundary conditions.

In his successive papers Chorin [123–126] improves his own method, and it is this more sophisticated technique that we shall discuss in detail. He designed an *auxiliary field* through which the velocity field and pressure values are projected from the old to the new time level. For this purpose, the pressure gradient does not appear explicitly in the ADI discrete simulation of the Navier–Stokes equations. At the new time level, the auxiliary field is decomposed into the velocity vector field and the pressure scalar field† using a set of algorithms in which his previous idea of an artificial compressibility has evolved into a formulation that is analogous to successive overrelaxation (SOR). The algorithm will be given in three dimensions.

Using $\partial_t = \partial/\partial t$, $\partial_i = \partial/\partial x_i$, the equations of motion of an incompressible fluid (cartesian coordinates) are

$$\partial_t u_i + u_j \partial_j u_i = -\rho_0^{-1} \partial_i p + \nu \nabla^2 u_i + E_i, \qquad \nabla^2 = \sum_j \partial_j^2, \qquad \partial_j u_j = 0. \tag{4.90}$$

With U as a reference velocity and d a reference length, a set of dimensionless variables

$$u_i' = u_i/U, \qquad x_i' = x_i/d, \qquad p' = p(d/\rho_0 \nu U),$$
$$E_i' = (\nu U/d^2)\, E_i, \qquad t' = t(\nu/d^2),$$

† This is based on writing the Navier–Stokes equations as

$$\partial_t u_i + \partial_i p = \mathscr{L}_i u, \qquad \partial_t = \partial/\partial t, \qquad \text{and} \qquad \partial_i = \partial/\partial x_i,$$

where the first term has zero divergence and the second has zero curl. This decomposition exists and is uniquely determined whenever the initial-value problem for the Navier–Stokes equation is well posed. The decomposition is extensively used in existence and uniqueness proofs for these equations (e.g., Fujita and Kato [127]).

when introduced into Eq. (4.90) transform that set into (the primes are dropped)

$$\partial_t u_i + R u_j \partial_j u_i = -\partial_i p + \nabla^2 u_i + E_i \tag{4.91}$$

$$\partial_j u_j = 0, \qquad R = Ud/\nu \tag{4.92}$$

(Padmanabhan [128] finds a modified set of dimensionless variables more useful in his problem with high Reynolds numbers (70,000).)

Equation (4.91) can be written as

$$\partial_t u_i + \partial_i p = \mathscr{L}_i u , \tag{4.93}$$

where $\mathscr{L}_i u$ depends on u_i and E_i, but not on p. Equation (4.92) when differentiated becomes

$$\partial_i(\partial_t u_i) = 0. \tag{4.94}$$

The method is summarized as follows: The time t is discretized; at every time step, $\mathscr{L}_i u$ is evaluated; it is then decomposed into the sum of a vector with zero divergence and a vector with zero curl. The component with zero divergence is $\partial_t u_i$, which can be used to obtain u_i at the next time level. The component with zero curl is $\partial_i p$.

In what follows u_i and p also denote the discrete approximation of Eqs. (4.91) and (4.92) and Du is a difference approximation to $\partial_j u_j$. At time $n\,\Delta t$ a velocity field $u_i{}^n$ is given satisfying $Du^n = 0$. Our job is to evaluate $u_i{}^{n+1}$ from Eq. (4.91), so that $Du^{n+1} = 0$.

Let $Tu_i = bu_i^{n+1} - Bu_i$ approximate $\partial_t u_i$, where b is a constant and Bu_i is a suitable linear combination of u_i^{n-j}, $j \geqslant 0$. First an auxiliary field, u_i^{aux}, is evaluated by means of

$$bu_i^{\text{aux}} - Bu_i = F_i u, \tag{4.95}$$

where $F_i u$ approximates $\mathscr{L}_i u$. Clearly u_i^{aux} differs from u_i^{n+1}, since the pressure and Eq. (4.92) are not incorporated. u_i^{aux} may be evaluated by an implicit scheme, that is, $F_i u$ may depend on $u_i{}^n$, u_i^{aux}, and intermediate fields, say $u_i{}^*$, u_i^{**}. $bu_i^{\text{aux}} - Bu_i$ now approximates $\mathscr{L}_i u$ to within an error which, generally, depends upon Δt.

Second, let $G_i p$ approximate $\partial_i p$. To obtain u_i^{n+1}, p^{n+1} it is necessary to perform the decomposition

$$F_i u = bu_i^{\text{aux}} - Bu_i = Tu_i + G_i p^{n+1},$$

$$D(Tu) = 0.$$

It is, however, assumed that $Du^{n-j} = 0, j \geqslant 0$, so it is only necessary to perform the decomposition

$$u_i^{\text{aux}} = u_i^{n+1} + b^{-1} G_i p^{n+1}, \tag{4.96}$$

where $Du_i^{n+1} = 0$ and u_i^{n+j} satisfies the prescribed boundary conditions. Since p^n is available and is a reasonable first guess for p^{n+1}, the decomposition, Eq. (4.96), is probably best done by iteration. The iteration scheme is (m is the iteration number)

$$u_i^{n+1,m+1} = u_i^{\text{aux}} - b^{-1} G_i{}^m p, \qquad m \geqslant 1, \tag{4.97a}$$

$$p^{n+1,m+1} = p^{n+1,m} - \lambda\, Du^{n+1,m+1}, \qquad m \geqslant 1, \tag{4.97b}$$

$$p^{n+1,1} = p^n.$$

Here λ is a parameter, $(\ )^{n+1,m+1}$ represent successive approximations to $(\ )^{n+1}$, and $G_i{}^m p$ is a function of $p^{n+1,m+1}$ and $p^{n+1,m}$ which converges to $G_i p$ as $|\, p^{n+1,m+1} - p^{n+1,m}\,| \to 0$. Equation (4.97a) is to be performed in the interior of the integration domain $\mathscr{D}$ and Eq. (4.97b) in $\mathscr{D}$ and on its boundary. Clearly Eq. (4.97a) tends to Eq. (4.96) if the iterations converge. $G_i{}^m p$ is used instead of $G_i p$ to enable improvement of the rate of convergence of the iterations. When for some l and a small predetermined constant $\epsilon > 0$

$$\max_{\mathscr{D}} |\, p^{n+1,l+1} - p^{n+1,l}\,| \leqslant \epsilon,$$

we set

$$u_i^{n+1} = u_i^{n+1,l+1}, \qquad p^{n+1} = p^{n+1,l+1}.$$

Chorin [124] conjectures that the over-all scheme is stable if the scheme $Tu_i = F_i u$ is stable. His experimental calculations and those of Padmanabhan [128] lend support to this conjecture, but proof is lacking.

Equation (4.95) represents one step in time for the solution of the Burgers-like equation

$$\partial_t u_i = \mathscr{L}_i u.$$

Schemes selected should be convenient to use, implicit, and accurate to $O(\Delta t) + O(\Delta x^2)$. Implicit schemes eliminate unduly restrictive conditions such as small Δt limitations ($\Delta t < \frac{1}{8}\Delta x^2$ is usual in three space dimensions). However, we do not select implicit schemes of higher accuracy to avoid the solution of nonlinear equations at each step. Two schemes, both variants of the alternating direction implicit method with

$$Tu_i = (u_i^{n+1} - u_i{}^n)/\Delta t, \qquad b^{-1} = \Delta t, \qquad Bu_i = u_i{}^n/\Delta t,$$

are chosen.

In two-dimensional problems, one can use a Peaceman–Rachford (ADI) scheme or in both two and three dimensions Samarskii's [129] variant of alternating direction. The latter takes the form (three dimensions)

$$
\begin{aligned}
u^{*}_{i(q,r,s)} &= u^{n}_{i(q,r,s)} - R(\Delta t/2\,\Delta x_1)\,u^{n}_{1(q,r,s)} \\
&\quad \cdot [u^{*}_{i(q+1,r,s)} - u^{*}_{i(q-1,r,s)}] + (\Delta t/\Delta x_1{}^2) \\
&\quad \cdot [u^{*}_{i(q+1,r,s)} + u^{*}_{i(q-1,r,s)} - 2u^{*}_{i(q,r,s)}], \\
u^{**}_{i(q,r,s)} &= u^{*}_{i(q,r,s)} - R(\Delta t/2\,\Delta x_2)\,u^{*}_{2(q,r,s)} \\
&\quad \cdot [u^{**}_{i(q,r+1,s)} - u^{**}_{i(q,r-1,s)}] + (\Delta t/\Delta x_2{}^2) \\
&\quad \cdot [u^{**}_{i(q,r+1,s)} + u^{**}_{i(q,r-1,s)} - 2u^{**}_{i(q,r,s)}], \\
\overset{\text{aux}}{u}_{i(q,r,s)} &= u^{**}_{i(q,r,s)} - R(\Delta t/2\,\Delta x_3)\,u^{**}_{3(q,r,s)} \\
&\quad \cdot [\overset{\text{aux}}{u}_{i(q,r,s+1)} - \overset{\text{aux}}{u}_{i(q,r,s-1)}] + (\Delta t/\Delta x_3{}^2) \\
&\quad \cdot [\overset{\text{aux}}{u}_{i(q,r,s+1)} + \overset{\text{aux}}{u}_{i(q,r,s-1)} - 2\overset{\text{aux}}{u}_{i(q,r,s)}] + \Delta t E_{i(q,r,s)}\,, \qquad (4.98\text{a})
\end{aligned}
$$

where $(\ )_{i(p,r,s)} = (\ )_i\,(q\,\Delta x_i\,,\, r\,\Delta x_2\,,\, s\,\Delta x_3)$. In symbolic form we can write Eq. (4.98a) as

$$
\begin{aligned}
(I - \Delta t Q_1)\,u_i{}^{*} &= u_i{}^{n}, \\
(I - \Delta t Q_2)\,u_i^{**} &= u_i{}^{*}, \\
(I - \Delta t Q_3)\,\overset{\text{aux}}{u_i} &= u_i^{**} + \Delta t E_i\,, \qquad (4.98\text{b})
\end{aligned}
$$

where I is the identity operator and Q_l involves differentiations with respect to x_l only. These equations require fewer arithmetic operations per time step than the Peaceman–Rachford method, are stable in three directions, and because of the simple right-hand side structure $u_i{}^{*}$ and u_i^{**} do not have to be stored separately.

If u_i^{n+1} are supplied as boundary conditions, one only needs to set

$$
\begin{aligned}
u_i{}^{*} &= u_i^{n+1} - \Delta t Q_2 u_i^{n+1} - \Delta t Q_3 u_i^{n+1} - \Delta t E_i + \Delta t \bar{G}_i p, \\
u_i^{**} &= u_i^{n+1} - \Delta t Q_3 u_i^{n+1} - \Delta t E_i + \Delta t \bar{G}_i p, \\
\overset{\text{aux}}{u_i} &= u_i^{n+1} + \Delta t \bar{G}_i p,
\end{aligned}
$$

where $\bar{G}_i p^n = G_i p^n + O(\Delta t)$.

For simplicity let $\mathscr{D}$ be two dimensional, $\mathscr{B}$ be its boundary, and $\mathscr{C}$ the set of mesh nodes with a neighbor in $\mathscr{B}$. In $\mathscr{D} - \mathscr{B}$ we set

$$Du = (2\,\Delta x_1)^{-1}[u_{1(q+1,r)} - u_{1(q-1,r)}] + (2\,\Delta x_2)^{-1}[u_{2(q,r+1)} - u_{2(q,r-1)}], \tag{4.99}$$

and at point of $\mathscr{B}$ we use a second-order, one-sided difference so that Du is accurate to $O(\Delta x^2)$ everywhere. Now define $G_i p$ at every point of $\mathscr{D} - \mathscr{B}$ by

$$G_1 p = (2\,\Delta x_1)^{-1}(p_{q+1,r} - p_{q-1,r}), \qquad G_2 p = (2\,\Delta x_2)^{-1}(p_{q,r+1} - p_{q,r-1}). \tag{4.100}$$

The operator G_i^m is defined at a point of $\mathscr{D} - \mathscr{B} - \mathscr{C}$ by implicit relations generated by selecting the value of p at (q, r) by

$$\tfrac{1}{2}(p_{q,r}^{n+1,m+1} + p_{q,r}^{n+1,m}),$$

while at other points we use $p^{n+1,m}$. This is crucial to the convergence of the method. With D expressed by Eq. (4.99) we obtain, after solving the implicit system,

$$p_{q,r}^{n+1,m+1} = (1 + \alpha_1 + \alpha_2)^{-1}\,[(1 - \alpha_1 - \alpha_2)\,p^{n+1,m} - \lambda D u^{\text{aux}}]$$
$$-\alpha_1[p_{q+2,r}^{n+1,m} + p_{q-2,r}^{n+1,m}] + \alpha_2[p_{q,r+2}^{n+1,m} + p_{q,r-2}^{n+1,m}], \tag{4.101}$$

where $\alpha_i = \lambda\,\Delta t/4\,\Delta x_i$. This is a Dufort–Frankel relaxation scheme (see Chorin [124] for choice of λ_{opt}). In $\mathscr{B}$ and $\mathscr{C}$ Eq. (4.101) must be modified by using the known values of u_i^{n+1} on the boundary. Chorin considers only rectangular domains.

Chorin [124] presents a numerical solution of the three-dimensional Bénard convection problem by his procedure. Reliable time-dependent results are asserted. Padmanabhan [128] and Padmanabhan *et al.* [130], by Chorin's method, investigated the wake collapse of a cylindrical fluid mass in a medium of changing viscosity. The collapse is, of necessity, accompanied by a horizontal elongation of the network. Coordinate stretching was periodically employed to convert the elongated network back into a nearly circular form. Nevertheless the advance of the solution for each time-step increased from an initial value of 6 sec to 10 min at the end. Wessel (in [113, p. 171]) used a straightforward primitive variable time-centered algorithm based on a DuFort–Frankel molecule and successive overrelaxation, to compute the collapse of a perturbation in an infinite density stratified fluid. He does not report any computational difficulties.

Pujol [131] compares a modified Chorin's method with that of Douglas–Aziz (Section 4.11) for Poiseuille flow of two-dimensional non-Newtonian fluids. While satisfactory results were obtained from both procedures, the Douglas–Aziz scheme proved more accurate at the eventual risk of numerical instability. The inclusion of a variable viscosity in Chorin's procedure did not impair its stability.

4.11 VECTOR POTENTIAL METHODS

Aziz [132] and Pearson [110] found that ADI methods of solving the parabolic vorticity transport equation were more accurate, converged faster, and were more stable than explicit methods such as that of DuFort–Frankel. The discretization error for these Peaceman–Rachford (cf. [77, p. 149]) ADI methods is locally second order in space and first order in time. Aziz and Hellums [133] in a study of three-dimensional laminar natural convection found that the Douglas [134] ADI scheme (Ames [77, p. 248]) was superior in the sense that both two- and three-dimensional forms were unconditionally stable for the linear diffusion equation. The Douglas scheme is a perturbation of the classical Crank–Nicolson molecule and as such is locally second order in space and time. While this procedure can and has been used in two-dimensional computation (cf., e.g., Aziz and Hellums [133] in convection and Pujol [131] in non-Newtonian problems) its major impact is felt in three-dimensional problems when the basic equations are recast in *vector potential form.*

The *vector potential method* of Aziz and Hellums [133] consists of a transformation of the complete Navier–Stokes equations in terms of a *vorticity* and a *vector potential.* These are discretized and solved using the Douglas [134] ADI method for the parabolic portion of the problem and SOR for the elliptic portion. The formulation of the equations of change in terms of the vector potential was found to be an essential ingredient in this analysis. The existence of the vector potential has been known for many years. Most of the early work is summarized by Jacob [135]. We shall briefly review the basic concepts herein.

A useful classification of vector fields[†] $\mathbf{E}$ is possible in terms of the divergence (div) and curl operators (cf., e.g., Moon and Spencer [136–138]. If div $\mathbf{E} = 0$ at every point of a region R, the field is said to be *solenoidal* in that region.[‡] Physically this means there are no sources or

[†] In this section, a vector is denoted by boldface, viz. $\mathbf{E}$.

[‡] Much of the foundation research has been with respect to electric and electromagnetic fields, especially with respect to the Maxwell equations. Thus the terms are reminiscent of that area.

sinks in R. If, at every point of R, curl $\mathbf{E} = 0$ the field is said to be *irrotational* in R. The classification of fields is as follows:

I. Solenoidal and irrotational:
$$\text{curl } \mathbf{E} = 0, \qquad \text{div } \mathbf{E} = 0;$$
II. Irrotational but not solenoidal:
$$\text{curl } \mathbf{E} = 0, \qquad \text{div } \mathbf{E} \neq 0;$$
III. Solenoidal but not irrotational:
$$\text{curl } \mathbf{E} \neq 0, \qquad \text{div } \mathbf{E} = 0;$$
IV. Neither solenoidal nor irrotational:
$$\text{curl } \mathbf{E} \neq 0, \qquad \text{div } \mathbf{E} \neq 0.$$

The velocity field of an incompressible fluid falls into Class III while that of a compressible fluid is of Class IV.

A necessary and sufficient condition for the existence of a *scalar potential* ϕ is that curl $\mathbf{E} = 0$, whereupon ϕ is defined by $\mathbf{E} = -\text{grad}\, \phi$ and div $\mathbf{E} = -\text{div grad}\, \phi = -\nabla^2\phi$. Thus, in fields of Class I and II it is always possible to introduce a scalar potential ϕ, defined by $\nabla^2\phi = K$ (Poisson's equation for a field of Class II).

For fields which are not irrotational (curl $\mathbf{E} \neq 0$), in some cases it is still possible to employ a scalar function and thus avoid the difficulties usually associated with the more difficult case, that of the vector potential. Suppose that curl $\mathbf{E} \neq 0$, but there exists a scalar function ρ such that curl$(\rho\mathbf{E}) = 0$. In this case a scalar *quasi potential* Γ exists, defined by the equation

$$\mathbf{E} = -\rho^{-1} \text{ grad } \Gamma. \tag{4.102}$$

Since $0 = \text{curl}(\rho\mathbf{E}) = \rho \text{ curl } \mathbf{E} + \text{grad}\, \rho \times \mathbf{E}$, it follows that

$$\begin{aligned} 0 = \mathbf{E} \cdot \text{curl}(\rho\mathbf{E}) &= \mathbf{E} \cdot \rho \text{ curl } \mathbf{E} + \mathbf{E} \cdot \text{grad}\, \rho \times \mathbf{E} \\ &= \rho(\mathbf{E} \cdot \text{curl } \mathbf{E}) + \text{grad}\, \rho \cdot \mathbf{E} \times \mathbf{E}. \end{aligned}$$

Consequently, *a necessary and sufficient condition for the existence of a quasi potential is that* $\mathbf{E} \cdot$*curl* $\mathbf{E} = 0$.

From Eq. (4.102),

$$\begin{aligned} K = \text{div } \mathbf{E} &= -\text{div}(\rho^{-1} \text{ grad } \Gamma) \\ &= -\rho^{-1}\nabla^2\Gamma - (\text{grad } \rho^{-1}) \cdot \text{grad } \Gamma, \end{aligned}$$

where K is zero for Class III and a function for Class IV. Thus a quasi potential can be determined by solving

$$\nabla^2\Gamma + (\text{grad } \rho^{-1}) \cdot \text{grad } \Gamma = -\rho K. \tag{4.103}$$

The latter equation occurs in field problems associated with inhomogeneous media, and a useful transformation is presented by Ames and de la Cuesta [139].

Most solenoidal fields do not admit a potential or quasi potential, but it is *always possible to introduce a vector potential* **A**, which, though not as simple as a scalar potential, nevertheless behaves in a somewhat similar fashion. If curl **E** = **B**, div **E** = 0, the vector potential is defined by the relation

$$\mathbf{E} = \operatorname{curl} \mathbf{A}. \tag{4.104}$$

Then

$$\operatorname{curl} \mathbf{E} = \operatorname{curl} \operatorname{curl} \mathbf{A} = \operatorname{grad} \operatorname{div} \mathbf{A} - \star\mathbf{A} = \mathbf{B}, \tag{4.105}$$

where ☆ is used to denote the *vector Laplacian*†

$$\star\mathbf{A} = \operatorname{grad} \operatorname{div} \mathbf{A} - \operatorname{curl} \operatorname{curl} \mathbf{A}. \tag{4.106}$$

We can assign any desired value to the divergence of our new vector **A**. If we take div **A** = 0, then Eq. (4.106) becomes

$$\star\mathbf{A} = -\mathbf{B}, \tag{4.107}$$

which is the vector form of Poisson's equation. In cartesian coordinates (x, y, z) it splits into three *scalar* Poisson equations,

$$\star_x\mathbf{A} = -\mathbf{B}_x, \qquad \star_y\mathbf{A} = -\mathbf{B}_y, \qquad \star_z\mathbf{A} = -\mathbf{B}_z. \tag{4.108}$$

In the case of a general orthogonal coordinate system (x^1, x^2, x^3), the vector Laplacian is given by Moon and Spencer [136–138]. To solve a problem using the vector potential, we first obtain **A** from Eq. (4.107) and then find **E** from Eq. (4.104).

For completeness we now present the scheme used by Helmholtz (cf., e.g., Wills [141]) in his classic study of vortex motion. In Class IV suppose **E** has known divergence and curl, neither of which is zero. If **U** is irrotational (curl **U** = 0) and **V** is solenoidal (div **V** = 0), we suppose

$$\mathbf{E} = \mathbf{U} + \mathbf{V}. \tag{4.109}$$

† The vector Laplacian is often written ∇^2, but Moon and Spencer [140] note that this is an ambiguous notation. Except in the special case of rectangular coordinates, where each component is similar to a scalar Laplacian the two operators are quite distinct!

Since curl $\mathbf{U} = 0$, a scalar potential ϕ can be introduced so that

$$\mathbf{U} = -\text{grad}\,\phi. \tag{4.110}$$

Likewise, since div $\mathbf{V} = 0$, there exists a vector potential $\mathbf{A}$ such that

$$\mathbf{V} = \text{curl}\,\mathbf{A}, \qquad \text{div}\,\mathbf{A} = 0. \tag{4.111}$$

From Eq. (4.109), we find

$$\text{div}\,\mathbf{E} = \text{div}\,\mathbf{U} + \text{div}\,\mathbf{V} = -\text{div grad}\,\phi = -\nabla^2\phi,$$

and

$$\text{curl}\,\mathbf{E} = \text{curl}\,\mathbf{U} + \text{curl}\,\mathbf{V} = \text{curl curl}\,\mathbf{A} = -☆\mathbf{A}.$$

Consequently fields of Class IV are evaluated by the solution of the scalar Poisson equation

$$\nabla^2\phi = -\text{div}\,\mathbf{E}, \tag{4.112}$$

and the vector Poisson equation

$$☆\mathbf{A} = -\text{curl}\,\mathbf{E}, \tag{4.113}$$

for the scalar potential ϕ and vector potential $\mathbf{A}$. Using these we calculate $\mathbf{U}$ and $\mathbf{V}$ by means of Eqs. (4.110) and (4.111). Finally $\mathbf{E}$ is calculated by means of Eq. (4.109). With this introduction, we are now in position to discuss the method of the *vector potential* (Aziz and Hellums [133], Douglas [134]).

The dimensionless equations describing the behavior of fluid layers heated from below are (cf., e.g., Chandrasekhar [142])

$$\text{div}\,\mathbf{V} = 0, \tag{4.114}$$

$$(\partial\mathbf{V}/\partial t) + \mathbf{V}\cdot\text{div}\,\mathbf{V} = -\text{div}\,P - \mathbf{Gr}\,T + ☆\mathbf{V}, \tag{4.115}$$

$$\partial T/\partial t + \mathbf{V}\cdot\text{grad}\,T + w = (\text{Pr})^{-1}\nabla^2 T, \tag{4.116}$$

where T and P denote deviations in the dimensionless variables from an initial condition of steady conduction, $\mathbf{Gr}$ denotes a vector with zero components in the x- and y-directions and with the Grashof number as its z-component, and Pr is the Prandtl number. (Again the reader is reminded that in cartesian coordinates the components of ☆ are the same as the corresponding scalar Laplace operators, ∇^2.) The boundary and initial conditions for the problem of convection in a cube of specified temperature on the upper and lower walls and with insulated side walls are

$$
\begin{aligned}
u = v = w &= 0, \qquad \text{on all solid boundaries,} \\
T(x, y, z, 0) &= \tilde{T}, \qquad \text{(temperature disturbance)} \qquad\qquad (4.117) \\
T &= 0, \qquad \text{at} \quad z = 0, 1, \qquad \partial T/\partial x = 0, \qquad \text{at} \quad x = 0, 1, \\
\partial T/\partial y &= 0, \qquad \text{at} \quad y = 0, 1.
\end{aligned}
$$

Elimination of the pressure from the vector equations, Eq. (4.115), results in the vorticity transport equation

$$\frac{\partial \mathbf{W}}{\partial t} + \mathbf{V} \cdot \operatorname{div} \mathbf{W} - \mathbf{W} \cdot \operatorname{div} \mathbf{V} = \begin{Bmatrix} -GrT_y \\ GrT_x \\ 0 \end{Bmatrix} + \star\mathbf{W}, \tag{4.118}$$

where $\mathbf{W} = \operatorname{curl} \mathbf{V}$ is the vorticity vector. To calculate the velocity from the vorticity, it is convenient to introduce a vector potential $\boldsymbol{\psi}$ as previously described. This may be viewed as the three-dimensional counterpart of the two-dimensional stream function. It is defined by

$$\mathbf{V} = \operatorname{curl} \boldsymbol{\psi},$$

and the degree of freedom in its selection permits the vector potential to be solenoidal, i.e., $\operatorname{div} \boldsymbol{\psi} = 0$. Thus we have

$$\star\boldsymbol{\psi} = -\mathbf{W}. \tag{4.119}$$

The set of Eqs. (4.116), (4.118), (4.119), and $\mathbf{W} = \operatorname{curl} \mathbf{V}$ were found to be a convenient form for numerical computation.

A word about the boundary conditions must be inserted here because of a minor controversy between Timman and Moreau which is discussed by Hirasaki and Hellums [143]. Moreau [144] and Hirasaki and Hellums [143] agree that on solid surfaces the proper boundary condition on the normal component of velocity is satisfied if the normal derivative of the normal component of the vector potential vanishes, and if the components of the vector potential tangential to the surface vanish.† The boundary conditions on the vector potential become

$$
\begin{aligned}
(\psi_1)_x = \psi_2 = \psi_3 &= 0, \qquad \text{at} \quad x = 0, 1, \\
\psi_1 = (\psi_2)_y = \psi_3 &= 0, \qquad \text{at} \quad y = 0, 1, \qquad\qquad (4.120) \\
\psi_1 = \psi_2 = (\psi_3)_z &= 0, \qquad \text{at} \quad z = 0, 1,
\end{aligned}
$$

† Difficulty in treatment of the boundary conditions is very often reported. Indeed stability problems are often a result of these difficulties.

and those on the vorticity are

$$\begin{aligned}
&\xi_1 = 0, &&\xi_2 = -\partial w/\partial x, &&\xi_3 = \partial v/\partial x, &&\text{at}\quad x = 0, 1,\\
&\xi_1 = \partial w/\partial y, &&\xi_2 = 0, &&\xi_3 = -\partial u/\partial y, &&\text{at}\quad y = 0, 1,\\
&\xi_1 = -\partial v/\partial z, &&\xi_2 = \partial u/\partial z, &&\xi_3 = 0, &&\text{at}\quad z = 0, 1.
\end{aligned} \tag{4.121}$$

Here we have written $\mathbf{W}^{\mathrm{T}} = \{\xi_1, \xi_2, \xi_3\}$ to denote the vorticity vector.

In the numerical computation we shall use the following notation:

$$\begin{aligned}
Q_n &= Q_n(i, j, k) = Q(i\,\Delta x, j\,\Delta y, k\,\Delta z, n\,\Delta t) = Q(x_i, y_j, z_k, t_n),\\
\nabla_x Q_n &= [Q_n(i+1, j, k) - Q_n(i-1, j, k)]/(2\,\Delta x),\\
\nabla_x{}^2 Q_n &\, [Q_n(i+1, j, k) - 2Q_n(i, j, k) + Q_n(i-1, j, k)]/(\Delta x)^2,\\
\delta_x Q_n &= (\nabla_x{}^2 - u\,\nabla_x)\,Q_n,\\
\delta_y Q_n &= (\nabla_y{}^2 - v\,\nabla_y)\,Q_n,\\
\delta_z Q_n &= (\nabla_z{}^2 - w\,\nabla_z)\,Q_n.
\end{aligned}$$

Corresponding operators ∇_y, ∇_z, etc., are defined in a similar manner.

Any one of the scalar parabolic equations (4.116) or (4.118), may be written as

$$S_t = (S_{xx} - uS_x) + (S_{yy} - vS_y) + (S_{zz} - wS_z) - \phi, \tag{4.122}$$

where S represents any of the dependent variables T, ξ_1, ξ_2, and ξ_3. The expression ϕ includes all of the remaining terms, e.g., $\phi = w$ in Eq. (4.116). The ADI method of Douglas [134], previously discussed, will now be described for Eq. (4.122).

Let us denote by Q the approximate value for S obtained by the finite difference procedure. From the known solution at t_n we find a first estimate of the solution Q^*_{n+1}, at time t_{n+1}, by†

$$\tfrac{1}{2}\delta_x(Q^*_{n+1} + Q_n) + \delta_y Q_n + \delta_z Q_n = (Q^*_{n+1} - Q_n)/\Delta t + \phi, \tag{4.123a}$$

where ϕ is evaluated at n, $n + \frac{1}{2}$, or $n + 1$ depending upon the method being used. This is followed by

$$\tfrac{1}{2}\delta_x(Q^*_{n+1} + Q_n) + \tfrac{1}{2}\delta_y(Q^{**}_{n+1} + Q_n) + \delta_z Q_n = (Q^{**}_{n+1} - Q_n)/\Delta t + \phi, \tag{4.123b}$$

† One asterisk denotes the first approximation with more asterisks for successive estimates. For the development of the method see Douglas [134] or Ames [77, p. 248]. This ADI scheme reduces to the one proposed by Douglas if we replace our operators δ_x, δ_y, and δ_z by $\nabla_x{}^2$, $\nabla_y{}^2$, and $\nabla_z{}^2$. The nonlinearities in the procedure of Douglas appear only through the term ϕ where $\phi = \phi(x, y, z, t, S)$.

and this by

$$\tfrac{1}{2}\delta_x(Q_{n+1}^{*} + Q_n) + \tfrac{1}{2}\delta_y(Q_{n+1}^{**} + Q_n) + \tfrac{1}{2}\delta_z(Q_{n+1} + Q_n) = (Q_{n+1} - Q_n)/\Delta t + \phi, \tag{4.124}$$

where Q_{n+1} is the accepted value. Equations (4.123) and (4.124) may be simplified by subtracting Eq. (4.123a) from Eq. (4.123b) and Eq. (4.123b) from Eq. (4.124), respectively. After some rearrangement, the new system becomes

$$[\delta_x - 2(\Delta t)^{-1}]\, Q_{n+1}^{*} = -[\delta_x + 2\delta_y + 2\delta_z + 2(\Delta t)^{-1}]\, Q_n + 2\phi, \tag{4.125}$$

$$[\delta_y - 2(\Delta t)^{-1}]\, Q_{n+1}^{**} = \delta_y Q_n - 2(\Delta t)^{-1}\, Q_{n+1}^{*}, \tag{4.126}$$

$$[\delta_z - 2(\Delta t)^{-1}]\, Q_{n+1} = \delta_z Q_n\,. \tag{4.127}$$

With u, v, w, and ϕ known in advance Eqs. (4.125)–(4.127) require the solution of a tridiagonal system of linear algebraic equations three times to find Q_{n+1}. If u, v, w, and ϕ are not evaluated at the old time step, it becomes necessary to perform several iterations at each time step.

The elliptic equations are all of the type

$$\nabla^2 S = -\phi,$$

where ϕ is again a scalar function. The resulting finite difference simulation may be solved by a direct application of the three-dimensional Douglas ADI scheme or by the SOR method of Young (see Ames [77]).

Lastly we remark on the four procedures tried in treating the velocity component coefficients of the nonlinear convective terms and the ϕ term of Eq. (4.122). These were:

(a) Advance values of the dependent variables to the $n + \frac{1}{2}$ level in time. From this calculate the velocity components u, v, and w, and the ϕ term at $n + \frac{1}{2}$ and use these to advance the values a full time step. This is essentially the "predictor–corrector" method of Douglas and Jones [145] found so useful by Miller [146] in obtaining numerical solutions of the Burgers equation $u_t + uu_x = \nu u_{xx}$ (see Part C).

(b) Evaluate the velocity components u, v, w, and ϕ entirely at the new level. This requires an iteration.

(c) Evaluate the velocity components u, v, w, and ϕ at the nth level. No iteration is required if the boundary vorticity is allowed to lag one step behind the inner vorticity. Stability difficulties can arise from this procedure.

(d) Use the average of the new and old values for determining u, v, w, and ϕ in the nonlinear terms.

In the trials, (d) was found most suitable from the point of view of stability. In two-dimensional problems, both (b) and (d) gave identical solutions, but (b) was preferred for storage reasons in three-dimensional problems.

C. SOME NEW DIRECTIONS

4.12 INTRODUCTORY REMARKS

In the remaining few pages we describe some of the directions that research in numerical analysis is taking. The *first* of these is the extension of *predictor–corrector methods* to nonlinear problems by Douglas and Jones. The *second*, promoted by Bellman and Protter and schools, we will call *functional methods*. The *third* direction involves a reformulation of the problem, usually in terms of *new dependent and independent variables*. Flugge-Lotz *et al.* (cf., Volume I) have had much success with such an approach in boundary-layer calculations.

We have selected these three areas for their future potential with the full realization that research and development in numerical analysis is one of the fastest growing areas of mathematical research. Many of the budding methods are marrying an approximate solution in one or more directions with a numerical scheme. In fluid mechanics one of the earliest computations, by Whitaker and Wendel ([147], also Volume I, p. 401), was of this type. Galerkin-type approximations and the associated computations for parabolic equations have been reported by Douglas and duPont [148]. The tremendous variety and modifications for individual problems is observable by Giese's [149] computer bibliography of numerical solutions of partial differential equations. The subsequent updating and computer search routine by Francis [150] provides improved accessibility to this file. Excellent summaries of computational methods in various fields of physics are published in the serial "Methods in Computational Physics." The issues through 1970 are summarized in reference [151].

Extensive generalizations of the basic ADI methods have been made with considerable activity by the Russian school. Yanenko's 1967 work, now translated [152], summarizes these generalizations in the so called *method of fractional steps* which subsumes the majority of the procedures.

4.13 PREDICTOR–CORRECTOR METHODS

Predictor–corrector methods have been successfully used by many in the numerical solution of *ordinary* differential equations. A discussion of some of these is to be found in the work of Hamming [153, 154] and Fox [155] variously labeled as the Adams–Bashforth method, methods of Milne, and so forth. The general approach starts from known or previously computed results at previous pivotal values, up to and including the point x_n, by "predicting" results at x_{n+1} with formulas which need no knowledge at x_{n+1}. These predicted results are relatively inaccurate. They are then improved by the use of more accurate "corrector" formulas which require information at x_{n+1}. Generally this amounts to computing results at x_{n+1} from a nonlinear algebraic equation, to the solution of which the "predictor" gives a first approximation and the "corrector" is used repeatedly, if necessary, to obtain the final result.

Douglas and Jones [145] have considered

$$u_{xx} = \psi(x, t, u, u_x, u_t), \qquad 0 < x < 1, \quad 0 < t \leqslant T, \tag{4.128}$$

with $u(x, 0)$, $u(0, t)$, and $u(1, t)$ as specified boundary conditions. If either

$$\psi = f_1(x, t, u)(\partial u/\partial t) + f_2(x, t, u)(\partial u/\partial x) + f_3(x, t, u), \tag{4.129}$$

or

$$\psi = g_1[x, t, u, (\partial u/\partial x)](\partial u/\partial t) + g_2[x, t, u, (\partial u/\partial x)], \tag{4.130}$$

a predictor–corrector modification of the Crank–Nicolson procedure is possible so that the resulting algebraic problem is linear. This is significant since the class given by Eq. (4.129) includes the Burgers equation

$$u_{xx} = uu_x + u_t,$$

of turbulence and suggests extension to higher-order systems in fluid mechanics. The class given by Eq. (4.130) includes the equation

$$(\partial/\partial x)(K(u)\,\partial u/\partial x) = \alpha(u)(\partial u/\partial t),$$

of nonlinear diffusion.

If ψ is of the form Eq. (4.129) the following predictor–corrector analog combined with the boundary data $u_{i,0}$, $u_{0,j}$, and $u_{M,j}$, leads to *linear* algebraic equations. The predictor is†

$$(h^2)^{-1}\delta_x^2 U_{i,j+1/2} = \psi[ih, (j+\tfrac{1}{2})k, U_{i,j}, (2h)^{-1}\mu\delta_x U_{i,j}, (2/k)(U_{i,j+1/2} - U_{i,j})], \qquad (4.131)$$

for $i = 1, 2,..., M-1$. This is followed by the corrector

$$\tfrac{1}{2}\delta_x^2[U_{i,j+1} + U_{i,j}] = \psi[ih, (j+\tfrac{1}{2})k, U_{i,j+1/2}, (4h)^{-1}\mu\delta_x(U_{i,j+1} + U_{i,j}), k^{-1}(U_{i,j+1} - U_{i,j})]. \qquad (4.132)$$

Equation (4.131) is a backward difference equation utilizing the intermediate time points $(j+\frac{1}{2})\,\Delta t$. Since Eq. (4.129) only involves $\partial u/\partial t$ linearly, the calculation into the $(j+\frac{1}{2})$ time row is a linear algebraic problem. To move up to the $(j+1)$st time, we use Eq. (4.132) and by virtue of the linearity of Eq. (4.129) in $\partial u/\partial x$, this problem is also a linear algebraic problem. As an alternate to Eq. (4.131) one may use the predictor

$$(2h^2)^{-1}\,\delta_x^2[U_{i,j+1/2} + U_{i,j}] = \psi[ih, (j+\tfrac{1}{2})k, U_{i,j}, (2h)^{-1}\mu\delta_x U_{i,j}, (2/k)(U_{i,j+1/2} - U_{i,j})]. \qquad (4.133)$$

When ψ is given by Eq. (4.130) and we replace the corrector by

$$(2h^2)^{-1}\delta_x^2[U_{i,j+1} + U_{i,j}] = \psi[ih, (j+\tfrac{1}{2})k, U_{i,j+1/2}, (2h)^{-1}\mu\delta_x U_{i,j+1/2}, k^{-1}(U_{i,j+1} - U_{i,j})], \qquad (4.134)$$

then the predictor–corrector system [Eqs. (4.131) and (4.134)] generates linear algebraic equations for the calculation of the finite difference approximation.

Douglas and Jones [145] have demonstrated that the predictor–corrector scheme defined by Eqs. (4.131) and (4.132) converges to the solution of Eq. (4.128) when ψ is specified by Eq. (4.129). The truncation error is $O[h^2 + k^2]$. When ψ is given by Eq. (4.130), convergence was also

† In this section we use the conventional definitions for δ_x^2 and μ, i.e.,

$$\delta_x^2 U_{i,j+1} = U_{i+1,j+1} - 2U_{i,j+1} + U_{i-1,j+1}, \qquad \mu_x U_{i,j} = \tfrac{1}{2}[U_{i+1/2,j} + U_{i-1/2,j}].$$

established when the corrector adopted is Eq. (4.134). In this case the error is $O[h^2 + k^{3/2}]$.

Miller [146] has investigated and compared this predictor–corrector method with the explicit method and the exact solution for a problem given by the Burgers equation

$$\begin{aligned} u_t + uu_x &= \nu u_{xx}, \qquad 0 < x < 1, \quad 0 < t \leqslant T, \\ u(0, t) &= u(1, t) = 0, \\ u(x, 0) &= \sin \pi x. \end{aligned} \tag{4.135}$$

The exact solution is obtained by transforming this problem into a linear diffusion problem. Solutions are obtained for $0 < \nu \leqslant 1.0$ with emphasis on small values of ν. For values of ν, $0.01 < \nu \leqslant 1.0$, all three solutions were in excellent agreement. For $\nu < 0.01$, computation by means of the exact solution is not practical because of the slow convergence of the Fourier series.

As ν decreases from 10^{-2} to 10^{-4}, a definite consistent pattern emerged on all grids when the explicit method was employed. A disturbance (shock ?) appears at $x = 0.5$ for small t and passes to the right with steepened front as t increases. After the disturbance reaches a maximum near $x = 1.0$, for some time t, all values of u tend to decrease in a uniform manner. As h is reduced, these disturbances increase always maximizing as close to $x = 1.0$ as the grid allows. The ripples become more exaggerated as $\nu \rightarrow 10^{-4}$. More ripples appear as ν decreases propagating back toward $x = 0$. As $\nu \rightarrow 10^{-4}$, the disturbance is larger and decays more slowly as is physically expected. Computer time limitations imposed by stability requirements prevented further use of the explicit method.

The known property of unconditional stability of the predictor–corrector method was employed to refine the grid in the x-direction while maintaining the same time step $k = 4 \times 10^{-4}$. The computation at $h = 5 \times 10^{-3}$ ($r = 16.00$, $r = k/h^2$) shows that the ripples are gone! The computed solutions, with the predictor–corrector, tend to the asymptotic approximate profile of Cole [156].

4.14 FUNCTIONAL METHODS

Promoted by Bellman *et al.* [157–159], Noh and Protter [160], and Protter [161] these procedures are based upon two themes. The first is that of using a more efficient way (than that of finite differences) of recreating a function than by storing its values at grid points. The second is that an approximating algorithm should as clearly as possible exhibit the properties of the actual solution. If the solution is non-negative, this fact should be evident from the relations used to obtain it computationally. Whether or not algorithms of the desired type always exist is an unsolved problem.

Bellman *et al.* [157–159] initiated their investigation with the equation

$$u_t = uu_x, \qquad u(x, 0) = g(x), \tag{4.136}$$

which has the merit of possessing an implicit general solution

$$u = g(x + ut). \tag{4.137}$$

In place of the usual finite difference approximation for the Burgers equation

$$u_t + uu_x = \nu u_{xx}, \qquad u(x, 0) = g(x), \quad 0 \leqslant x \leqslant 1, \quad t > 0, \tag{4.138}$$

an alternative is adopted by Bellman *et al.* [158]. Suppose $g(x + \pi) = g(x)$ and let the approximating algorithm be

$$u(x, t + \Delta) = \lambda u[x - au(x, t)\Delta, t] + [(1 - \lambda)/2][u(x + b\Delta^{1/2}, t) + u(x - b\Delta^{1/2}, t)], \tag{4.139}$$

where Δ is the (time) integration step size, and λ, a, and b are constants which will be determined so that Eq. (4.139) is consistent with Eq. (4.138) to $O(\Delta^2)$. To evaluate λ, a, and b, we expand both sides of Eq. (4.139) in a Taylor series to the term in Δ^2, obtaining the equation

$$u_t = -\lambda a u u_x + \tfrac{1}{2}(1 - \lambda)\, b^2 u_{xx}\,. \tag{4.140}$$

For Eq. (4.140) to approximate Eq. (4.138) to $O(\Delta^2)$, we must have

$$a = 1/\lambda, \qquad b = [2\nu/(1 - \lambda)]^{1/2}.$$

If ν is fixed, a and b are functions of λ, and Eq. (4.139) becomes

$$u(x, t + \Delta) = \lambda u[x - u(x, t)(\Delta/\lambda), t] + \frac{1 - \lambda}{2}\left\{u\left[x + \left(\frac{2\nu\Delta}{1 - \lambda}\right)^{1/2}, t\right] + u\left[x - \left(\frac{2\nu\Delta}{1 - \lambda}\right)^{1/2}, t\right]\right\}. \tag{4.141}$$

Let $t = 0, \Delta, 2\Delta, \ldots$, and at each stage of the calculation store $u(x, t)$ by means of the finite sum

$$u(x, t) = \sum_{n=1}^{M} u_n(t) \sin n\pi x,$$

where the coefficients $u_n(t)$ are obtained by

$$u_n(t) = 2 \int_0^1 u(x, t) \sin n\pi x \, dx$$

$$\cong (2/R) \sum_{k=1}^{R-1} u(k/R, t) \sin(n\pi k/R).$$

Thus the values $u(k/R, t)$, $k = 1, 2, \ldots, R - 1$, store $u(x, t)$ at time t, and by means of Eq. (4.141), $u(x, t + \Delta)$ can be obtained.

Various computational results for changes in M, R, and λ are discussed by Bellman *et al.* [158]. A higher-order approximation would take the general form

$$u(x, t + \Delta) = \lambda u[x - au(x, t)\Delta, t]$$
$$+ \sum_{i=1}^{N} a_i[u(x + b_i\Delta^{1/2}, t) + u(x - b_i\Delta^{1/2}, t)].$$

Noh and Protter [160] and Protter [161] use the term "soft solution" as follows: Any function u satisfying the relation $u = f[x - t\phi(u)]$ is a solution of $u_t + \phi(u)u_x = 0$. For an arbitrary f (not necessarily differentiable), the expression $u = f[x - t\phi(u)]$, when it can be solved for u, is called a soft solution. In transport with chemical reaction,

$$u_t + uu_x = u^n, \qquad u(x, 0) = f(x),$$

has a soft solution

$$u^{1-n} = (1 - n)t + f\{x + (n - 2)^{-1} u^{2-n} + (2 - n)^{-1}[u^{1-n} + (n - 1)t]^{(n-2)/(n-1)}\},$$

for $n \neq 1$, $n \neq 2$. For $n = 1, 2$, we find

$$u(x, t) = e^t f[x - u + ue^t], \qquad n = 1,$$

$$u(x, t) = \frac{f[x + \ln(1 - tu)]}{1 + tf[x + \ln(1 - tu)]}, \qquad n = 2.$$

Soft solutions (Protter [161]) can be obtained for a variety of more complicated problems. The existence of soft solutions is useful for testing computational techniques and in the work of Protter suggests natural numerical schemes.

As an alternative to the Bellman *et al.* procedure, the "exact" difference methods of Noh and Protter [160] are available. To illustrate their process, we briefly discuss how to obtain various difference methods from the soft solution

$$u = f(x - ut), \tag{4.142}$$

of the equation

$$u_t + uu_x = 0, \qquad u(x, 0) = f(x), \quad -\infty < x < \infty. \tag{4.143}$$

In the upper half-plane, select a fixed rectangular grid with mesh sizes Δx and Δt, and let $u^n(x) = u(x, t_n)$ denote the solution of Eq. (4.143) at time $t_n = n\,\Delta t$. Because of the soft solution Eq. (4.142), we have

$$u^n(x) = f[x - t_n u(x, t_n)] = f[x - t_n u^n(x)]. \tag{4.144}$$

Consequently, at time $t_{n+1} = (n + 1)\,\Delta t$, we have from Eq. (4.144)

$$u^{n+1}(x) = f\{x - t_n u^n[x - u^{n+1}(x)\,\Delta t] - \Delta t u^{n+1}(x)\}. \tag{4.145}$$

On the other hand, using Eq. (4.142) directly, we have

$$u^{n+1}(x) = f[x - t_{n+1}u^{n+1}(x)] = f[x - t_n u^{n+1}(x) - \Delta t u^{n+1}]. \tag{4.146}$$

Since Eqs. (4.145) and (4.146) must coincide, we obtain

$$u^{n+1}(x) = u^n[x - \Delta t u^{n+1}(x)], \tag{4.147}$$

as the basic functional relation.

Suppose we have a difference method which determines a solution of a difference equation corresponding to Eq. (4.143). At the mesh points $(k\,\Delta x, n\,\Delta t)$, $k = 0, \pm 1, \pm 2,\ldots$, $n = 0, 1, 2,\ldots$, we denote that solution by $\{u_k^n\}$. If this solution satisfies

$$u_k^{n+1} = u^n(x_k - u_k^{n+1}\,\Delta t), \tag{4.148}$$

with $x_k = k\,\Delta x$, then $\{u_k^{n+1}\}$ will precisely coincide with the soft solution given by Eq. (4.147). A difference method which, at the mesh points where it is defined, coincides with the corresponding soft solution is called an *exact difference method.*

A soft solution is prescribed when the initial function $f(x)$ is prescribed. Suppose u_k^0, $k = 0, \pm 1, \pm 2,\ldots$, are given and f is determined by linear interpolation between mesh points on the initial line. Thus for $x_{k-1} \leqslant x \leqslant x_k$,

$$u^0(x) = u(x, 0) = f(x) = u_k^0 + (x - x_k)[(u_k^0 - u_{k-1}^0)/\Delta x]. \tag{4.149}$$

Since Eq. (4.148) is to be satisfied, we have

$$u_k^{\,1} = u^0[x_k - u_k^{\,1}\,\Delta t] = u_k^{\,0} - u_k^{\,1}(\Delta t/\Delta x)(u_k^{\,0} - u_{k-1}^0),$$

where the second step follows from Eq. (4.149). Solving for $u_k^{\,1}$, we have

$$u_k^{\,1} = u_k^{\,0}/[1 + r(u_k^{\,0} - u_{k-1}^0)], \qquad \text{if} \quad u_k^{\,1} \geqslant 0,$$

$$u_k^{\,1} = u_k^{\,0}/[1 + r(u_{k+1}^0 - u_k^{\,0})], \qquad \text{if} \quad u_k^{\,1} < 0,$$

where $r = \Delta t/\Delta x$. Upon replacing 1 by $n + 1$ and 0 by n, an exact difference method is obtained.

Each interpolation method gives rise to a corresponding exact difference method. If quadratic interpolation is applied at the nth step, the corresponding exact method for Eq. (4.143), at step $n + 1$, is

$$u_k^{n+1} = \frac{1 + \frac{1}{2}r\,\Delta u_k^n - [(1 + \frac{1}{2}r\,\Delta u_k^n)^2 - 2u_k^n r^2\,\Delta^2 u_k^n]^{1/2}}{r^2\,\Delta^2 u_k^n},$$

where

$$\Delta u_k^n = u_{k+1}^n - u_{k-1}^n\,, \qquad \Delta^2 u_k^n = u_{k-1}^n - 2u_k^n + u_{k-1}^n\,.$$

Other examples are also discussed by Noh and Protter [160], and calculations are detailed for problems in gas dynamics.

4.15 REFORMULATION IN NEW INDEPENDENT VARIABLES

In analytic studies of nonlinear equations, we do not hesitate to employ a great variety of complicated transformations. Quite the opposite appears to be the case in numerical computation. The boundary-layer equations have been intensively studied and transformed. Flugge-Lotz (Volume I, p. 349) has, in her numerical studies, demonstrated the importance of other independent variables than the primitive ones. A typical example of such behavior is described by Laganelli *et al.* [162].

We suppose that a flat plate of nonporous length l is fixed in a steady incompressible laminar stream followed downstream by a porous transpiration region. The boundary-layer equations with zero pressure gradient in Cartesian coordinates are

$$u_x + v_y = 0, \tag{4.150a}$$

$$uu_x + vu_y = \nu u_{yy}\,. \tag{4.150b}$$

The classical Blasius solution describes the boundary-layer growth over the nonporous section. This solution evaluated at l provides the "initial" condition, at $x = 0$, for continuation of the solution into the transpiration region. The boundary and initial conditions for that region are

$$y = 0\colon u = 0,\ v = v_c\,, \qquad \text{for} \quad x > 0,$$

$$y \to \infty\colon u \to u_1\,, \qquad \text{for} \quad x > 0,$$

$$x = 0\colon u = \text{Blasius solution.}$$

A stream function, ψ, defined through the relations

$$u = \partial\psi/\partial y, \qquad v = -(\partial\psi/\partial x), \tag{4.151}$$

guarantees that the continuity Eq. (4.150a) is satisfied. Upon introduction of these relations into Eq. (4.150b) we obtain the third-order equation

$$\psi_y\psi_{xy} - \psi_x\psi_{yy} = \nu\psi_{yyy}\,. \tag{4.152}$$

Such a form is undesirable for several reasons, including our relative lack of information regarding their numerical analysis. Thus a way is sought to maintain the second-order nature of the equations. This can be accomplished by application of the von Mises transformation (see Volume I). The goal in this transformation is to change from (x, y) to (x, ψ) coordinates. To accomplish this, we find

$$\left.\frac{\partial u}{\partial x}\right|_y = \left.\frac{\partial u}{\partial x}\right|_\psi + \frac{\partial u}{\partial \psi}\frac{\partial \psi}{\partial x} = \left.\frac{\partial u}{\partial x}\right|_\psi - v\,\frac{\partial u}{\partial \psi},$$

$$\left.\frac{\partial u}{\partial y}\right|_x = \frac{\partial u}{\partial \psi}\frac{\partial \psi}{\partial y} = u\,\frac{\partial u}{\partial \psi}, \qquad \left.\frac{\partial^2 u}{\partial y^2}\right|_x = u\,\frac{\partial}{\partial \psi}\left(u\,\frac{\partial u}{\partial \psi}\right),$$

whereupon Eq. (4.150b) becomes

$$u\left.\frac{\partial u}{\partial x}\right|_\psi = \nu u\,\frac{\partial}{\partial \psi}\left(u\,\frac{\partial u}{\partial \psi}\right)\Big|_x\,. \tag{4.153}$$

With the understanding that $u = u(x, \psi)$, we drop the subscripts in Eq. (4.153) and treat

$$u_x = \nu(uu_\psi)_\psi\,, \tag{4.154}$$

which is second order!

The boundary and initial conditions must now be transformed to (x, ψ) coordinates. Upon integrating Eq. (4.151), we have

$$\psi = -\int v(x, y)\,dx + f(y), \qquad \psi = \int u(x, y)\,dy + g(x), \tag{4.155}$$

where f and g are arbitrary functions. Upon employing the boundary conditions at $y = 0$, there results

$$\psi(x, 0) = -v_c x + f(0), \qquad \psi(x, 0) = g(x).$$

The equality of these two relations implies that $g(x) = -v_c x + f(0)$, thereby satisfying the stream function at $x = 0$. [If $v_c = F(x)$, the method is easily generalized. Throughout our discussion v_c is held constant.] The value $f(0)$ is chosen to be zero establishing the reference $\psi = 0$.

As a consequence of the foregoing arguments, Eq. (4.154) is subject to the following boundary and initial conditions:

$$\begin{aligned} &\psi = -v_c x\colon u = 0,\ v = v_c, \qquad x > 0,\\ &\psi \to \infty\colon u \to u_1, \qquad x > 0,\\ &x = 0\colon u = \text{Blasius solution.} \end{aligned} \tag{4.156}$$

Finally, the dimensionless variables

$$\eta = x/l, \qquad \xi = (\psi/u_1 l)(u_1 l/\nu)^{1/2}, \qquad \Gamma = 1 - (u/u_1)^2,$$

are introduced, whereupon Eq. (4.154) becomes

$$\partial\Gamma/\partial\eta = (1 - \Gamma)^{1/2}(\partial^2\Gamma/\partial\xi^2),$$

subject to the auxiliary data

$$\begin{aligned} &\xi \to \infty\colon \Gamma \to 0,\\ &\xi = -(v_c/u_1)(u_1 l/\nu)^{1/2}\,\eta = -\beta\eta\colon \Gamma = 1.\end{aligned}$$

$\Gamma(0, \xi) =$ Blasius solution (see Laganelli *et al.* [162] for data).

The integration domain does not have orthogonal boundaries (for $v_c \neq 0$), since the second boundary condition is applied along the sloping boundary $\xi = -\beta\eta$. Irregular mesh point techniques must be employed near that boundary.

Since the integration domain is infinite, with propagation in the η direction, an explicit method is employed. Thus setting $G_{ij} = G(i\,\Delta\eta, j\,\Delta\xi)$, we have the finite difference equation

$$G_{i+1,j} = G_{ij} + r(1 - G_{ij})^{1/2}\{G_{i,j+1} - 2G_{i,j} + G_{i,j-1}\} \tag{4.157}$$

[$G_{i,j}$ is the approximate solution, obtained from Eq. (4.157) for $\Gamma_{i,j}$]. A heuristic stability argument may be employed here. Thus we find that the quantity

$$(1 - G_{i,j})^{1/2}\,\Delta\eta/(\Delta\xi)^2 \leqslant \tfrac{1}{2},$$

should suffice for the computational stability of the finite difference approximation. Now initially, and in fact everywhere except at (0, 0), $0 \leqslant G \leqslant 1$, so that a choice of $\Delta\eta/(\Delta\xi)^2 = \frac{1}{2}$ should be safe. This was found, in actual practice, to be satisfactory.

Beginning with the initial values along the ξ-axis, there is no difficulty in obtaining that part of the solution above a 45° line, beginning at $\xi = \eta = 0$, using the explicit algorithm, Eq. (4.157). The calculation below the line must proceed by employing the boundary values on $\xi = -\beta\eta$, but the explicit molecule cannot be employed, since there is insufficient information on the previous line.

By performing the von Mises transformation, one can bring to bear a mass of information and methodology for second-order equations. Similar knowledge is not available for the third-order equation (4.152).

References

1. Hrenikoff, A., *J. Appl. Mech.* **8**, 169 (1941).
2. McHenry, D., *J. Inst. Civil Eng.* **21**, 59 (1943).
3. Courant, R., *Bull. Amer. Math. Soc.* **49**, 1 (1943).
4. Newmark, N. M., "Numerical Methods of Analysis in Bars, Plates and Elastic Bodies," *in* "Numerical Methods of Analysis in Engineering" (L. E. Grinter, ed.). Macmillan, New York, 1949.
5. Prager, W., and Synge, J. L., *Quart. Appl. Math.* **5**, 241 (1947).
6. Synge, J. L., "The Hypercircle in Mathematical Physics." Cambridge Univ. Press, London and New York, 1957.
7. Turner, M. J., Clough, R. W., Martin, H. C., and Topp, L. J., *J. Aeronaut. Sci.* **23**, 805 (1956).
8. Argyris, J. H., "Energy Theorems and Structural Analysis." Butterworth, London, 1960; (reprinted from Aircraft Eng. 1954-1955).
9. Clough, R. W., "The Finite Element in Plane Stress Analysis." Proc. 2nd ASCE Conf. on Electronic Computation, Pittsburgh, Pennsylvania, September (1960).
10. Clough, R. W., "The Finite Element Method in Structural Mechanics," Chapter 7 *in* "Stress Analysis" (O. C. Zienkiewicz and G. S. Holister, eds.). Wiley, New York, 1965.
11. Zienkiewicz, O. C., and Cheung, Y. K., *Eng.* **200**, 507, September (1965).
12. Zienkiewicz, O. C., and Cheung, Y. K., "The Finite Element Method in Structural and Continuum Mechanics." McGraw-Hill, New York, 1967: Second Edition,1971.
13. Wilson, E. L., and Nickell, R. E., *Nucl. Eng. Des.* **3**, 1 (1966).
14. Herrmann, L. R., *J. Eng. Mech. Div., Proc. ASCE* **91**, 11 (1965).
15. Zienkiewicz, O. C., Arlett, P. L., and Bahrani, A. K., *Eng.* **224**, 547 October (1967).
16. Winslow, A. M., *J. Comput. Phys.* **1**, 149 (1967).
17. Pian, T. H. H., *AIAA J.* **2**, 576 (1964).
18. Zienkiewicz, O. C., *Appl. Mech. Rev.* **23**, 249 (1970).
19. de Veubeke, B., "Displacement and Equilibrium Models in the Finite Element

Method,' Chapter 9 *in* "Stress Analysis" (O. C. Zienkiewicz and G. S. Holister, eds.). Wiley, New York, 1965.

20. Argyris, J. H., *J. Roy. Aeronaut. Soc.* **69**, 711 (1965).
21. Irons, B. M., "Numerical Integration Applied to Finite Element Methods." Conf. on use of digital computers in Structural Engineering, Univ. of Newcastle upon Tyne, England (1966).
22. Gallagher, R. H., Padlog, J., and Bijlaard, P. P., *J. Aerosp. Sci.* 700 (1962).
23. Melosh, R. J., *Proc. ASCE* **4**, 205 (1963).
24. Argyris, J. H., *AIAA J.* **3**, 45 (1965).
25. Argyris, J. H., *Ing. Ark.* **34**, 33 (1965).
26. Irons, B., "Stress Analysis by Stiffnesses using Numerical Integration." *Int. Rep. Rolls Royce* (1963); see Zienkiewicz [12].
27. Argyris, J. H., "Continua and Discontinua," *Proc. Conf. Matrix Meth. Struct. Mech.*, Wright Patterson Air Force Base, Ohio, October (1965).
28. de Veubeke, B., "Bending and Stretching of Plates." *Proc. Conf. Matrix Meth. Struct. Mech.*, Wright Patterson Air Force Base, Ohio (1965).
29. Herrmann, L. R., "A Bending Analysis of Plates." *Proc. Conf. Matrix Meth. Struct. Mech.*, Wright Patterson Air Force Base, Ohio (1965).
30. Pian, T. H. H., and Tong, P., *Int. J. Num. Meth. Eng.* **1**, 3 (1969).
31. Pian, T. H. H., *AIAA J.* **2**, 1232 (1964).
32. Severn, R. T., and Taylor, D. R., *Proc. Inst. Civil Eng.* **34**, 153 (1966).
33. Visser, W., "A Finite Element Method for the Determination of Non-Stationary Temperature Distribution and Thermal Deformations." *Proc. Conf. Matrix Meth. Struct. Mech.*, Wright Patterson Air Force Base, Ohio (1965).
34. Zienkiewicz, O. C., Mayer, P., and Cheung, Y. K., *Proc. ASCE* **92**, EMI, 111 (1966).
35. Adler, A., and Gallagher, R. H., "The Use of Finite Element Methods in Heat Flow Analysis." Bell Aerosystems Co. Rep. No. 9500-920134 (1968).
36. Zienkiewicz, O. C., *Appl. Mech. Rev.* **23**, 249 (1970).
37. Proc. of First Air Force Conf. on Matrix Meth. Struct. Mech. (1965). Wright-Patterson Air Force Base, Ohio AFFDL-TR-66-80 (1966).
38. Proc. of Second Air Force Conf. on Matrix Meth. Struct. Mech., Wright-Patterson Air Force Base, Ohio (1968).
39. Gallagher, R. H., Yamada, Y., and Oden, J. T. (eds.), "Recent Advances on Matrix Methods of Structural Analysis and Design." Proc. U.S.–Japan Seminar (Tokyo August 1969) Univ. of Alabama Press, University, Alabama, 1971.
40. Turner, M. J., Dill, E. H., Martin, H. C., and Melosh, R. J., *J. Aeronaut. Sci.* **27**, 97 (1960).
41. Gallagher, R. H., "The Finite Element Method in Elastic Instability Analysis." IDS/ISSC Symposium of Finite Element Techniques, Stuttgart, West Germany, 1969.
42. Mendelson, A., and Manson, S. S., "Practical Solution of Plastic Deformation Problems in the Elastic-Plastic Range." NASA TR R28 (1959).
43. Padlog, J., Huff, R. D., and Holloway, G. F., "The Unelastic Behavior of Structures Subjected to Cyclic, Thermal and Mechanical Stressing Conditions." Bell Aerosystems Co., Report WPADD TR 60-271 (1960).
44. Argyris, J. H., *J. Roy. Aeronaut. Soc.* **69**, 633 (1965).
45. Percy, J. H., Loden, W. A., and Navaratna, D. R., "A Study of Matrix Analysis Methods for Inelastic Structures." RTD-TDR-63-4032 (1963).
46. Argyris, J. H., Kelsey, S., and Kamel, W. H., "Matrix Methods of Structural

Analysis. A Precis of Recent Developments." Proc. 14th Mtg. Struct. and Mat. Panel, AGARD (1963).

47. Jensen, W. R., Falby, W. E., and Prince, N., "Matrix Analysis Methods for Anisotropic Inelastic Structures." AFFDL-TR-65-220 (1966).
48. Marcal, P. V., and Mallett, R. H., "Elastic-Plastic Analysis of Flat Plates by the Finite Element Method." Proc. ASME Winter Ann. Mtg. Paper 68-WA/PVP-10 December (1968).
49. Greenbaum, G. A., and Rubinstein, M. F., *Nucl. Eng. Des.* **7**, 379 (1968).
50. Drucker, D. C., "Basic Concepts of Plasticity," *in* "Handbook of Engineering Mechanics" (W. Flugge, ed.). McGraw-Hill, New York, 1962.
51. Purdy, M. D., and Przemieniecki, J. S., "Influence of Higher-Order Terms in the Large Deflection Analysis of Frameworks." Air Force Inst. Tech. Rep., Wright Patterson Air Force Base, Ohio.
52. Oden, J. T., *Proc. ASCE, J. Struct. Div.* **93**, 235 (1967).
53. Mallett, R. H., and Berke, L., "Automated Method for the Large Deflection and Instability Analysis of Three Dimensional Truss and Frame Assemblies." AFFDL-TR-66-102 (1966).
54. Bogner, F., Fox, R., and Schmit, L., *AIAA J.* **6**, 1251 (1968).
55. Murray, D. W., and Wilson, E. L., *Proc. ASCE, J. Mech. Div.* **95**, EM1 (1969).
56. Grafton, P. E., and Strome, D. R., *AIAA J.* **1**, 2343 (1963).
57. Navaratna, D. R., "Elastic Stability of Shells of Revolution by the Variational Approach using Discrete Elements," ASRL-TR-139-1. MIT Press, Cambridge, Massachusetts, 1966.
58. Navaratna, D. R., Pian, T. H. H., and Witmer, E. A., "Analysis of Elastic Stability of Shells of Revolution by the Finite Element Method." Proc. AIAA/ASME Eighth Struct., Struct. Dyn., Mat. Conf., Palm Springs, California, March (1967).
59. Pope, G., "A Discrete Element Method for Analysis of Plane Elastic-Plastic Stress Problems." *Roy. Aeronaut. Est.*, TR65028 (1965).
60. Swedlow, J. L., and Yang, W. H., "Stiffness Analysis of Elastic-Plastic Plates." *Grad. Aero. Lab. Calif. Inst. Tech. Rep.* SM 65-10 (1965).
61. Marcal, P. V., and King, I. P., *Int. J. Mech. Sci.* **9**, 143 (1967).
62. Yamada, Y., Yoshimura, N., and Sakurai, T., "Plastic Stress-Strain Matrix and its Application for the Solution of Elastic-Plastic Problems by the Finite Element Method." *Rep. Inst. Ind. Sci., Univ. Tokyo* August (1967).
63. Levy, N. J., Application of Finite Element Methods to Large Scale Elastic-Plastic Problems of Fracture Mechanics, Ph.D. Dissertation, Brown Univ., Providence, Rhode Island (1969).
64. Marcal, P. V., *AIAA J.* **6**, 157 (1967).
65. Hayes, D. J., and Marcal, P. V., *Int. J. Mech. Sci.* **9**, 245 (1967).
66. Oden, J. T., and Kubitza, W. K., "Numerical Analysis of Nonlinear Pneumatic Structures." *Proc. First Int. Coll. Pneumatic Struct.*, Stuttgart, May (1967).
67. Marcal, P. V., "Large Deflection Analysis of Elastic-Plastic Shells of Revolution." *Proc. Tenth ASME/AIAA Struct., Struct. Dyn. Mat. Conf.*, New Orleans, Louisiana, March (1969).
68. Marcal, P. V., "Large Deflection Analysis of Elastic-Plastic Plates and Shells." *Proc. First Int. Conf. Press. Vessel Tech. ASME/Roy. Neth. Eng. Soc.*, Delft, September (1969).
69. Marcal, P. V., "Finite Element Analysis of Combined Problems of Nonlinear Material and Geometric Behavior." *Joint ASME Comp. Conf. Comp. Appr. Applied Mech.*, Chicago, Illinois, June (1969).

70. Popov, E. P., and Yaghmai, S., "Linear and Nonlinear Static Analysis of Axisymmetrically Loaded Thin Shells of Revolution." *Proc. First Int. Conf. Press. Vessel Tech., ASME/Roy. Neth. Eng. Soc.*, Delft, September (1969).
71. Szabo, B. A., and Lee, G. C., "Derivation of the Stiffness Matrix for Plates by Galerkin's Method." *SUNY Buffalo, Civ. Eng. Rep.* 17.1 (1968).
72. Szabo, B. A., and Lee, G. C., "Derivation of Stiffness Matrices for Problems in Plane Elasticity." *SUNY Buffalo, Civ. Eng. Rep.* 17.2 (1968).
73. Langhaar, H. L., and Chu. S. C., *Develop. Theo. Appl. Mech.* **4**, Pergamon Press, New York, 553 (1970).
74. Leonard, J. W., Bramlette, T. T., *Proc. ASCE, J. Mech. Div.* **96**, EM6 1277 (1970).
75. Thom, A., *Proc. Roy. Soc. (London)* **A141**, 651 (1933).
76. Richtmyer, R. D., and Morton, K. W., "Difference Methods for Initial Value Problems," (2nd ed.). Wiley (Interscience), New York, 1967.
77. Ames, W. F., "Numerical Methods for Partial Differential Equations." Nelson, London; Barnes & Noble, New York, 1969.
78. Argyris, J. H., Mareczek, G., and Scharpf, D. W., *Aero. J. Roy. Aero. Soc.* **73**, 961 (1969).
79. Doctors, L. J., *Int. J. Num. Meth. Eng.* **2**, 243 (1970).
80. Argyris, J. H., *Aero. J. Roy. Aero. Soc.* **74**, 13 (1970), 111 (1970).
81. deVries, G., and Norrie, D. H., "Applications of the Finite Element Technique to Potential Flow Problems." Reps. 7 and 8, Dept. of Mechanical Engineering, Univ. of Calgary, Alberta, Canada (1969).
82. Javandel, I., and Witherspoon, P. A., *Soc. Petrol. Eng. J.* 241, (1968).
83. Taylor, R. L., and Brown, C. B., *J. Hyd. Div., ASCE* **92**, 25 (1967).
84. Sandhu, R. S., and Wilson, E. L., *J. Eng. Mech. Div., ASCE* **95**, 641 (1969).
85. Volker, R. E., *J. Hyd. Div., ASCE* **95**, 2093 (1969).
86. Tong, P., and Fung, Y. C., "The Effect of Wall Elasticity on a Liquid in a Cylindrical Container." Rep. SM64-40, Calif. Inst. Tech., Pasadena, California, 1965.
87. Tong, P., Liquid Sloshing in a Container, Ph.D. Dissertation, Calif. Inst. Tech. Pasadena, California (1966); Also issued as AFOSR 66-0943.
88. Luk, C. H., "Finite Element Analysis for Liquid Sloshing Problems," SM Thesis. MIT, Cambridge, Massachusetts, 1969.
89. Argyris, J. H., and Scharpf, D. W., *Aero. J. Roy Aero. Soc.* **73**, 1044 (1969).
90. Reddi, M. M., *J. Lab. Tech., Trans. ASME* **91**, 524 (1969).
91. Reddi, M. M., and Chu, T. Y., "Finite Element Solution of the Steady State Compressible Lubrication Problem." Paper 69-LUB-12 Oct. ASME Mtg. (1969).
92. Skiba, E., "A Finite Element Solution of General Fluid Dynamic Problems—Natural Convection in Rectangular Cavities." M. App. Sci. Thesis, Univ. of Waterloo, Ontario, Canada (1970).
93. Oden, J. T., and Somogyi, D., *J. Eng. Mech. Div., ASCE* **95**, 821 (1964).
94. Tong, P., "Finite Element Method for Low Reynolds Number Flow and its Application to Biomechanics." 2nd Canad. Cong. Appl. Mech., Waterloo, Canada, 1969.
95. Atkinson, B., Brocklebank, M. P., Card, C. C. M., and Smith, J. M., *AIChE J.* **15**, 548 (1969).
96. Zienkiewicz, O. C., and Newton, R., "Coupled Vibrations of a Structure Submerged in a Compressible Fluid." *Int. Symp. Fin. Element Tech. Shipbuilding*, Stuttgart, Germany, 1969.
97. Zienkiewicz, O. C., Irons, B., and Nath, P., "Natural Frequencies of Complex

Free or Submerged Structures by the Finite Element Method." Symp. on Vibration in Civil Eng., Inst. Civ. Eng., Butterworth (1965).

98. Olson, M. D., *AIAA J.* **8**, 747 (1970).

99. Norrie, D. H., and deVries, G., "Finite Elements and Fluid Mechanics." Academic Press, New York (in preparation).

100. Kuethe, A. M., and Schetzer, J. D., "Foundations of Aerodynamics," (2nd ed). Wiley, New York, 1967.

101. Emmons, H. W., "The Numerical Solution of the Turbulence Problem." *Proc.* 1947 *Symp. Appl. Math., Amer. Math. Soc. I*, 67 (1949); See also "A New Approach to the Turbulence Problem," Vol. 16. Annals of the Computation Laboratory. Harvard Univ. Press, Cambridge, Massachusetts, 1947.

102. Payne, R. B., Rep. No. 3407. *Aero. Res. Council*, London (1956).

103. Payne, R. B., *J. Fluid Mech.* **4**, 81 (1958).

104. Fromm, J. E., "A Method for Computing Nonsteady Incompressible Viscous Fluid Flows," Rep. No. 2910. Los Alamos Scientific Laboratory, Los Alamos, New Mexico (1963); See also Fromm, J. E., and Harlow, F. H., *Phys. Fluids* **6**, 975 (1963).

105. Fromm, J. E., "The Time Dependent Flow of an Incompressible Viscous Fluid," *in "Methods in Computational Physics"* **3**, 345. Academic Press, New York, 1964.

106. DuFort, E. C., and Frankel, S. P., *Math. Tab. Nat. Res. Coun.*, Washington, D. C. **7**, 135 (1953).

107. Harlow, F. H., "Stability of Difference Equations; Selected Topics," Rep. No. 2452. Los Alamos Scientific Laboratory, Los Alamos, New Mexico (1960).

108. Harlow, F. H., and Fromm, J. E., *Phys. Fluids* **8**, 1147 (1964).

109. Wilkes, J. O., The Finite Difference Computation of Natural Convection in an Enclosed Rectangular Cavity, Ph.D. Dissertation, Univ. of Michigan, Ann Arbor, Michigan, 1963.

110. Pearson, E. E., *J. Fluid Mech.* **21**, 611 (1965).

111. Keller, H. B., and Takami, H., "Numerical Studies of Steady Viscous Flow about Cylinders," p. 115 *in* "Numerical Solutions of Nonlinear Differential Equations" (D. Greenspan, ed.). Wiley, New York, 1966; See also Takami, H., and Keller, H. B., p. 51 of Ref. 113.

112. Keller, H. B., *Quart. Appl. Math.* **16**, 209 (1958).

113. Int. Union Theo. Appl. Mech. Symp. on High Speed Computing in Fluid Dynamics, *Phys. Fluids* **12**, December (1969).

114. Jenson, V. G., *Proc. Roy. Soc., London* **A249**, 346 (1959).

115. Hamielec, A. E., Hoffman, T. W., and Ross, L. L., *AIChE J.* **13**, 14 (1967).

116. Rimon, Y., and Cheng, S. I., *Phys. Fluids* **12**, 949 (1969).

117. Alonso C. V., Time Dependent Confined Rotating Flow of an Incompressible Viscous Fluid, Ph.D. Dissertation, Univ. of Iowa, Iowa City, Iowa, 1971.

118. Harlow F. H., and Welch, J. E., *Phys. Fluids* **8**, 2182 (1964).

119. Welch J. E., Harlow F. H., Shannon, J. P., and Daly, B. J., "The MAC Method. A Computing Technique for Solving Viscous Incompressible Transient Fluid-Flow Problems Involving Free Surface," Rep. 3425, Los Alamos Scientific Laboratory, Los Alamos, New Mexico, 1966; see also *Phys. Fluids* **10**, 927 (1967).

120. Chorin, A. J., *Bull. Amer. Math. Soc.* **73**, 918 (1967).

121. Chorin, A. J., "Numerical Study of Thermal Convection in a Fluid Layer Heated from Below," Rep. 1480-61. Courant Institute, New York Univ., New York, 1966.

122. Ames, W. F., "Recent Developments in the Nonlinear Equations of Transport Processes." *Ind. Eng. Chem. Fund.* **8**, 522 (1969).

123. Chorin, A. J., *J. Comput. Phys.* **2**, 12 (1967).
124. Chorin, A. J., *Math. Comp.* **22**, 745 (1968).
125. Chorin, A. J., "On the Convergence of Discrete Approximations to the Navier–Stokes Equations," Rep. 1430-106. Courant Institute, New York Univ., New York, 1968.
126. Chorin, A. J., "Numerical Solution of Incompressible Flow Problems." Invited paper SIAM Meeting, Philadelphia, Pennsylvania, October (1968).
127. Fujita, H., and Kato, T., *Arch. Rational Mech. Anal.* **16**, 269 (1964).
128. Padmanabhan, H., Wake Deformation in Density-Stratified Fluids, Ph.D. Dissertation, Univ. of Iowa, Iowa City, Iowa, 1969.
129. Samarskii, A. A., *USSR Comput. Math. Math. Phys.* **3**, 894 (1963).
130. Padmanabhan, H., Ames, W. F., Kennedy, J. F., and Hung, T. K., *J. Eng. Math.* **4**, 229 (1970).
131. Pujol, A., Numerical Experiments on the Stability of Poiseuille Flows of Non-Newtonian Fluids, Ph.D. Dissertation, University of Iowa, Iowa City, Iowa, 1971.
132. Aziz, K., A Numerical Study of Cellular Convection, Ph.D. Dissertation, Rice Univ., Houston, Texas (1965).
133. Aziz, K., and Hellums, J. D., *Phys. Fluids* **10**, 314 (1967).
134. Douglas, J., Jr., *Numer. Math.* **4**, 41 (1962).
135. Jacob, C., "Introduction Mathematique a la Mecanique des Fluids." Gauthier-Villars, Paris, 1959.
136. Moon, P., and Spencer, D. E., "Foundations of Electrodynamics." Van Nostrand–Reinhold, Princeton, New Jersey, 1960.
137. Moon, P., and Spencer, D. E., "Field Theory for Engineers." Van Nostrand–Reinhold, Princeton, New Jersey, 1961.
138. Moon, P., and Spencer, D. E., "Field Theory Handbook." Springer-Verlag, Berlin and New York, 1961.
139. Ames, W. F., and de la Cuesta, H., *J. Math. and Phys.* **42**, 301 (1963).
140. Moon, P., and Spencer, D. E., *J. Franklin Inst.* **256**, 551 (1953).
141. Wills, A. P., "Vector Analysis." Prentice-Hall, Englewood Cliffs, New Jersey, 1931.
142. Chandrasekhar, S., "Hydrodynamic and Hydromagnetic Stability." Oxford Univ. Press (Clarendon), London and New York, 1961.
143. Hirasaki, G. J., and Hellums, J. D., *Quart. Appl. Math.* **26**, 331 (1968).
144. Moreau, J., *Comp. Rend.* **248**, 3406 (1959).
145. Douglas, J., Jr., and Jones, B. F., *J. Soc. Ind. Appl. Math.* **11**, 195 (1963).
146. Miller, E. L., Predictor-Corrector Studies of Burgers Model of Turbulent Flow, MS thesis, Univ. of Delaware, Newark, Delaware, 1966.
147. Whitaker, S., and Wendel, M. M., *Appl. Sci. Res.* **12**, 91 (1963).
148. Douglas, J., Jr., and duPont, T., *J. Num. Anal. Soc. Ind. Appl. Math.* **1**, 575 (1970).
149. Giese, J. H., "A Bibliography for the Numerical Solution of Partial Differential Equations," BRL Mem. Rep. No. 1991. Ballistic Research Laboratories, Aberdeen Proving Ground, Maryland, 1969; updated in BRL Mem. Rep. No. 2114, 1971.
150. Francis, G. C., "A Computer Based Searchable File of Journal References in the Field of Partial Differential Equations," BRL Mem. Rep. No. 2025. Ballistic Research Laboratories, Aberdeen Proving Ground, Maryland, 1970.
151. "Methods in Computational Physics," (B. Alder, and S. Fernbach, eds.). Academic Press, New York.
    Vol. 1: "Statistical Physics" (1963)
    2: "Quantum Mechanics" (1964)
    3: "Fundamental Methods in Hydrodynamics" (1964)

4: "Applications in Hydrodynamics" (1965)
5: "Nuclear Particle Kinematics" (1966)
6: "Nuclear Physics" (1966)
7: "Astrophysics" (1967)
8: "Energy Bands of Solids" (1968)
9: "Plasma Physics" (1970)

152. Yanenko, N. N., "The Method of Fractional Steps." Springer-Verlag, Berlin and New York, 1971.
153. Hamming, R. W., "Numerical Methods for Scientists and Engineers." McGraw-Hill, New York, 1962.
154. Hamming, R. W., *J. Ass. Comput. Mach.* **6**, 37 (1959).
155. Fox, L. (ed.), "Numerical Solution of Ordinary and Partial Differential Equations." Macmillan, New York, 1962.
156. Cole, J. D., *Quart. Appl. Math.* **9**, 225 (1951).
157. Bellman, R., Kalaba, R., and Kotkin, B., *Proc. Nat. Acad. Sci.* USA **48**, 1325 (1962).
158. Bellman, R., Azen, S., and Richardson, J. M., *Quart. J. Appl. Math.* **23**, 55 (1965).
159. Bellman, R., and Kalaba, R., "New Methods for the Solution of Partial Differential Equations," chapter *in* "Nonlinear Partial Differential Equations—Methods of Solutions," p. 43, (W. F. Ames, ed.). Academic Press, New York, 1967.
160. Noh, W. F., and Protter, M., *J. Math. Mech.* **12**, 149 (1963).
161. Protter, M. H., "Difference Methods and Soft Solutions," chapter *in* "Nonlinear Partial Differential Equations—Methods of Solution," p. 161, (W. F. Ames, ed.). Academic Press, New York, 1967.
162. Laganelli, A. L., Ames, W. F., and Hartnett, J. P., *AIAA J.* **6**, 193 (1968).

Author Index

Numbers in parentheses are reference numbers and indicate that an author's work is referred to, although his name is not cited in the text. Numbers in italics show the page on which the complete reference is listed.

O

P

R

S

Subject Index

W